T0317611

Semiconductor Microchips and Fabrication

IEEE Press
445 Hoes Lane
Piscataway, NJ 08854

Semiconductor Microchips and Fabrication

A Practical Guide to Theory and Manufacturing

Yaguang Lian
University of Illinois
Urbana, USA

Published by John Wiley & Sons, Inc., Hoboken, New Jersey.
Published simultaneously in Canada.

Library of Congress Cataloging-in-Publication Data Applied for:
Hardback ISBN: 9781119867784

Cover Design: Wiley
Cover Image: © Den Rise/Shutterstock

Set in 9.5/12.5pt STIXTwoText by Straive, Chennai, India

Contents

Author Biography

Yaguang Lian is a research engineer in Holonyak Micro & Nanotechnology Lab at the University of Illinois at Urbana-Champaign (UIUC). In 1979, he studied at the Department of Electronics, Hebei University, China. In 1983, with a bachelor's degree, he worked on silicon epitaxy for a semiconductor company for two years. Yaguang returned to Hebei University as a graduate student in 1985. In 1988, he got his master's degree. From 1988 to 2001, he worked with Hebei Semiconductor Research Institute (HSRI). In 2001, Yaguang joined a start-up company at California as a senior semiconductor process engineer. Two years later, he worked in the University of California, Los Angeles (UCLA). From November 2004 till now, Yaguang works as research engineer at the University of Illinois.

With over 35 years of experience in the semiconductor field, Yaguang has dealt with different processes for different kinds of semiconductors, such as Si and GaAs. When he worked in HSRI in China, he was manufacturing GaAs MESFET digital IC and designing circuits, and also managed a process flow from implantation to packaging. In the company at California, he worked mainly in the manufacture of silicon optical waveguide. At UCLA, Yaguang started the equipment repair and maintenance, and users' training, which he further evolved at UIUC. He is also responsible for designing different recipes to meet the needs of professors in different projects, such as dry etching of silicon, III–V materials, dielectric films and metals, and deposition of silicon nitride film with different stresses. He solves different types of process problems for students paving way for their research.

Yaguang believes that to be a good engineer, one must have an important trait, which is "it is not enough for you to know how to do it, you must know why to do it."

Preface

"Semiconductor process engineer" is the label of my whole career. As a research engineer, I have been working in Holonyak Micro & Nanotechnology Laboratory (HMNTL) at the University of Illinois for nearly two decades. The core part of HMNTL is the clean room with different types of equipment, which are used to fabricate various semiconductor devices. The lab is open to the campus and society. Up to now, I have trained thousands of users in using the lab machines. Most users are doctoral students. These training experiences have given me the opportunities to encounter various issues and solve different technical problems with different kinds of users.

During my years of training, I have met students with different backgrounds. Most of them have EE or ECE backgrounds, and some of them don't have these backgrounds. The student without EE or ECE backgrounds lack basic knowledge of semiconductors and processing, while students majoring in EE or ECE need to make up the working principles of manufacturing processes and the basic structures of equipment. Lack of adequate understanding of process and equipment not only occurs in many doctoral students with backgrounds in EE or ECE, but also in some postdoctoral researchers. An important reason for this phenomenon is that they do not understand the process and equipment from the perspective of physical and chemical principles. These issues and the encouragement from students drove me to write this book.

We are using microchips to describe semiconductor devices and integrated circuits (ICs). To meet the needs of readers without EE or ECE backgrounds, this book includes the knowledge of semiconductor concepts, theories, histories, and basic structures of microchips, which help to lay the foundation for them to understand semiconductor manufacturing processes. To help the readers with EE or ECE backgrounds better understand the process and equipment, this book strives to clarify the principles of processes, the basic structures of equipment, and the design of process recipes based on the physical laws, chemical reactions, and electrical circuit theories. It not only shows readers how to do the process, but also explains why the process is designed in such a way. It combines the processes with the actual machines used in the clean room at the University. Therefore, the book will be very useful for readers with different backgrounds.

This is a handy book for many audiences as it starts with the basic concepts and daily examples. It uses simple language to explain complicated concepts and theories. The needs of various levels of audiences will be satisfied, such as undergraduates, graduates, researchers, engineers, and professors. The book will pave the ways for readers in their semiconductor research, process, and manufacture. From this book, readers can find many valuable suggestions and solutions to the problems that students or engineers often encounter in semiconductor processing. These suggestions and solutions are based on my years of working experience. Moreover, readers can also find some useful experimental results in the book, which will help them in their processing work.

Nowadays, semiconductors are widely used in many fields. This book is also written for those who are not majoring but interested in the research and production of semiconductor microchips. Even those who do not have enough knowledge of semiconductor and processes, as long as they have basic knowledge of physics, chemistry, and circuit, after reading this book, can easily learn and quickly grasp the principles, knowledge, and manufacturing processes of semiconductors. From this book, readers can obtain the fundamental concepts and skills, which will be a necessity in the development of semiconductor processing. They can apply all these concepts and skills in semiconductor technology to improve their product quality or the project research.

I could not accomplish my book without HMNTL. In HMNTL, I have gotten a good time with my colleagues. I would like to express my appreciation to my excellent colleagues: Mr. John Hughes, Dr. Mark McCollum, Dr. Edmond Chow, Dr. Glennys Mensing, Ken Tarman, Hal Romans, Michael Hansen, Lavendra Mandyam, Karthick Jeganathan, Paul DiPippo. From them, I have learned more about the layout and management of the clean room at the university. From them, I also have gotten a lot of help in the equipment troubleshooting, recipe design, and parameter testing. I am so glad that I have worked with them for so many years.

In writing this book, many friends have offered their support. Dr. Ruijie Zhao gave me good advice at the beginning. Dr. Wenjuan Zhu reviewed the draft. Dr. Anming Gao pushed me to write the book. Mr. Raman Kumar and Alvin Flores provided me good suggestions on some theoretical issues, and some students supplied me with nice images. I especially thank Tianyi Bai, my nephew, a graduate student at the University of Pennsylvania, who has given me valuable opinions in some areas. Finally, I would like to thank the Internet, which was born from semiconductor technology. From the Internet, I can easily find the information I need. Here, I deeply appreciate the companies and individuals who agreed to let me use their images in the book.

February 12, 2022

Yaguang Lian
University of Illinois at Urbana-Champaign
Champaign, IL

1

Introduction to the Basic Concepts

1.1 What Is a Microchip?

Looking back on the development history of human society, it has gone through different stages of civilization, from the primitive stone age to the modern information age. The material used to support stone age was stone. The material used to support information age is semiconductor. So contemporary society is essentially a semiconductor era represented by silicon. This era began in the late 1950s and early 1960s in the Bay Area of Northern California near San Francisco in the United States. Later, people called this area "Silicon Valley," the sign of high technology (abbreviated to high-tech), which have brought us into the information age. Silicon and other semiconductors are the cornerstone of this era. If petroleum is thought as the blood of modern society, semiconductor microchips can be regarded as the brain. Semiconductor technology has been integrated by different industries to enhance their technical level, and it has also come into our households. A microchip is a semiconductor device or an integrated circuit (IC). An IC is to make a lot of tiny semiconductor devices onto a small flat piece of semiconductor (a die).

1.2 Ohm's Law and Resistivity

Due to microchips are operated by electricity, so first, let us get to know what electricity is, and how electricity works. See Figure 1.1. It is a voltage converter for a small household electronic product. The explanation of some technical words on the converter are listed as follows:

- "VAC" means "Volts Alternating Current"
- "Hz-Hertz" is the frequency unit
- "W-Watt" is the unit of power
- "mA" means "milli-Ampere." "Ampere-A" is the unit of electric current (abbreviated to current)

Semiconductor Microchips and Fabrication: A Practical Guide to Theory and Manufacturing, First Edition. Yaguang Lian.

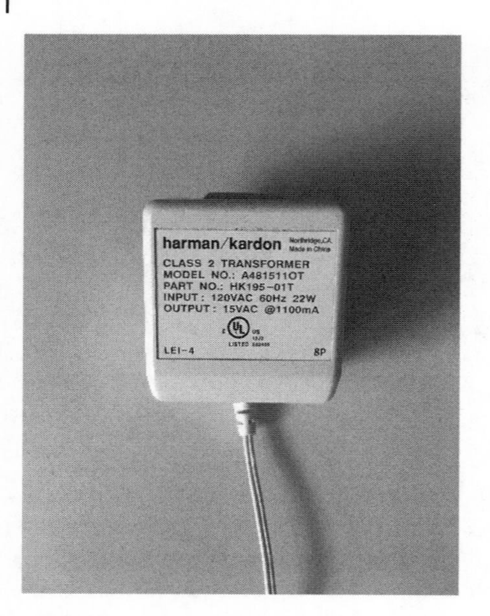

Figure 1.1 A voltage converter for an electronic product. Source: Harman International Industries, Incorporated.

Among them, voltage, current, and watt are the three basic parameters used to express the characteristics of electricity. Another basic parameter is resistance, we will talk about it in Ohm's law below.

Electricity is the set of physical phenomena. A physical phenomenon refers to a process that does not produce new substances, such as the movement of objects, the freezing and boiling of water, and so on. Corresponding to this is a chemical phenomenon, which refers to a process that can produce new substances. We call this process a chemical reaction, for example, oxygen and hydrogen generate water through a chemical reaction. There is also a nuclear phenomenon, which is beyond the scope of this book and will not be discussed. Now, let us return to the topic of electricity. Electricity is generated by the motion of matter that has a property of electric charge. An electric charge can be positive or negative. Positive one is called "positive charge" and represented by "+." Negative one is called "negative charge" and represented by "−." The movement of electric charges is an electric current. In most cases, the current is produced by the movement of electrons, which are negative charges. The unit of electric current, Ampere, is named in honor of a French mathematician and physicist André-Marie Ampère (1775–1836), who is considered the father of electrodynamics.

Electricity comes into our homes through electrical wires and then goes to various electrical appliances. There are usually two kinds of electrical wires (abbreviated to wire) for household use. They are two-core wires, just like the wire attached to the converter in Figure 1.1, and three-core wires. If we strip a three-core wire, its structure is shown in Figure 1.2. "Cable Jacket" is the insulating sheath, "Wire Insulation" is the insulating layer, and "Stripped Wire" is the wire exposed after

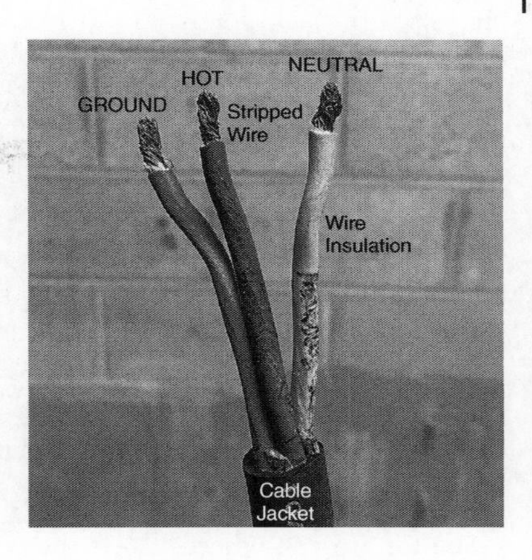

Figure 1.2 The basic structure of three-core wire.

the insulating layer is stripped. The three wires are hot, neutral, and ground. Most of the insulating sheathes and layers are made of rubber and plastic; the wire is made of metal. In most cases, the metal is aluminum or copper. The kind of materials such as rubber and plastic are called insulators, in which, electrons cannot move. Thus, current cannot flow in the insulators. We call aluminum and copper, the type of materials, conductors, because electrons can move in them. Thus, electrical current can flow in the conductors.

All materials are composed of atoms, and atoms will hinder the movement of electrons, which means that all materials have resistance in the opposite direction of the current flow. This resistance is called electrical resistance (abbreviated to resistance), and the unit of the resistance is Ohm, which is represented by "Ω." Ω is named in honor of the German physicist Georg Simon Ohm (March 16, 1789–July 6, 1854). How do electrons move in a conductor to generate the current? They are driven by the pressure of electricity. This special pressure is called voltage, and the unit is volt, which is expressed by "V." The voltage unit volt is to commemorate the Italian physicist Alessandro Volta (1745–1827), who invented voltaic pile, the first electrical battery in the world. Now, we have three parameters related to electricity: current, represented by "I"; resistance, represented by "R"; voltage, represented by "V." Sometimes lowercase letters are also used to indicate current, resistance, and voltage. The relationship between them is the famous Ohm's law:

$$I = \frac{V}{R} \tag{1.1}$$

Why can electrons flow in metal? It is because the metal has little resistance; electrons cannot flow in the insulator because the insulator has large resistance. Scientists use resistivity to express the resistance of a material per unit length. "ρ" is used to represent resistivity, its unit is ohm · cm (Ω · cm). The relationship between resistance and resistivity is expressed by the following formula:

$$R = \rho\frac{L}{A} \tag{1.2}$$

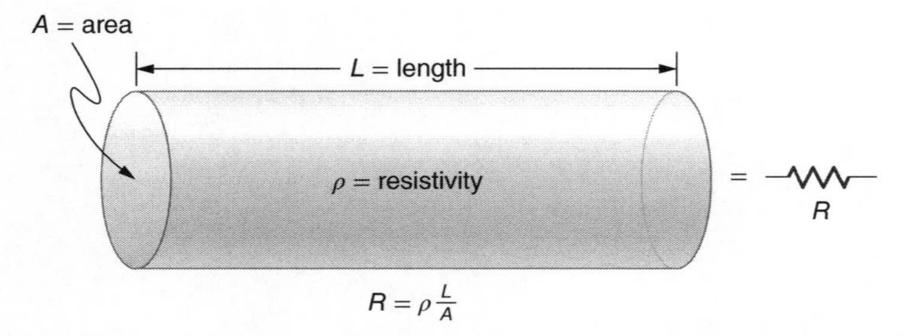

Figure 1.3 A diagram of a section of resistive material, current flows along the length direction. Source: Reproduced with permission of Physics LibreTexts.

See Figure 1.3, where "A" is the cross-sectional area, and "L" is the length of a piece of material. To make more convenient for practical use, we introduce the concept of conductance. The symbol of conductance is G, and the relationship with resistance R is as follows:

$$G = \frac{1}{R} \tag{1.3}$$

The unit of G is Siemens and is represented by the letter S to commemorate Werner von Siemens (December 13, 1816–December 6, 1892). He was a German scientist and founder of Siemens. Correspondingly, there is conductivity, expressed by σ, and the relationship with resistivity ρ is

$$\sigma = \frac{1}{\rho} \tag{1.4}$$

In practical application, we often use devices called resistors, which are shown in the left of Figure 1.4, and the symbol is shown in the right of Figure 1.4.

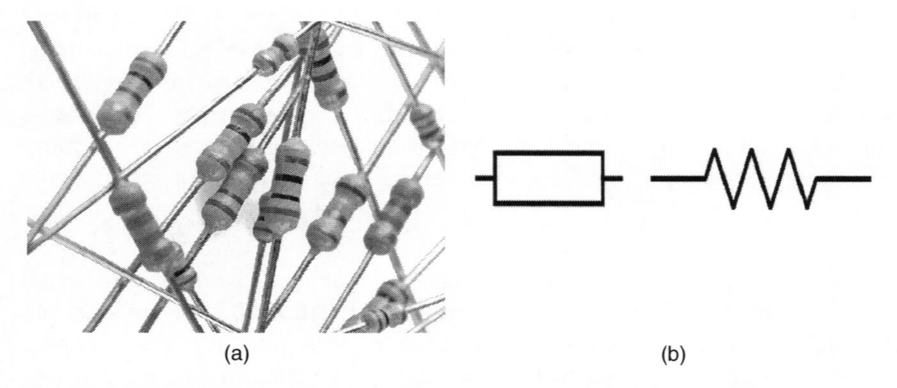

Figure 1.4 The picture of resistors (a) and the symbol (b). Source: Reproduced with permission of SparkFun Electronics.

1.3 Conductor, Insulator, and Semiconductor

Now, we use resistivity to distinguish conductors and insulators. In general, the resistivity of a conductor is very low, and the resistivity of an insulator is very high. For example, the resistivity of copper is $1.55 \times 10^{-6}\ \Omega \cdot \text{cm}$, and for aluminum is $2.5 \times 10^{-6}\ \Omega \cdot \text{cm}$ [1], where 10^{-6} is one millionth. Polyvinyl chloride (PVC) is a kind of plastic commonly used for making insulating materials, its resistivity is 2×10^{12}–$2 \times 10^{14}\ \Omega \cdot \text{cm}$, nylon is $4.56 \times 10^{16}\ \Omega \cdot \text{cm}$ [2]. Mathematically, 10^2 means there is a zero after 10, 10^3 means there are two zeros after 10, and so on. From which, we can know that the resistivity of the above two insulating materials has 11–15 zeros after 10. Conductors and insulators have the resistivities either extremely small or extremely large. Does a material with resistivity between them exist? Yes, this material exists, we call it a semiconductor. At room temperature, the resistivity of silicon is $6.3 \times 10^4\ \Omega \cdot \text{cm}$, and the resistivity of germanium is $46\ \Omega \cdot \text{cm}$ [1]. Silicon with the symbol Si is the most important material in the modern semiconductor industry. That is why the Bay Area of Northern California is called Silicon Valley. Germanium with the symbol Ge was used to make the first transistor in the world. Silicon and germanium are single-element semiconductors. Another kind of semiconductors are compound semiconductors. The most used one is gallium arsenide with symbol GaAs, which has a resistivity of 10^7–$10^9\ \Omega \cdot \text{cm}$ [3]. With such a high resistivity, we call this material a semi-insulator. Due to the high resistivity, pure GaAs cannot be used to make a device. It must be changed to a semiconductor by a process of doping. In fact, silicon also needs to be doped to make devices. We will discuss doping process in Chapter 17.

So far, by using the resistivity, we have distinguished between conductors, semiconductors, semi-insulators, and insulators. Generally, semi-insulating materials represented by GaAs need to be converted into semiconductors before they can be used to make devices. So, in the following discussion, we will classify semi-insulating materials as semiconductors. It is too simple to distinguish materials from resistivity. To really understand them, especially semiconductors, we have to use quantum mechanics and energy band theory. It is necessary for us to give a brief introduction of quantum mechanics and energy band theory.

References

1 饭田修一等, (1979). 物理学常用数表, [日]. 科学出版社, 133–135.
2 Fink, D.G. and Beaty, H.W. (1987). *Standard Handbook for Electrical Engineers*, 12e, 4–153. McGraw-Hill Companies.
3 Soares, R., Graffeuil, J., and Obrégon, J. (1983). *Applications of GaAs MESFETs*, 17. Artech House.

2

Brief Introduction of Theories

This chapter is a brief introduction of quantum mechanics, and then energy band theory. By using these theories, we can easily understand what is a conductor, an insulator, and a semiconductor.

2.1 The Birth of Quantum Mechanics

At the end of the nineteenth century and the beginning of the twentieth century, Newtonian mechanics, Maxwell's theory of the electromagnetic field, and Maxwell–Boltzmann statistics constituted what is now called classical physics that ruled the physical world at that time. The physical quantities discussed in classical physics have two characteristics: continuity and controllability. However, there were two problems that could not be solved by using the theory of classical physics. One was blackbody radiation, and the other was Michelson–Morley experiment. In 1900, Max Planck (April 23, 1858–October 4, 1947), a German theoretical physicist proposed that in the radiation and absorption of electromagnetic field, the energy appeared in a discrete rather than continuous form. This discrete energy is called the quantization of energy. This assumption explains blackbody radiation very well and is considered as the beginning of quantum mechanics. In 1905, Albert Einstein (March 14, 1879–April 18, 1955) published the theory of special relativity to explain Michelson–Morley experiment. Since then, physics has entered the post-Newtonian era of modern physics.

According to Planck's assumption, each part of energy is proportional to the frequency of electromagnetic radiation. We use E to represent the energy and ν to represent the frequency. The Plank's equation is

$$E = h\nu \tag{2.1}$$

The "h" here is called Planck's constant. The frequency is the number of occurrences of a repeating event per unit of time. In most cases, frequency is represented

Semiconductor Microchips and Fabrication: A Practical Guide to Theory and Manufacturing, First Edition. Yaguang Lian.

by the letter "f." However, in quantum mechanics it is represented by "ν." "T" is used to represent the period. It is the duration of time of one cycle in a repeating event. The relationship between f and T is as follows:

$$f = \frac{1}{T} \tag{2.2}$$

If we use seconds to express time, then unit of frequency is Hertz (Hz), named after German physicist Heinrich Hertz (February 22, 1857–January 1, 1894). He used the experiment to confirm the existence of electromagnetic waves. Electromagnetic waves were theoretically predicted by James Clerk Maxwell (June 13, 1831–November 5, 1879). The theory is well-known as Maxwell's equations. The experiment also proved that the light is electromagnetic waves that were predicted by Maxwell.

Planck's Eq. (2.1) plays a very important role in physics. It is one of key differences between classic physics and modern physics. In classic physics, the energy is supposed to exist in the continuous form and is valid at a large (macroscopic) scale. The Planck equation points out that at small (microscopic) scale, energy exists in the discrete (quantum) form, which is one of the basic characteristics of quantum mechanics. Therefore, when dealing with microscopic worlds such as atoms and subatomic particles, we must use quantum mechanics.

In 1905, Einstein published four papers – photoelectric effect, Brownian motion, special theory of relativity, and mass–energy equivalence. These four articles contributed substantially to the foundation of modern physics and changed people's views from beginning of history on space, time, mass, and energy. So this year is also called the "miracle year" of physics. In the paper of mass–energy equivalence, Einstein wrote down a well-known equation:

$$E = mc^2 \tag{2.3}$$

In this equation, E is energy, m is mass, c is the speed of light, $c = 300\,000$ km/s.

Now, let us talk about photoelectric effect. The effect means that when light beams shine on the surface of an object (mostly metal), and if the light frequency is higher than a certain number, the electrons on the surface will be excited and escape from the object. This phenomenon was first discovered by Hertz. The escaped electrons are called photoelectrons. In the paper of photoelectric effect, Einstein assumed that light travel through space, not in the form of waves as described in the classical theory of electromagnetic field, but in discrete "wave packets." A wave packet is called a "photon." A photon obeys Planck's equation and has energy of $h\nu$. When the frequency (energy) of photons that are illuminating the object reaches or exceeds a certain threshold frequency, electrons will be emitted out of the surface of the object (see Figure 2.1).

In 1913, Niels Bohr (October 7, 1885–November 18, 1962), a Danish physicist, and Ernest Rutherford (August 30, 1871–October 19, 1937), a British physicist born

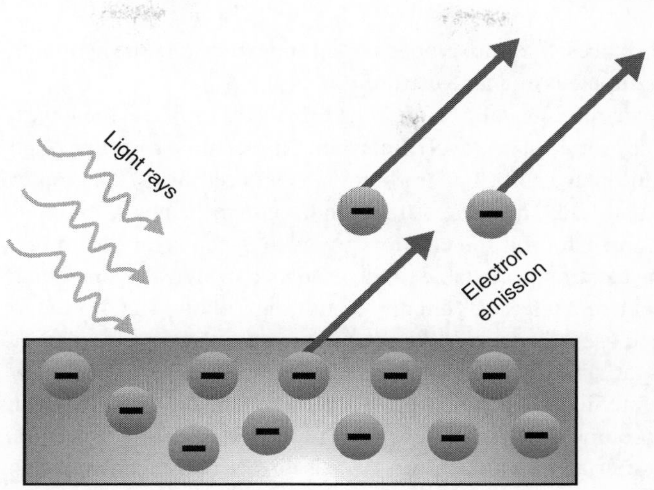

Figure 2.1 The schematic diagram of photoelectric effect. Source: Reproduced with permission of Science ABC. What is the Photoelectric Effect? ≫ Science ABC.

in New Zealand, together proposed a model to describe atoms. This model states that an atom contains small, high-density nucleus, surrounded by electrons. This is like the structure of the solar system, except that this attraction comes from electromagnetic force rather than gravity. We call this model Rutherford–Bohr model, or simply Bohr model. Figure 2.2 is Bohr model of a hydrogen atom. In this figure, the nucleus is in the center, which is composed of a neutron and a proton. An electron rotates in outer orbits. Neutron is uncharged and proton is positively charged. Since the number of protons in an atom is the same as the number of electrons,

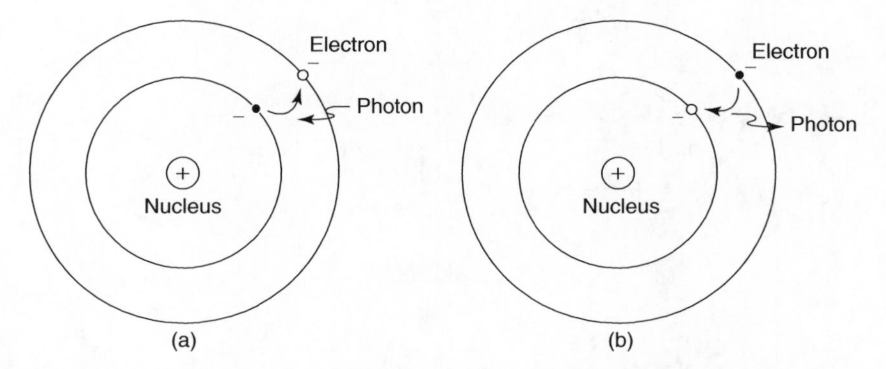

Figure 2.2 Bohr model of a hydrogen atom. (a) A photon is absorbed and an electron jumps from an inner orbit to an outer orbit. (b) An electron jumps from an outer orbit to an inner orbit and a photon is emitted. Source: [1] Bertolotti / Taylor & Francis.

under normal circumstances, the atom is not charged and is neutral. The hydrogen atom is composed of a nucleus and an electron.

In this model, the electron is usually on an inner orbit. This orbit has the lowest energy and is called the ground state (level). But when the electron absorbs enough energy, it jumps to an outer orbit (higher energy), as showed in (a). The outer orbit is called an excited state. The excited state can have many states with different energy orbits, and which orbit the electron jumps to depends on how much energy it absorbs. The electron is unstable in the excited states. It will jump back to the lower energy orbit and release its energy by emitting a photon, as showed in (b). Sometimes, we can use $\Delta E = h\nu$ to represent the emitted photon. "Δ" usually means difference in mathematics. Figure 2.3 is a schematic diagram of the energy levels of a silicon atom. Silicon is composed of 1 nucleus and 14 electrons. The nucleus contains 14 protons with positive charge. The "shell" in the figure means that the electrons move so fast that an electron cloud is formed around the nucleus, just like a shell. Valence electrons are the electrons located in outermost orbit. Ionized level refers to a state where an electron has absorbed enough energy to get rid of the bondage of the nucleus and become a free electron. In this case, the neutrality of an atom is broken. The remaining atom becomes positive charge. Such atoms are called positively charged ions (positive ions). The electrons in photoelectric effect are ionized electrons. As mentioned above, electrons tend to occupy the low-energy levels, which are the shell 1 and shell 2 as shown in the figure. They are the inner shells. The electrons fully occupy the states in these two shells. The

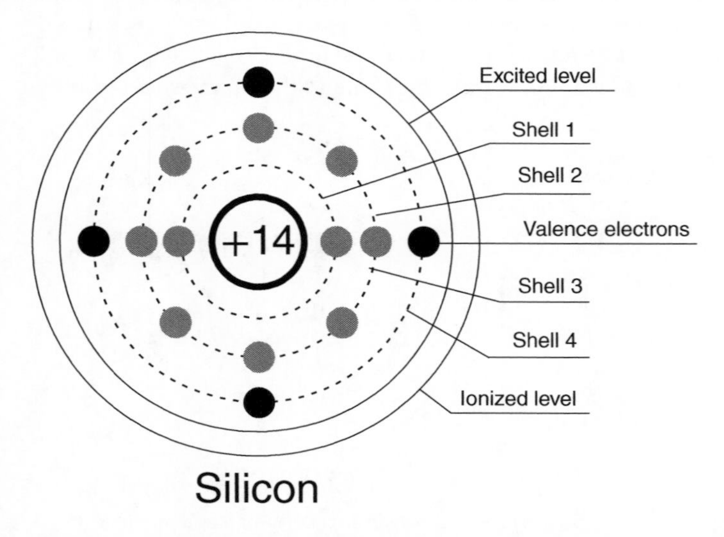

Figure 2.3 Bohr model of a silicon atom. Source: Reprinted from "HyperPhysics" of Georgia State University.

electrons at these two levels are stable. In shell 3, the number of electrons is less than the number of states in this level, so electrons cannot completely occupy the states. The electrons in this level are valence electrons that determine the chemical properties of substance. They participate in chemical reaction and are easily excited to higher-energy states. Silicon has four valence electrons, and germanium also has four.

Based on the works of Planck, Einstein, and Bohr, Erwin Schrödinger (August 12, 1887–January 4, 1961), an Austrian physicist, published Schrödinger equation in 1926. So far, quantum mechanics had been initially established.

2.2 Energy Band (Band)

The photoelectric effect and Bohr model imply an important characteristic of electrons, that is, they only occupy some special energy levels. In the case of a single atom, these energy levels are discrete. But in a crystal material, such as silicon, discrete energy levels become energy bands. Matter usually has three states, solid, liquid, and gas. The branch of physics that is the study of solids is called solid-state physics. If a solid has a periodic and repeating structure, it is a crystalline solid (single crystal material). The semiconductors we are using to make microchips are mainly crystalline solids. Silicon has a single-atom structure. In silicon crystals, atoms are arranged periodically and orderly. Figure 2.4 is a schematic diagram of silicon crystal structure. The small balls in the figure represent silicon atoms, and X–Y–Z is the coordinate system. Because the atoms are very close in the crystal, the valence electrons in an atom appear to be shared by other atoms. Therefore, the discrete energy levels of electrons in a single atom become energy

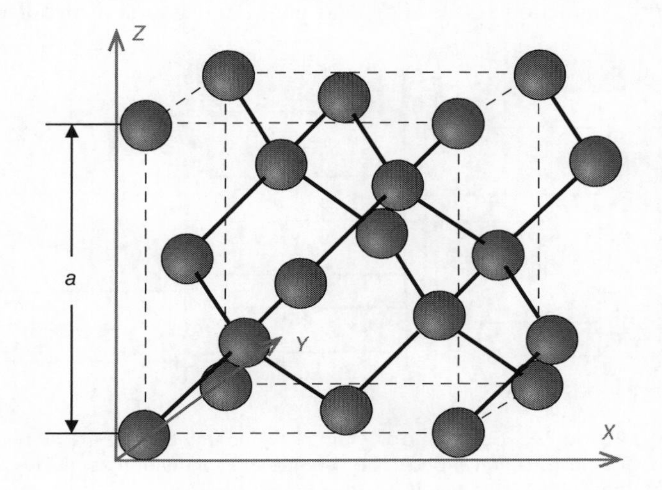

Figure 2.4 Schematic diagram of a silicon crystal.

bands in a crystal material. Using this view and extending quantum mechanics to solid-state physics, one of the results obtained is energy band theory. By solving the Schrödinger equation, the band structures of different crystalline solids can be obtained.

Different crystal materials have different energy band structures, but they all have one thing in common. Some energy bands allow electrons to occupy, and some energy bands prohibit electrons to occupy. Energy bands that prohibit electronic occupation are called forbidden bands. Energy bands that allow electronic occupation are divided into two categories: valance band (full band) and conduction band (empty band). According to the theory of solid-state physics, adjacent atoms share valence electrons (Figure 2.3). A pair of valence electrons form a bond. This is a covalent bond. As mentioned above, in a solid, the discrete energy levels become energy bands. The energy band is composed of massive energy levels with subtle differences. For valence electrons, this change is caused by the energy level splitting of the valence electrons. In the structure of energy band, the band generated by the energy level splitting of the valence electrons is called the valence band. If the valence band is filled with electrons, this band is called full band. Similarly, an excited state energy level will split to form an excited state energy band. If there are no electrons in this excitation band, the band is called empty band. Under certain conditions, some valence electrons will be excited to transit into this band, and these electrons will generate electric current. At this time, the band is called conduction band. In most cases, we do not mention full band and empty band separately but classify them as valence band and conduction band.

Band theory and band structure clearly show the difference between conductors, insulators, and semiconductors. Figure 2.5 is a schematic diagram of the energy band structure of these materials. In this figure, we can see valence band, conduction band, and band gap. Band gap is also called the forbidden band.

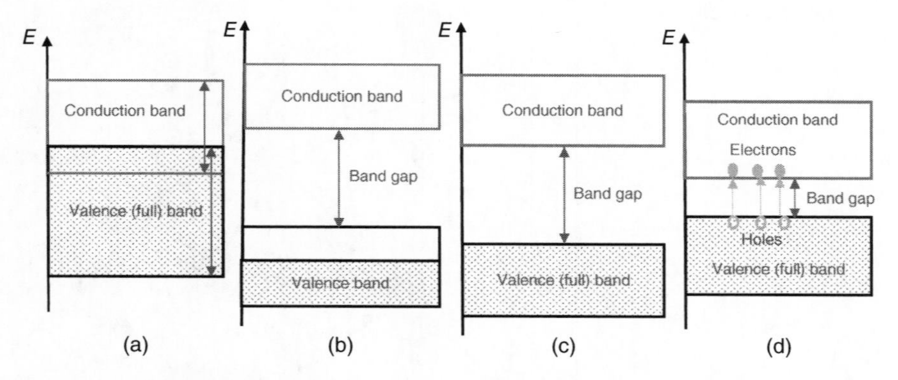

Figure 2.5 Schematic diagram of the energy band structure of a solid [2]. Source: Chinese Technical Books. (a) Valence and conduction bands overlap (b) Valence band is not full (c) Band gap is big (d) Band gap is small.

The figure shows that electrons are in the conduction band and holes are in valence band. In Figure 2.5a, the top of valence band overlaps with the bottom of conduction band. There is no forbidden band. Many electrons automatically reach the conduction band and participate in conduction. This is a type of conductor. Calcium is an example of such conductor. In Figure 2.5b, although there is a large forbidden band, the valence band is not full, and electrons can easily flow in the valence band to participate in conduction. This is another type of conductor, and copper is one example of this kind of conductor. So according to energy band theory, there are two types of metals. Figure 2.5c is a schematic diagram of the energy band of an insulator. This structure has a very large forbidden band. Under normal circumstances, electrons cannot reach the conduction band. The conduction band is basically an empty band, while the valence band is full. No electronic flow, no electric current is generated. Figure 2.5d is a schematic diagram of the semiconductor energy band. This energy band has a forbidden band, but the forbidden band is very narrow. At room temperature, some of the electrons in the valence band will transit from the valence band to the conduction band through thermal excitation. The transition of electrons will leave some vacancies in the valence band, which are holes mentioned above. A hole is positive charge. The electrons entering the conduction band will participate in conduction, and the holes remaining in the valence band will also participate in conduction.

From the energy band theory, we can clearly see the difference between conductors, insulators, and semiconductors, as well as the difference of electric conduction between conductors and semiconductors. In a conductor, only electrons contribute electric current. In a semiconductor, electrons and holes contribute electric current. Although semiconductors have two types of charged particles (positive and negative) that participate in conduction at the same time, since the total charged particle concentration is less than the concentration of electrons in the conductor, the conductivity of the semiconductor is smaller than that of the conductor. In other words, the resistivity of the semiconductor is bigger than that of the conductor.

We know that electric current is generated by the movement of electric charges driven by voltage. In a semiconductor, the movement of electrons and holes create the current. However, their movement velocities are different. We use mobility to further describe this velocity. Mobility is defined as how fast charge carriers like electrons move in a semiconductor driven by voltage. μ is used to express mobility. We will discuss it more in Chapter 5.

In Chapter 1, we said that atoms would hinder the movement of electrons and generate resistance. We also said before that if the structure of a solid is periodically and orderly, this solid is a crystalline material. A solid is composed of atoms. In order to visualize the arrangement of atoms in a crystalline material, an atom can be simplified into one point, and these points can be connected by imaginary lines to form a spatial framework with obvious regularity. Figure 2.4 illustrates the schematic diagram of silicon crystalline structure. This kind of spatial framework representing the regular arrangement of atoms in a crystalline material is called a lattice. Figure 2.4 shows the smallest structure of crystalline silicon (c-Si). This smallest structure is called unit cell. c-Si is composed of this repeating unit cell. The dimension of the unit cell is lattice constant labeled "a" in the figure. In a conductor, as the temperature increases, the lattice vibration intensifies, the obstacle to the movement of electrons increases, and the resistance increases. In semiconductors, as the temperature increases, more electrons enter the conduction band, which increases conductance and decreases resistance. This is in addition to the difference in resistance magnitude, another difference between conductor and semiconductor resistance.

Although Stephen Gray (December 1666–February 1736), a British scientist, discovered conductors and insulators in 1729 and another British scientist Michael Faraday (September 22, 1791–August 25, 1867) discovered semiconductor in 1833. The real understanding of the mechanism and difference of these materials was after the appearance of band theory.

In quantum mechanics, the unit of energy E is electron volt, and the symbol is eV. In most cases, the unit of energy is Joule (symbol is J), which is a commemoration of British physicist James Prescott Joule (1818–1889). $1\,\text{eV} = 1.6 \times 10^{-19}$ J. In semiconductor manufacturing, SiO_2 and Si_3N_4 are the two most used dielectrics (insulators). At room temperature, the band gap of SiO_2 is 9 eV, and that of Si_3N_4 is 5 eV. The band gap for Ge is 0.66 eV, for Si is 1.12 eV, and for GaAs is 1.42 eV [3].

In 1947, three American scientists, John Bardeen (May 23, 1908–January 30, 1991), Walter Brattain (February 10, 1902–October 13, 1987), and William Shockley (February 13, 1910–August 12, 1989) used Ge to invent the world's first semiconductor transistor. The transistor is the cornerstone of modern electronics. The invention led the world to enter the era of the IT – Information Technology. Figure 2.6 shows the photos of the transistors we are using and the first transistor.

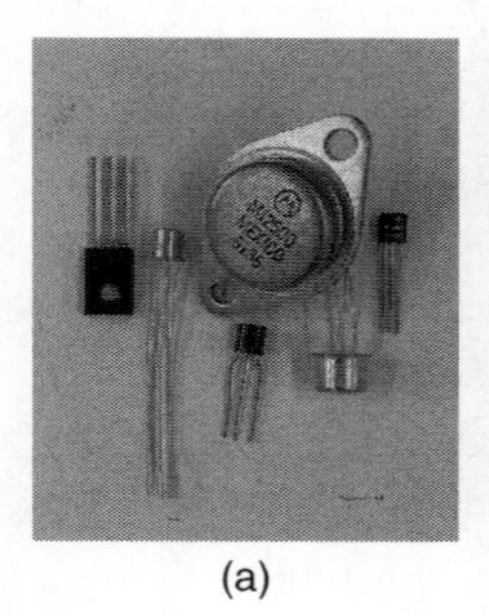

(a)

(b)

Figure 2.6 Transistors (a), the first transistor and John Bardeen (b). Source: Alchetron; Lucent Technologies Inc.

Why did the scientists invent the transistor? To know the reason, we need to have a preliminary understanding of early radio communication technology.

References

1 Bertolotti, M. (2004). *The History of the Laser*, 1e, 72. CRC Press.
2 童诗白主编. (1980). 模拟电子技术基础, 上册. 人民教育出版社, 4页。.
3 Sze, S.M. (1985). *Physics of Semiconductor Devices*, 2e, 850–852. Wiley.

3

Early Radio Communication

This chapter briefly introduces the early radio communication technology represented by wireless telegraphy and the invention of electron tube.

3.1 Telegraph Technology

In 1864, Maxwell solved the equations named after him – Maxwell's equations. He found that electromagnetic waves could propagate at the speed of light in free space. One function of electromagnetic waves is to produce electric sparks in the manner of "action at a distance." The physical meaning of action at a distance is that one object can affect another object through nonphysical contact (just like mechanical contact), changing its motion or other characteristics. Beginning in 1887, Hertz began a series of experiments. He not only experimentally confirmed the existence of electromagnetic waves but also verified that electromagnetic waves propagate at the speed of light. He made oscillators (discussion later) by using capacitor and inductor (discussion later) to generate and receive electric sparks. One was used as a transmitter, and another was used as a receiver. This experiment generated and received what we now call radio waves (see Figure 3.1). The transmitter is made of a pair of copper wires with a small gap between them and two hollow zinc spheres located at the ends. The spheres act as capacitors. A battery connects to an induction coil to power the transmitter and create electric sparks in the gap. The sparks cause the current pulses in the wires. The pulses reach the zinc spheres and generate electromagnetic radiation (waves)-radio waves. The transmitter is used as an "antenna." After careful adjustment, the oscillation frequency of the receiver is the same as that of the radio waves. This phenomenon is called resonance. So the waves can create big current pulses and the sparks jump across the small gap in the receiver. With this apparatus, Hertz produced and detected radio waves for the first time in the world.

Semiconductor Microchips and Fabrication: A Practical Guide to Theory and Manufacturing, First Edition. Yaguang Lian.

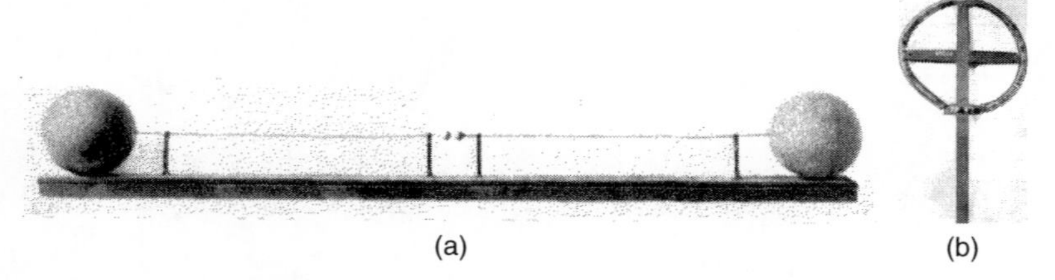

Figure 3.1 Hertz's first radio transmitter (a) and one of Hertz's radio receiver (b). Source: [1], Rollo Appleyard (1927) / ITT Inc.

In Hertz's experiment, the transmitter and receiver did not contact each other and did not have interconnect wires between them. After Hertz demonstrated the method of generating and detecting radio waves, inspired by this experiment, an Italian inventor Guglielmo Giovanni Maria Marconi (April 25, 1874–July 20, 1937) started research on wireless telegraphy from the early 1890s. At that time, there was already the technology of electric telegraph (wired telegraphy). By improving the design of the antenna, using a new device (coherer) to replace the gap of the receiver, in December 1894, Marconi developed a wireless telegraphy (radio telegraphy) system based on radio waves, as shown in Figure 3.2. The transmitter includes the induction coil, the antenna in the form of a copper sheet, a spark gap, and a telegraph key. The era of radio communication began with wireless telegraphy.

Figure 3.2 Marconi's first radio transmitter (telegraph) [2].

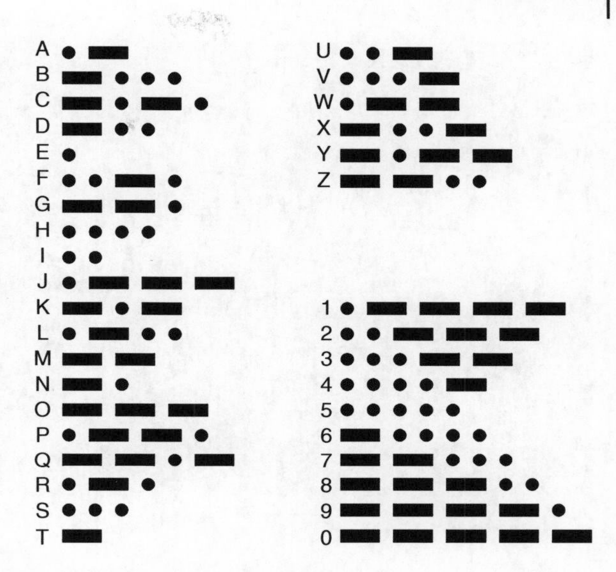

Figure 3.3 International Morse Code.

To transmit the messages in the wires of electric telegraph, from 1836 to 1844, Samuel Finley Breese Morse (April 27, 1791–April 2, 1872), an American inventor and others jointly developed Morse code, as shown in Figure 3.3. Morse code arranges two signals of different lengths into a standard sequence to compile and decipher the text used. These two signals of different lengths are called "di" and "da" (dits and dahs), and they can be used in the telegram. When Morse code was accepted by the radio communication technology, "di" and "da" were transmitted and received as pulse signals of different lengths. Telegram using Morse code had been widely used all over the world until the invention of the Internet.

3.2 Electron Tube

In the early times, it was difficult to use Marconi's telegraph for long-distance communication such as transoceanic communication. An important reason was that the original technology did not have the function of amplifying the transmission signal. In 1904 and 1906, John Ambrose Fleming (November 29, 1849–April 18, 1945), a British inventor, and Lee de Forest (August 26, 1873–June 30, 1961), an American inventor, created the electron diode and the electron triode. As they are made by putting electrodes in a vacuum glass tube, they are also called vacuum diode and vacuum triode. Mostly, we name this kind device as electron tube or vacuum tube. The invention of electron tube started a new field of electronics.

The vacuum tube has an amplifying function, which can amplify the power of radio signals to realize transoceanic radio communication (see Figure 3.4).

Figure 3.4 Vacuum tubes.

Vacuum tubes are a very important invention. Replacing the primitive spark-gap technology with vacuum tubes, we not only achieved the long-distance telegram but also achieved transoceanic telephone. Vacuum tubes also brought radio and television into the home. Another breakthrough was the realization of digital computing. The computers had been made with vacuum tubes. These computers

Figure 3.5 First vacuum tube made general-purpose digital computer (ENIAC). Source: Everett Collection Historical / Alamy Stock Photo.

Figure 3.6 Jack Kilby and the invention of integrated circuit.

are called vacuum tube computers. This kind of computer is the first generation of computers (see Figure 3.5). It is Electronic Numerical Integrator and Computer (ENIAC), the first programmable, electronic, general-purpose digital computer.

Vacuum tubes have two major problems: large size and high-power consumption. Since the 1920s, many scientists had tried to replace vacuum tubes with a new type of device. They wanted to develop this new type of device as a solid device. After the invention of the semiconductor transistor in 1947, in 1958, Jack St. Clair Kilby (November 8, 1923–June 20, 2005) and Robert Noyce (December 12, 1927–June 3, 1990) invented the integrated circuit, referred to as IC (see the Figure 3.6). Packaged integrated circuit products are what we often call "microchips" or "chips" now. Another well-known role of Robert Noyce was as the cofounder of the two companies – Fairchild Semiconductor in 1957 and Intel Corporation in 1968.

Since most of the electronic devices used today are made of silicon semiconductors, which is a solid, modern electronics is also called solid-state electronics. The use of solid-state devices instead of vacuum tubes has achieved two breakthroughs – small size and low-power consumption. These two breakthroughs have realized the miniaturization and household using of electronic products. PC – personal computers and mobile phones have become our daily ordinary consumer goods. To know about electronics and semiconductor devices, we need to understand the electric circuits and the components used in the circuits.

References

1 Appleyard, R. (1927). Pioneers of electrical communication-Heinrich Rudolph Hertz-V. *International Standard Electric Corporation* 63–77.
2 Marconi, G. (1926). Looking back over thirty years of radio. *Radio Broadcast Magazine*, Doubleday, Page, and Co., New York, Vol. 10, No. 1, (November 1926), p. 31.

4

Basic Knowledge of Electric Circuits (Circuits)

This chapter will give a brief introduction of electric circuits and the components used in the circuits, electric, and magnetic fields.

4.1 Electric Circuits and the Components

An electric circuit (circuit) usually includes a power supply (also called a supply or a power source), such as a battery, wall power, etc.; a load, such as a mobile phone, an electric light, etc.; electrical wires that are used to connect the power supply and load; a switch used to turn on or off the connection between the power supply and load. Figure 4.1 is a typical circuit diagram. In a circuit, the power supply can be direct current – DC – or alternating current – AC. The frequency of DC power source is zero, and the battery is DC supply. The frequency of AC power source is not zero, and the wall power is AC supply. If the power supply is a DC one, the electrical circuit is a DC circuit. If the power supply is an AC one, the electrical circuit is an AC circuit. Figure 4.2 are the symbols of DC and AC supplies. The long line of the DC supply is the anode "+," which can be simply considered as high-voltage side; and the short line is the cathode "−," which can be simply considered as low-voltage side. This voltage difference drives the current to flow in the circuit and power the load. At the beginning, people thought that the current was generated by the flow of positive charges, so the current flowed from anode to cathode. Later, scientists discovered that in most cases, the current was generated by the flow of electrons, and the direction of the electron flow was from cathode to anode. But from the historical perspective, we still say that the electrical current flows from the anode to the cathode. The anode and cathode in a DC supply do not alternate, so the current flows in only one direction. The anode and cathode in an AC power alternate periodically over time, the current periodically reverse flow direction, so the anode and cathode are not marked in the AC power. Figure 4.3 is the schematic diagram of the difference

Semiconductor Microchips and Fabrication: A Practical Guide to Theory and Manufacturing, First Edition. Yaguang Lian.

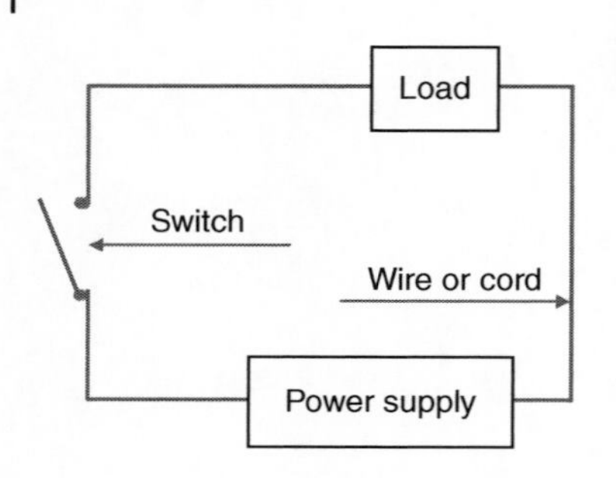

Figure 4.1 A simple circuit schematic.

Figure 4.2 DC power supply (a) and AC power supply (b).

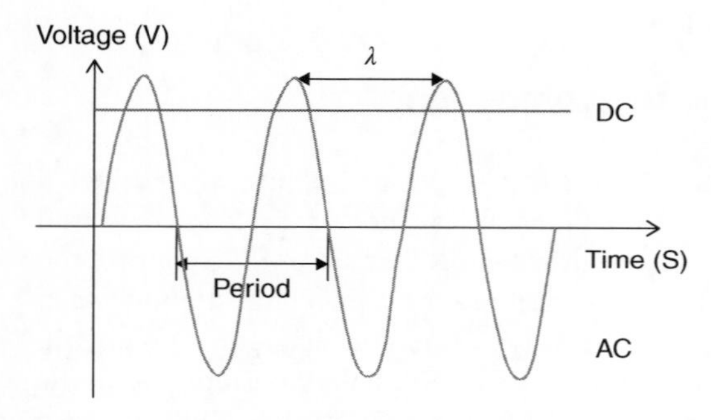

Figure 4.3 The difference between DC and AC power supplies. Source: Reprinted with permission of Electrical Academia.

between DC and AC supplies. AC is transmitted in the form of wave. The two parameters of the wave are also marked in the figure, period (T – Eq. (2.2)) and wavelength (λ). In a DC circuit, we consider the resistors (see Figure 1.4) only. In an AC circuit, in addition to resistors, two other components also need to be taken into consideration. They are capacitors and inductors. Resistors, capacitors, and inductors are the most basic components that make up a circuit. They are passive components. This corresponds to active components, which we will discuss later. Figure 4.4 is the picture of capacitors and the symbol, and Figure 4.5 is the picture of inductors and the symbol. As shown in Eq. (1.1), we use "R" to represent resistance and resistor, while capacitance and capacitor are represented by "C," and inductance and inductor are represented by "L." Resistors are power consumption devices, capacitors, and inductors are power storage devices. Capacitors store electric field energy and inductors store magnetic field energy.

The electric energy we often talk about is expressed in the form of an electric field, so when a capacitor stores electric field energy, it stores electric energy.

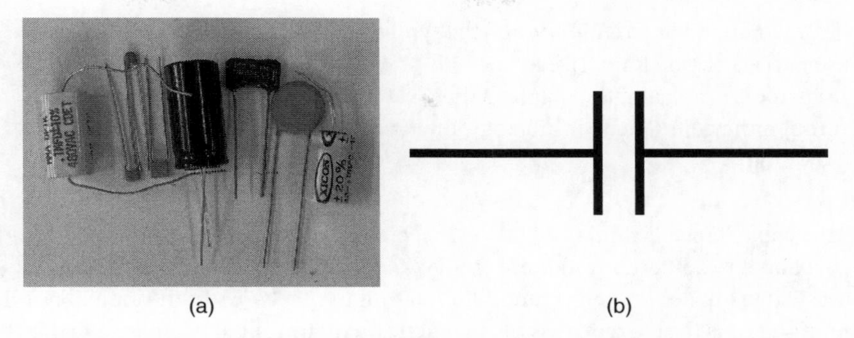

Figure 4.4 The picture of capacitors (a) and the symbol (b).

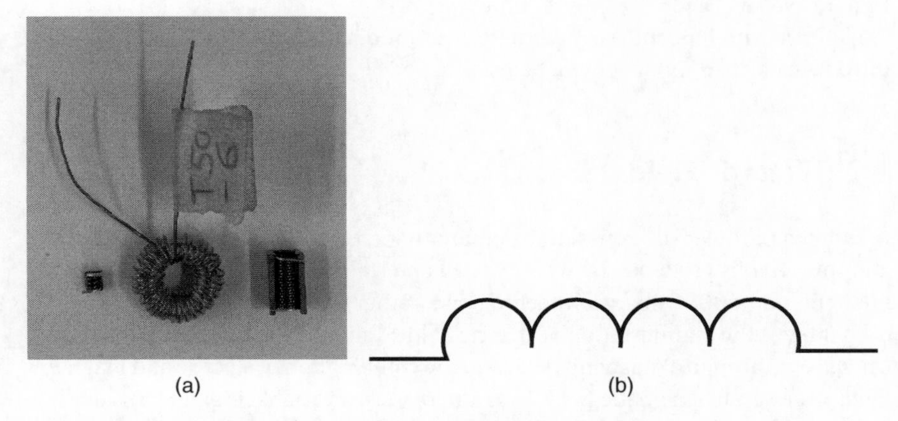

Figure 4.5 The picture of inductors (a) and the symbol (b).

Magnetic energy is manifested in the form of a magnetic field, so when an inductor stores magnetic field energy, it stores magnetic energy. The unit of capacitance is farad, and the symbol is F to commemorate Faraday (Chapter 2). His greatest contribution is the discovery of the law of electromagnetic induction (Faraday's law of induction) – the change in magnetic flux can generate electromotive force. This law can be simply demonstrated as follows: when a closed wire moves along the vertical direction of the magnetic field lines (cutting the magnetic field lines), the electric current will be created in the closed wire. This is how modern power generators work. The structure of a capacitor is two conductive plates are placed close to each other with a nonconductive dielectric in between. The definition of capacitance is:

$$C = \frac{Q}{V} \tag{4.1}$$

The unit of inductance is Henry, and the symbol is H to commemorate the American scientist Joseph Henry (December 17, 1797–May 13, 1878). The structure of the inductor is a coil, and the magnetic field can be generated by the current passing through the coil. The definition of inductance is:

$$L = \frac{\Phi}{I} \tag{4.2}$$

In the capacitance formula (4.1), V is the voltage, Q is the electric charge. The unit of charge is Coulomb, and the symbol is C, to commemorate French scientist Charles-Augustin de Coulomb (June 14, 1736–August 23, 1806). In the inductance formula (4.2), I is the current, Φ is the magnetic flux. The unit of the magnetic flux is Weber, and the symbol is "Wb," to commemorate the German physicist Wilhelm Eduard Weber (October 24, 1804–June 23, 1891).

To further understand how capacitors and inductors work, we need to briefly introduce electric and magnetic fields.

4.2 Electric Field

A charged particle will exert a force on another charged particle without contact, and this force is produced by what we call an electric field. We often say that the phenomenon of like charges repelling each other and unlike charges attracting each other is the manifestation of electric field. Similarly, the force of magnetism is carried out through a magnetic field, and the combination of electric and magnetic fields is an electromagnetic field. In mathematics, a variable is called a scalar if it has only a quantity, without a change in direction. The mass is a scalar. If this variable not only has a magnitude but also a change in direction, then it is called a vector. Since the electric field has both a magnitude and a change in direction, the electric field is a vector field. Similarly, the magnetic field is also a vector field.

Coulomb's law describes the interaction of two stationary charged particles. If the charges of the two particles are q_1 and q_2, and the distance between them is d, then the interaction force F between them is:

$$F = \frac{1}{4\pi\varepsilon_o}\frac{q_1 q_2}{d_2} \tag{4.3}$$

The unit of force F in the formula is Newton, and the symbol is N, to commemorate the founder of classical mechanics, British scientist Isaac Newton (1642–1727). The unit of charge is Coulomb (C) and the unit of distance is meter (m). ε_o is the permittivity of vacuum (free space), also called the absolute permittivity, and its unit is farad/m (F/m). The force of formula (4.3) is called Coulomb force or electrostatic force. Charged particles can be positively charged or negatively charged. As we said earlier, like charges repel each other and unlike charges attract each other. Therefore, atomic nuclei can attract electrons because the protons in the nucleus are positively charged, and electrons are negatively charged. The attractive force in an atom can be thought of as electrostatic force.

For convenience, we divide by q_2 in formula 4.3, and the expression will depend only on electric charge. If we remove the number of this electric charge and replace d in the formula with x according to the convention of mathematics, then we can get the expression for electric field $\mathcal{E}$:

$$\mathcal{E} = \frac{1}{4\pi\varepsilon_o} \frac{q}{x^2} \tag{4.4}$$

We now connect a capacitor to a DC power supply, as shown in Figure 4.6. In the figure, "Positive electrode" is the plate of the capacitor that connects with the anode of the supply. "Negative electrode" is the plate of the capacitor that connects with the cathode of the supply. In this circuit, the anode of the power supply drives the positive charges to the positive plate of the capacitor, and the cathode of the power supply drives the negative charges to the negative plate of the capacitor. The actual situation is that electrons are repelled by the cathode of the power supply to the negative plate of the capacitor, causing the other plate to become positive due to lack of electrons. In this way, an electric field directed from the positive plate of the capacitor to the negative plate is established. The electric field cannot be stabilized until the voltage between the two electrodes of the capacitor is equal to the voltage of the power supply. The relationship between electric field and voltage is:

$$\mathcal{E} = \frac{V}{d} \tag{4.5}$$

In the formula, V is the voltage and d is the distance between the two electrode plates of the capacitor, please see Figure 4.7. Under the DC power supply, no

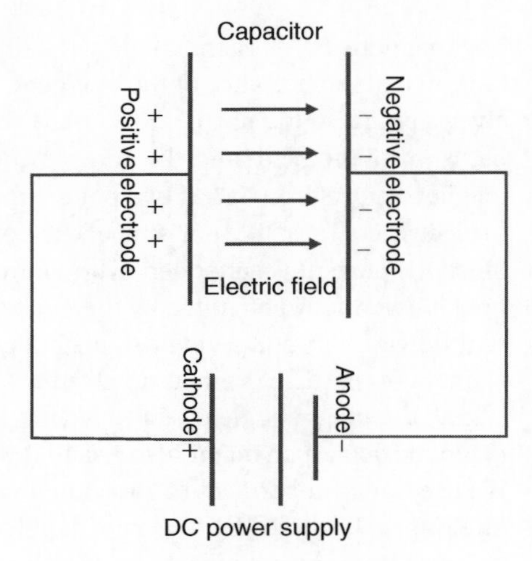

Figure 4.6 A capacitor connects with a DC power supply.

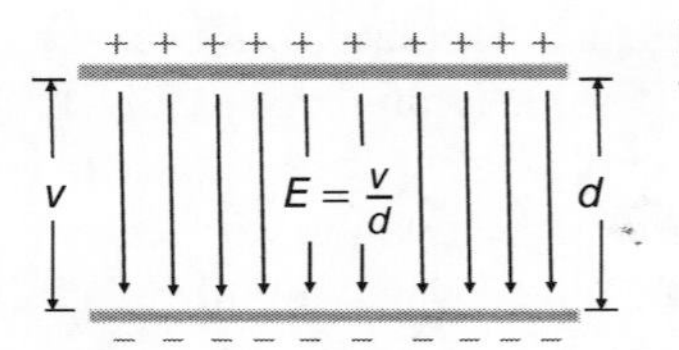

Figure 4.7 The relationship of electric field and voltage.

electric current flows through the capacitor until the voltage of the power supply is very high, so that a breakdown occurs between the two plates, and then there will be the current through the capacitor. This is how lightning occurs in a thunderstorm. We can imagine that when an AC power supply is connected, the polarity of the power supply periodically changes, causing electrons to flow back and forth in the connected wires, as if there is electric current flowing through the capacitor. The higher the frequency of the power supply, the bigger the current flow. Maxwell named this kind of the current as displacement current. The essence of displacement current is the change of electric field, which can exist in conductors, dielectrics, and vacuum. Corresponding to displacement current is conduction current which is what we are commonly talking about. Conduction current can only flow in conductors. As the frequency of the power supply is higher, the conductance of a capacitor becomes stronger. That is, at low frequencies, a capacitor is open (circuit disconnected); at high frequencies, a capacitor is short (circuit connected).

From the above discussion, we can see that a capacitor is a device associated with an electric field, and an electric field energy is stored between two plates.

4.3 Magnetic Field

The magnetic field is generated by the flow of electric current. It is a vector field. We can feel the presence of the magnetic field in our daily life. A compass needle always points to the north and south poles of the earth. This is because the earth has a magnetic field, and the compass needle is made of magnetic materials. A magnet attracts iron. It has two poles – the south pole and the north pole. Similar to electric charges, like poles repel each other and unlike poles attract each other. Since the current is generated by the movement of electric charges, the magnetic field is essentially generated by the movement of electric charges. An electric current flowing through a wire can generate a magnetic field. The spin of a charged elementary particle (such as an electron) can also generate a magnetic field. This is how the magnetic field of a magnet is generated. Since a magnet always has a magnetic field, such materials are called permanent magnets in the academic field. We use fine iron particles to show the magnetic field lines of a magnet, as shown in Figure 4.8, where "N" is the north pole and "S" is the south pole. The magnetic field generated by the current passing through the wire obeys the right-hand rule.

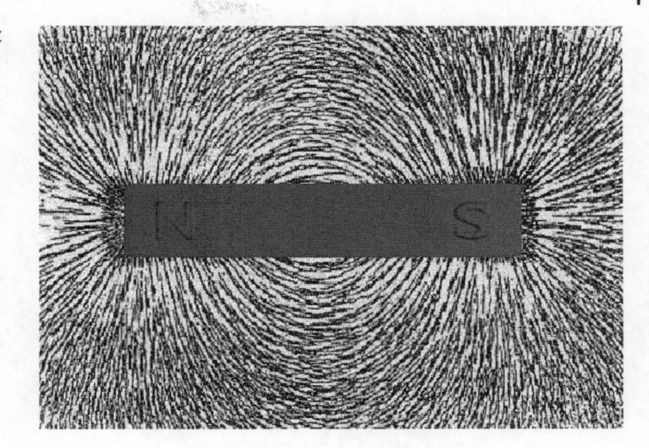

Figure 4.8 The magnetic field lines. Source: Berndt Meyer / Wikimedia Commons.

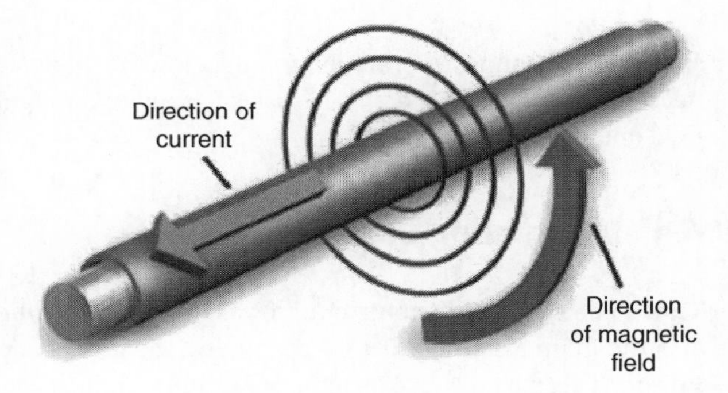

Figure 4.9 The current creates magnetic field. Source: MolecularExpressions.com at Florida State University.

The rule is: when you extend the right hand, the thumb points to the direction of the current, and the rotation direction of the remaining four fingers is the direction of the magnetic field. Please see Figure 4.9, where the current direction is from the anode to the cathode, that is the direction of positive charge flow. The right-hand rule is also used in our daily life. For example, the switch faucet is designed according to the right-hand rule. A conductor coil with current passing through can also generate a magnetic field, see Figure 4.10. The B in Figure 4.10 refers to the magnetic field. The unit of magnetic field strength is Tesla, and the symbol is T to commemorate the American Serbian scientist Nikola Tesla (July 10, 1856–January 7, 1943).

When connected to an AC power supply, the current passing through the coil changes, and the generated magnetic field also changes. This changed magnetic field generates induced current. The magnetic field created by the induced current always hinders the change of the flux of the original magnetic field that causes the induced current. This is Lenz's law, named after Russian physicist Emil Lenz (February 12, 1804–February 10, 1865). As the frequency of the power supply is higher, the hindering becomes stronger. That is, at low frequencies, the inductor is

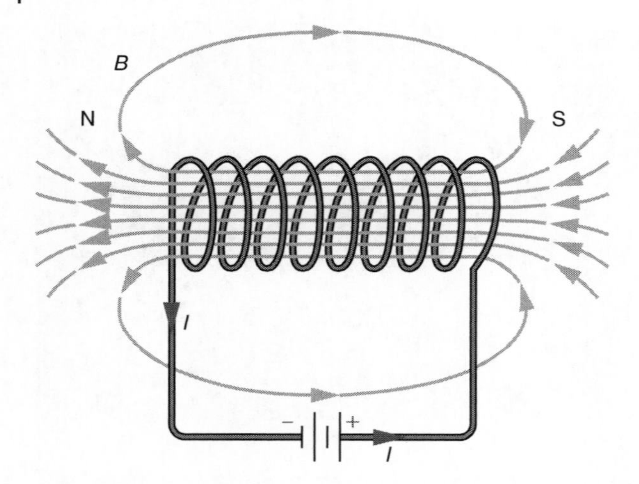

Figure 4.10 The magnetic field generated by a coil. Source: Reprinted with permission of Brilliant.

short (circuit connected); at high frequencies, the inductor is open (circuit disconnected). An inductor is a device associating with a magnetic field, and magnetic field energy is stored in a coil.

4.4 Alternating Current

Tesla's most important contribution was the invention of AC technology. This was also the main argument between him and the great American inventor Thomas Edison (February 11, 1847–October 18, 1931). Edison promoted DC while Tesla advocated AC. To understand this controversy, let us first talk about the power of the current. Power is represented by P. Its definition is energy per unit time. The unit of power is watt, and its symbol is W, to commemorate the British scientist James Watt (January 19, 1736–August 19, 1819). His invention of the steam engine started the industrial revolution. The relationship between power unit and energy unit is:

$$W = \frac{J}{S} \tag{4.6}$$

The "S" in the formula is time (seconds). From the definition of power and Ohm's law, we can get the following formula:

$$P = \frac{E}{t} = IV = I^2R = \frac{\mathrm{V}^2}{R} \tag{4.7}$$

In the formula, P is power, E is energy, t is time, I is current, V is voltage, and R is resistance. It can be seen from the above equation that when the power is constant, the higher the voltage, the smaller the current. A piece of wire with resistance R will consume power and generate heat when the current is passing through. Therefore, if we want to reduce the power consumption of the wire, we must decrease the current. Due to geographical or other constraints, power plants can only be built in certain places. Most power plants produce AC electricity. When a power plant generates electricity, it must be transmitted to other places over long distances through transmission lines made of Al or Cu wires. In the case of a certain power generation, in order to carry out long-distance power transmission, it is necessary to try to reduce the power loss of the transmission lines. It is an effective means to increase the transmission voltage and decrease the transmission current. An important feature of AC is that the voltage can be changed by using a transformer, through which a low voltage can be turned into a high voltage, or a high voltage can be turned into a low voltage. In this way, after a power plant generates electricity, the transformer is used to increase the voltage for long-distance transmission, and then the transformer is used to decrease the voltage before reaching the user. This is the AC high-voltage line transmission technology. The DC did not have this technology at that time. In the battle between DC and AC, Tesla defeated Edison, and AC technology won. Figure 4.11 is a schematic diagram of high-voltage power transmission technology.

The transformer is very important for the long-distance transmission of AC. It is made of conductive coils. Please see the schematic diagram in Figure 4.12 and the actual photo of the transformer.

In an AC circuit, Ohm's law is:

$$I = \frac{V}{Z} \tag{4.8}$$

where Z is the impedance of the AC circuit:

$$Z = R + j(XL - XC) \tag{4.9}$$

In the above formula, R is resistance; XL is inductive reactance, which corresponds to inductance; XC is capacitive reactance, which corresponds to capacitance; j is an imaginary unit:

$$j = \sqrt{-1} \tag{4.10}$$

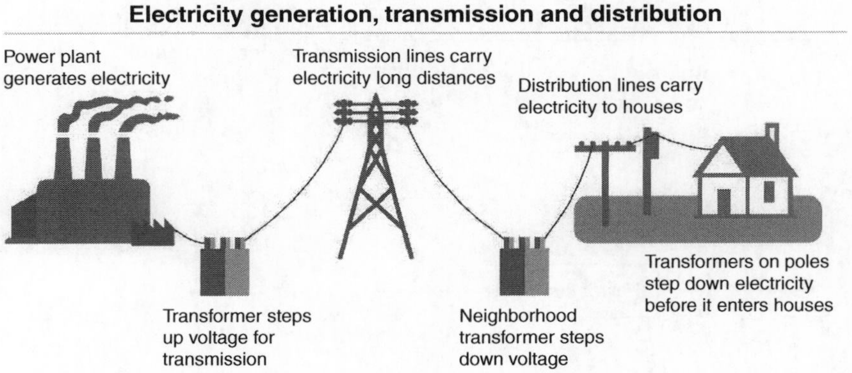

Figure 4.11 The electricity from a power plant to a house. Source: EIA / Public Domain.

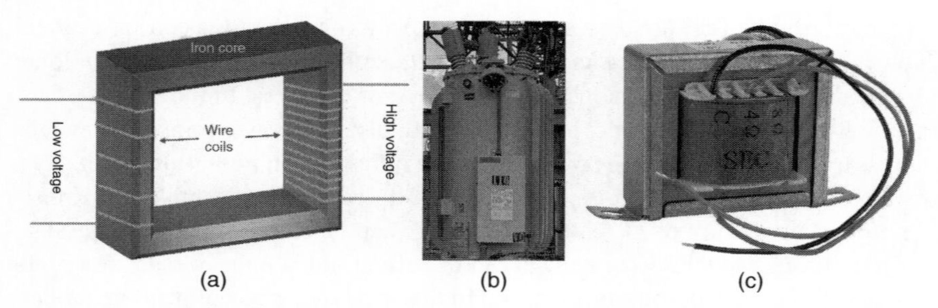

Figure 4.12 Transformers. Source: The image c is reprinted with permission of Parts Express.

The imaginary unit is usually represented by i in mathematics. But in the circuit, i is commonly used to represent the current. So, j is used to represent the imaginary unit in electricity and electronics. In the DC circuit, inductance and capacitance are not considered, and $Z = R$, at this time.

In an electric circuit, besides a power supply, it usually includes a resistor, a capacitor, and an inductor. In fact, most of the circuit loads can be expressed by these three basic components. Please see Figure 4.13. Because the circuit contains an inductor and a capacitor, under the action of AC supply, the circuit will resonate when certain conditions are reached. The resonant frequency is:

$$f = \frac{1}{2\pi\sqrt{LC}} \tag{4.11}$$

In the experiments of Hertz and Marconi, the oscillators they made were based on this formula.

Above, we have discussed some basic knowledges of the electric circuits, the components, and basic concepts of electric and magnetic fields. In Chapter 5, we will discuss semiconductors in depth and introduce the basic structure and one of the preliminary applications of diodes.

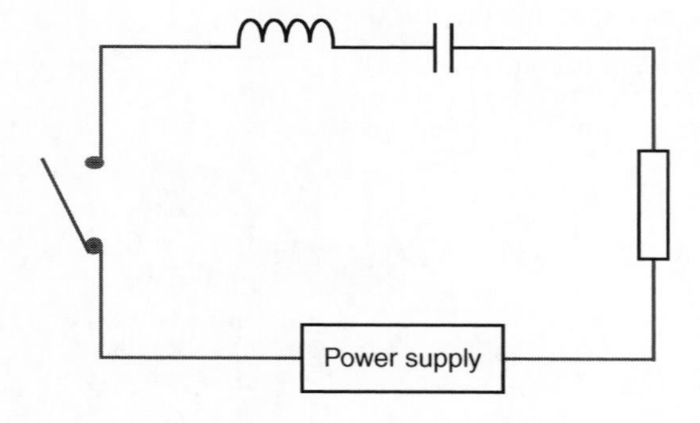

Figure 4.13 A typical structure of an electric circuit.

5

Further Discussion of Semiconductors and Diodes

Below we will discuss the basic structure of actual semiconductor energy bands, doping, and diodes.

5.1 Semiconductor Energy Band

Silicon is the most important material in the semiconductor industry and applications. Gallium arsenide is another widely used semiconductor material, which is mainly used to make optoelectronic devices. Please look at Figure 5.1 of the periodic table of elements. The table was invented by Russian scientist Dmitri Mendeleev (February 8, 1834–February 2, 1907). Elements of silicon and germanium used to make transistor are in group 14. They all have four valence electrons. Arsenic is in group 15 and has five valence electrons; gallium is in group 13 and has three valence electrons. So, GaAs is also known as the three–five (III–V) semiconductor material. Because it is a compound, it is sometimes called a compound semiconductor material.

Now we further discuss the energy band structure of semiconductors. Figure 5.2 is a schematic diagram of the basic structure of semiconductor energy bands. The actual energy band is much more complicated than this. In general, the valence band of a semiconductor is full, and no current will be generated. If the electrons absorb enough energy, this energy may come from photons or heat. Some electrons on the top of the valence band jump to the bottom of the conduction band. These negatively charged electrons are driven by the electric field (voltage). It will flow in the conduction band, thereby generating the current. The positively charged holes on the top of the valence band will also flow in the valence band and contribute to the current. In this case, electrons and holes are also called charge carriers.

We introduced mobility μ in Chapter 2, electric field $\mathcal{E}$ in Chapter 4, and charge carrier here. Let us give a complete description of mobility. In Chapter 4, we knew

Semiconductor Microchips and Fabrication: A Practical Guide to Theory and Manufacturing, First Edition. Yaguang Lian.

	group 1 Ia	2 IIa	3 IIIb	4 IVb	5 Vb	6 VIb	7 VIIb	8	9 VIII	10	11 Ib	12 IIb	13 IIIa	14 IVa	15 Va	16 VIa	17 VIIa	18 0
1s	1 H																	2 He
2s, 2p	3 Li	4 Be											5 B	6 C	7 N	8 O	9 F	10 Ne
3s, 3p	11 Na	12 Mg											13 Al	14 Si	15 P	16 S	17 Cl	18 Ar
4s 3d, 4p	19 K	20 Ca	21 Sc	22 Ti	23 V	24 Cr	25 Mn	26 Fe	27 Co	28 Ni	29 Cu	30 Zn	31 Ga	32 Ge	33 As	34 Se	35 Br	36 Kr
5s 4d, 5p	37 Rb	38 Sr	39 Y	40 Zr	41 Nb	42 Mo	43 Tc	44 Ru	45 Rh	46 Pd	47 Ag	48 Cd	49 In	50 Sn	51 Sb	52 Te	53 I	54 Xe
6s, 4f 5d, 6p	55 Cs	56 Ba	57 La *	72 Hf	73 Ta	74 W	75 Re	76 Os	77 Ir	78 Pt	79 Au	80 Hg	81 Tl	82 Pb	83 Bi	84 Po	85 At	86 Rn
7s, 5f 6d, 7p	87 Fr	88 Ra	89 Ac **	104 Rf	105 Db	106 Sg	107 Bh	108 Hs	109 Mt	110 Ds	111 Rg	112 Cn	113 (Uut)	114 (Uuq)	115 (Uup)	116 (Uuh)	117 (Uus)	118 (Uuo)

* 4f, 5d	58 Ce	59 Pr	60 Nd	61 Pm	62 Sm	63 Eu	64 Gd	65 Tb	66 Dy	67 Ho	68 Er	69 Tm	70 Yb	71 Lu
** 5f, 6d	90 Th	91 Pa	92 U	93 Np	94 Pu	95 Am	96 Cm	97 Bk	98 Cf	99 Es	100 Fm	101 Md	102 No	103 Lr

Figure 5.1 Periodic table of elements.

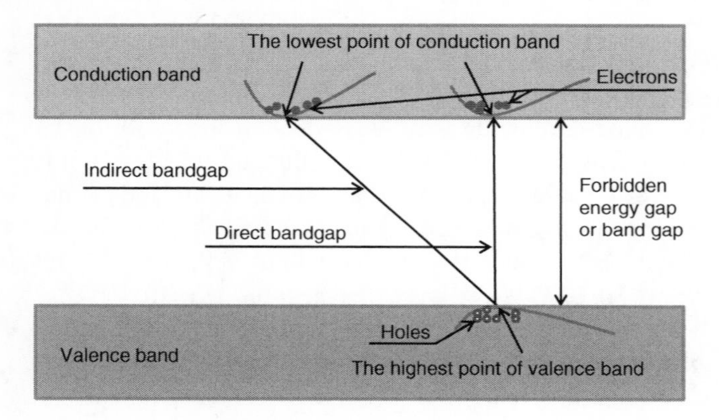

Figure 5.2 Schematic diagram of semiconductor energy band.

that voltage is related to electric field. The movement of charges driven by voltage (refer to Chapter 2) is actually pulled by an electric field. The charge carriers move in a semiconductor when pulled by an electric field. The term carrier mobility refers in general to both electron and hole. The average velocity of the carriers in a semiconductor is called the drift velocity v_{d}. The mobility is defined as follows

$$v_{\mathrm{d}} = \mu \mathcal{E} \tag{5.1}$$

The unit of mobility is cm^2/(V·s). Here, V is voltage and s is second. Through mobility, we establish the relationship between the carrier drift velocity and electric field. Mostly, the electron mobility is greater than hole mobility in semiconductors. For example, the electron mobility in silicon is 1500 $cm^2/(V \cdot s)$ and the hole mobility is 450 $cm^2/(V \cdot s)$ [1]. This means that under the action of an electric field, electrons can get a faster drift velocity than holes.

If electrons jump straight up and down between the valence band and the conduction band, such a semiconductor is a direct band gap semiconductor, otherwise, it is an indirect band gap semiconductor. Please see Figure 5.2. "The highest point of valence band" is the top of the valence band. When light hits the semiconductor surface, the electrons in the direct band gap directly interact with photons, and the photoelectric conversion efficiency is high. While the electrons in the indirect band gap, in addition to interacting with photons, also interact with the crystal lattice of the material, loss of some energy and low photoelectric conversion efficiency. Figure 5.3 is the energy band diagrams of germanium, silicon, and gallium arsenide. It is obvious from the figure that germanium and silicon are indirect band gaps, while gallium arsenide is a direct band gap. Therefore, gallium arsenide is widely used in the manufacture of optoelectronic devices. In addition, this figure shows that the real energy band structure is much more complicated than the band in Figure 5.2. The real energy band also contains many sublevel bands.

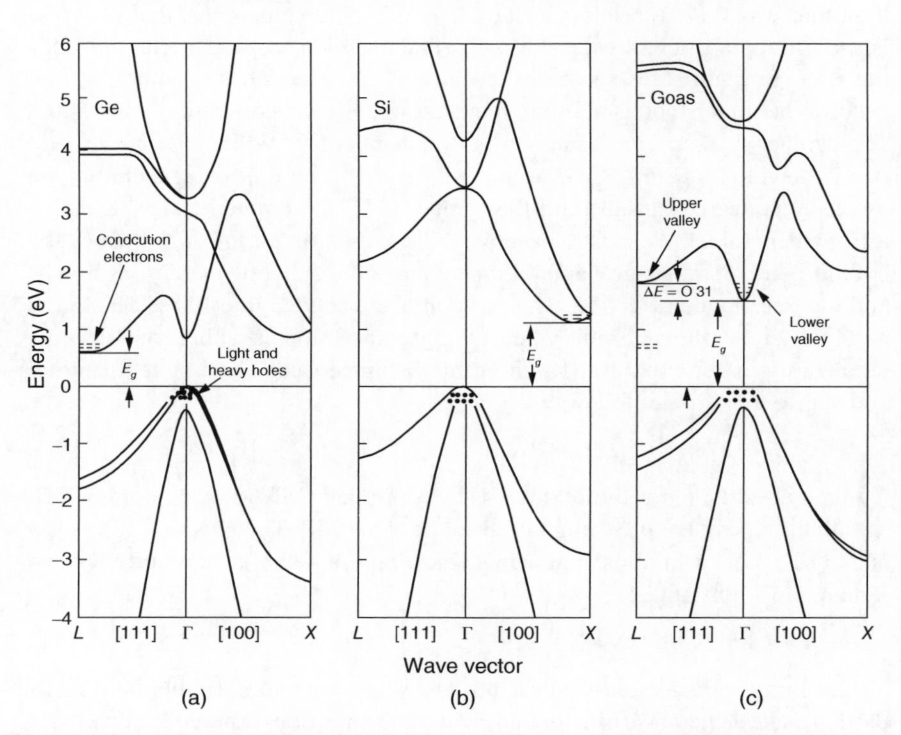

Figure 5.3 The energy bands of Ge, Si, and GaAs. Source: [1] Sze / John Wiley & Sons.

5.2 Semiconductor Doping

In most cases, pure semiconductor materials cannot be used to make devices and must be doped. To understand the principle of doping, it is necessary to introduce the basic concepts and constants of statistical physics. Statistical physics is a branch of physics that uses the method of probability and statistics to study the physical properties and laws of macroscopic objects composed of a large number of microscopic particles. Classical statistical physics was developed by Maxwell and Boltzmann. Maxwell was the physicist who proposed the electromagnetic wave equations. Ludwig Edward Boltzmann (February 20, 1844–September 5, 1906) was an Austrian physicist. Based on Maxwell's work, he proposed the famous Boltzmann equation, which is widely used in thermodynamics, statistical mechanics, and many other fields. Because of his important contribution to classical statistical physics, an important constant in thermodynamics is named after him-Boltzmann's constant, with the symbol k_{B}. After the birth of quantum mechanics, according to the different properties of microscopic particles, two statistical theories appeared. One was Fermi–Dirac statistics and the other was Bose–Einstein statistics. The microscopic particles that obey the Fermi–Dirac statistics are called fermions, and the microscopic particles that obey the Bose–Einstein statistics are called bosons. Another physical quantity that needs to be understood is absolute temperature, also called Kelvin temperature. The symbol is K to commemorate a British scientist, William Thomson, 1st Baron Kelvin (June 26, 1824–December 17, 1907). The unit of temperature we use daily is degree Celsius, and the symbol is °C to commemorate a Swedish astronomer, Anders Celsius (November 27, 1701–April 25, 1744). It defines the boiling point of water at 1 atm (the standard atmosphere) pressure as 100 °C and the freezing point as 0 °C. The two points are divided into 100 scales. In the absolute temperature system, when the substance stops shaking, the absolute temperature is zero, that is, 0 K. The relationship between absolute temperature and degree Celsius is as follows:

$$0\,\mathrm{K} = -273.15°\mathrm{C}, \quad 0°\mathrm{C} = 273.15\,\mathrm{K} \tag{5.2}$$

About absolute temperature, there is the third law of thermodynamics, which says that the temperature cannot be absolute zero (0 K). The meaning of this law is that substance will not stop moving. Knowing the Kelvin temperature K, then Boltzmann's constant is:

$$k_{\mathrm{B}} = 8.62 \times 10^{-5}\,\mathrm{eV/K} \tag{5.3}$$

Pure semiconductors, are called intrinsic semiconductors, cannot be directly used to make devices. We intentionally add some impurities (known as dopants) to the semiconductor crystals. The atoms of dopants can replace the semiconductor atoms on the crystal lattice, change the resistivity, and other characteristics. The process of adding impurities is called doping, and the semiconductors after doping are called doped semiconductors. Commonly used methods are high temperature (900–1200 °C) diffusion and ion implantation, which we will discuss later. In silicon, the most used dopants are phosphorus with the symbol P and boron with the symbol B.

Let us take another look at the periodic table of the elements in Figure 5.1. We can see that P is in group 15 and has five valence electrons. Boron is in group 13 and has three valence electrons. Silicon has four valence electrons and phosphorus has five. After phosphorus is doped into Si, a phosphorus atom that replaces a silicon atom only needs to take out four electrons to form a covalent bond with the surrounding silicon atoms (see Chapter 2). The fifth electron of P will be free from the bondage and become a free electron. This kind of negatively charged electron conduction is called n-type semiconductor. Sometimes capital N is used. n is the first letter of "negative." Boron has only three valence electrons. When a boron enters the silicon crystal to replace a silicon atom, the covalent bond formed with silicon lacks one electron, and one adjacent valence electron can be attracted by this incomplete covalent bond. This attraction procedure grabs one electron from a neighbor Si atom and leaves a hole in the original position. The hole is positively charged. This kind of semiconductor is called a p-type semiconductor. Sometimes capitalized P is used. p is the first letter of "positive." Phosphorus is to provide electrons in silicon, it is called donor impurity. Boron is to accept electrons in silicon, it is called acceptor impurity. Please see Figure 5.4.

Now let us discuss the band diagram of silicon after doping, please see Figure 5.5. To understand this picture, we should introduce an important concept, Fermi energy. Fermi energy is a concept of energy in solid-state physics and to commemorate the American Italian physicist Enrico Fermi (September 29, 1901–November 28, 1954). In semiconductors, Fermi energy is used to describe

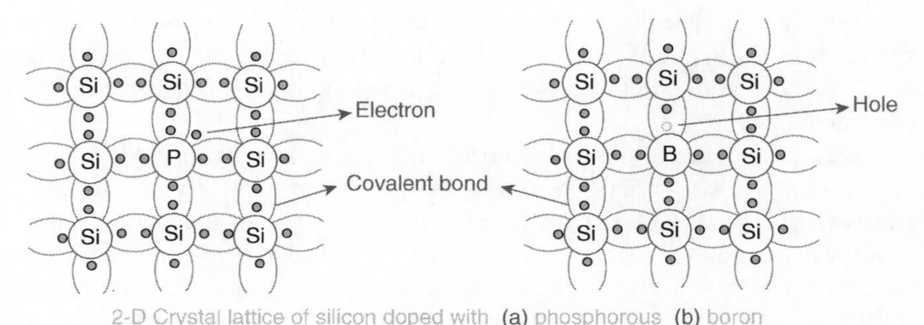

Figure 5.4 Schematic diagram of the impurities of P and B in Si. Source: Reprinted from Electrical 4U.

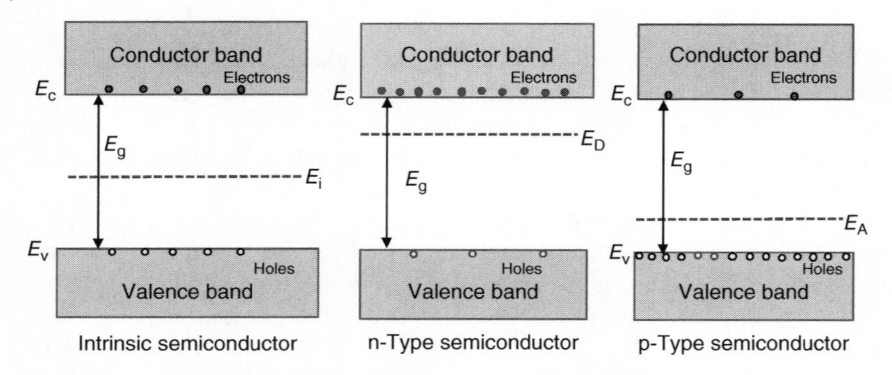

Figure 5.5 Schematic diagram of intrinsic, n-type and p-type semiconductor energy bands.

the energy level of electrons or holes and can be determined from the charge neutrality condition, so it is more commonly called the Fermi energy level (Fermi level) with the symbol E_f. In the energy band diagram of Figure 5.5, "E_c" is the bottom energy of the conduction band, "E_v" is the energy at the top of the valence band, and "E_g" is the forbidden band width. In intrinsic semiconductors, the number of electrons and holes is equal, and they are called electron–hole pair (EHP). The Fermi level is almost in the center of the forbidden band, which is represented by E_i. In n-type semiconductors, E_D represents the ionization energy of donor doping. The number of electrons is more than holes, Fermi level is close to the bottom of the conduction band. The higher the donor concentration, the closer the Fermi level is to the bottom of the conduction band. In p-type semiconductors, E_A is the ionization energy of acceptor doping. There are more holes than electrons, and the Fermi level is near the top of the valence band. The higher the acceptor concentration, the closer the Fermi level is to the top of the valence band. Figure 5.6 is the ionization energy level distributions of impurities in germanium, silicon, and gallium arsenide. From the figure, we can see that in silicon, the ionization energy of phosphorus is 0.046 eV and that of boron is 0.044 eV. The doped materials are called extrinsic semiconductors [2].

To further understand the characteristics of intrinsic and extrinsic silicon, we briefly introduce the distribution of electrons and holes in silicon crystal. The electron is a fermion, and it obeys the distribution of Fermi–Dirac statistics. A hole also beys the distribution of Fermi–Dirac statistics. Let's use n_i to represent the intrinsic carrier density, n to represent the electron density in n-type and p to represent the hole density in p-type. By solving Fermi–Dirac statistics, it can be obtained that

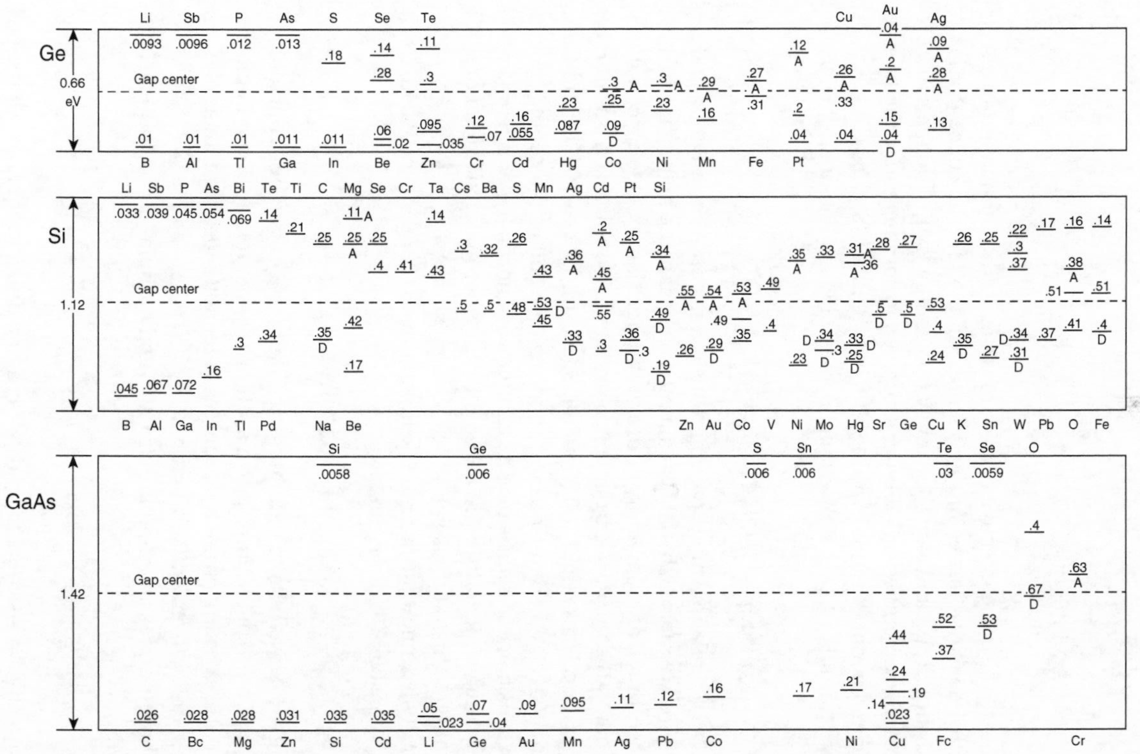

Figure 5.6 Ionization energies for various impurities in Ge, Si, and GaAs. The levels below the gap centers are from the top of the valence band and are acceptor levels unless indicated by D for donor level. The levels above the gap centers are from the bottom of the conduction band and are donor levels unless indicated by A for acceptor level. Source: [1] Sze / John Wiley & Sons.

the carrier concentrations of intrinsic and n-type are as follows. The concentration of p follows a similar rule as n [1]:

$$n_i = (N_C N_V)^{1/2} e^{-E_g/2k_B T} \tag{5.4}$$

$$n_i \approx \frac{1}{\sqrt{2}} (N_D N_C)^{1/2} e^{-(E_C - E_D)/2k_B T} \tag{5.5}$$

In the formula, N_C is the effective density of states in the conduction band. N_V is the effective density of states in the valence band. N_D is the concentration of donor impurities. E_g is band gap, E_D is ionization energy level of donor. $e \approx 2.72$ is natural constant, sometimes called Euler's number, to commemorate the Russian mathematician Leonhard Euler (April 15, 1707–September 18, 1783).

Sometimes formula (5.4) is also written as the following formula, same as formula (5.5):

$$n_i = (N_C N_V)^{1/2} \exp(-E_g/2k_B T) \tag{5.6}$$

Figure 5.7 is the relationship between intrinsic concentration and temperature. From the figure, we can see that, the smaller the band gap, the higher the intrinsic carrier concentration at room temperature. Assuming that the doping concentration is 10^{15} cm^{-3}, the intrinsic carrier concentration in Ge will reach this number when temperature is less than 100 °C, while Si must be about 300 °C, and GaAs will be about 500 °C. When the intrinsic concentration equals the doping concentration in extrinsic materials, the devices cannot work normally. This temperature is called operating temperature. However, due the other material limitations, the actual operating temperature is below 300 °C in most cases.

We generally assume that the room temperature is $T = 27\,°\text{C} \approx 300\,\text{K}$. Using Boltzmann's constant (5.2), we can have:

$$k_B T = 8.62 \times 10^{-5} \times 300\ \text{eV} = 0.02586\ \text{eV} \approx 0.026\ \text{eV} \tag{5.7}$$

According to Maxwell–Boltzmann statistics, the average energy of a microscopic particle is $\frac{3}{2}k_B$T. So, to put it simply, we may think the electrons in donor impurities can obtain 0.026 eV energy from heat sources at room temperature. In a solid, the prime heat source is from lattice vibration.

Taking phosphorus as an example, add 0.046 and formula (5.7) in formula (5.5), we can have:

$$e^{-(E_C - E_D)/2k_B T} = 2.72^{-0.046/2\times 0.026} \approx 0.41 \tag{5.8}$$

Mathematically, "=" is equal to, and "≈" is approximately equal to. This shows that at room temperature, we can simply understand that more than 40% phosphorus doping are activated by *thermal lattice vibration*. These excess electrons are ionized into the conduction band to become free electrons. The situation is

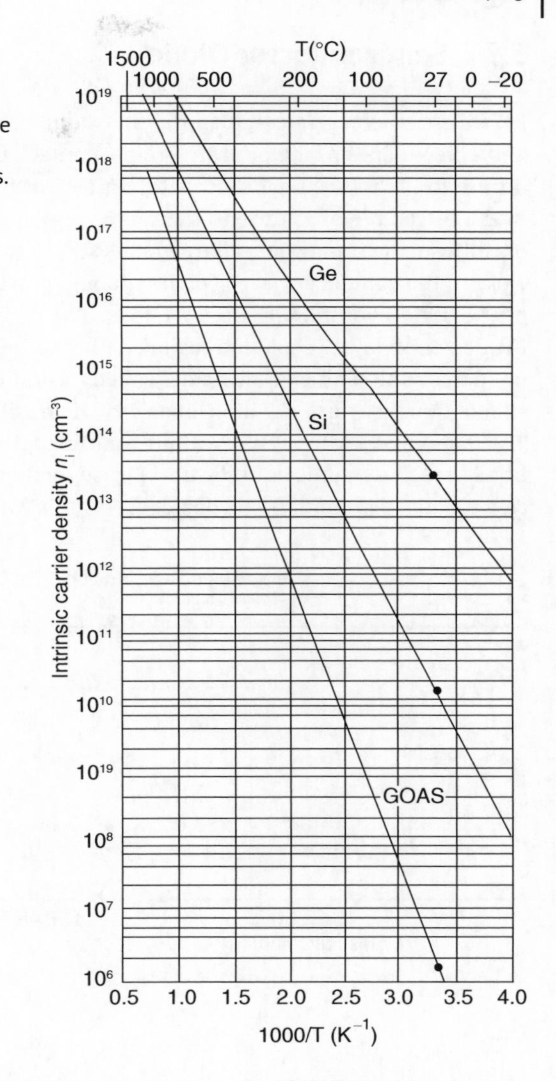

Figure 5.7 Intrinsic carrier concentration for Ge, Si, and GaAs as a function of inverse temperature. The room temperature values are marked for reference. Source: [1] Sze / John Wiley & Sons.

similar for boron. This kind of impurity that can be ionized at room temperature is called shallow level doping, which corresponds to deep level doping. In GaAs, we generally dope silicon. Si replaces gallium, becoming a shallow donor impurity, making GaAs an n-type semiconductor. In semiconductors, these impurities greatly change the characteristics of the material, for example, making GaAs from semi-insulating to a real semiconductor material. According to needs, we can change the doping concentration to make different semiconductors for manufacturing different devices.

5.3 Semiconductor Diode

Let us use silicon semiconductor as an example. When a piece of n-type material and a piece of p-type material are in close contact, the electrons in the n-type region diffuse to the p-type region near the contact surface. At the same time, the holes in the p-type region diffuse to the n-type region. After the movable charge carriers diffuse, the immovable charged ions stay in the corresponding lattice position close to the contact surface. The n-type region will remain positive ions, and the p-type region will remain negative ions. The region lacking the carriers is called the depletion layer (depletion region, depletion zone), and an electric field is built up in this zone. This electric field is called the built-in electric field. With the diffusion of more carriers, the built-in electric field will become stronger and stronger. This enhanced electric field will prevent further diffusion of charge carriers and finally reach dynamic equilibrium. Please see Figure 5.8. The V_{bi} in the figure is called built-in potential. The depletion layer is called p–n junction.

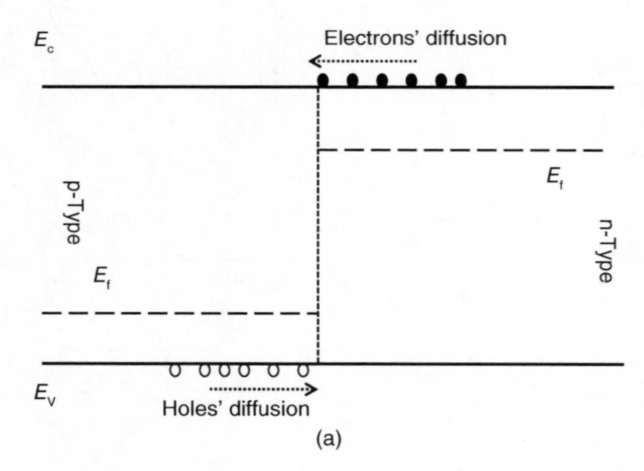

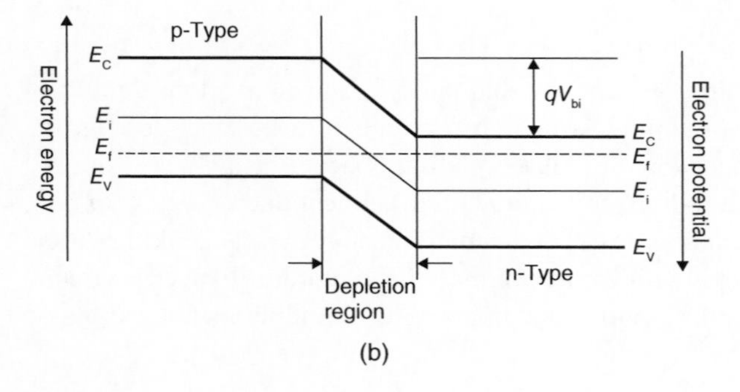

Figure 5.8 Schematic diagram of the energy band structure of p–n junction. (a) In the vicinity of the interface between p-type Si and n-type Si, electrons diffuse from n region to p region, and holes diffuse from p region to n region. (b) Finally, the Fermi levels of the p and n regions are equal, and the depletion region and built-in potential are established.

During the establishment of p–n junction, due to the mutual diffusion of electrons and holes, the Fermi level of the n-type semiconductor and the Fermi level of the p-type semiconductor eventually tend to be the same, and a built-in potential is established. For convenience, E_f is used to denote the Fermi levels in the n-type region and p-type region in the figure. The potential energy of the diffusion zone is the energy of a charge at a certain point in the electric field. Like the introduction of an electric field, the potential is the potential energy divided by the charge. The voltage between two points is the difference in potential between the two points. In other words, the potential difference is the voltage. The establishment of electric potential in the depletion layer will hinder the diffusion of the carriers, which is called a potential barrier. The p–n junction is vital to our understanding of modern electronics and semiconductor microchips. Metal electrodes (metal–semiconductor contact) are placed on both ends of the p–n junction, and then the device is packaged. This packaged diode is a product of a diode. We will discuss metal–semiconductor contact later.

In the p–n junction, there is another type of charge movement called charge carrier drift. Close to the depletion layer, a small number of electrons in the p-type region will be attracted by the built-in electric field and drift from the p-type region to the n-type region. Conversely, a small number of holes will be attracted by the built-in electric field in the n-type region and drift from the n-type region to the p-type region. If the diode is not connected to the battery, all the carriers' movement reaches dynamic balance, and no current flows in the p–n junction. When the diode is connected to a battery, there are two situations:

(1) The anode of a battery is connected to the p-type zone, and the cathode is connected to the n-type zone. In this scenario, the direction of electric field of the external battery is opposite to the built-in electric field. The built-in electric field is weakened, and the depletion layer becomes thinner, the diffusion of charge carriers is not offset by the built-in electric field. The diffused carriers flow in the p–n junction, the electrical current will flow through the diode at this time. Please look at the top image of Figure 5.9. This is called forward voltage.
(2) The anode of the battery is connected to the n-type zone, and the cathode is connected to the p-type zone. In this scenario, the electric field of the external battery is in the same direction as the built-in electric field. The built-in electric field is strengthened, and the depletion layer becomes thicker. There are no diffused carriers flowing through the p–n junction, only a small amount of drift carriers flows through the p–n junction. At this time, only a small amount of electrical current flows through the diode, as shown in the bottom image of Figure 5.9. This is called reverse voltage.

The charge carriers that we often refer to are also divided into two categories: majority charge carriers and minority charge carriers. In an n-semiconductor,

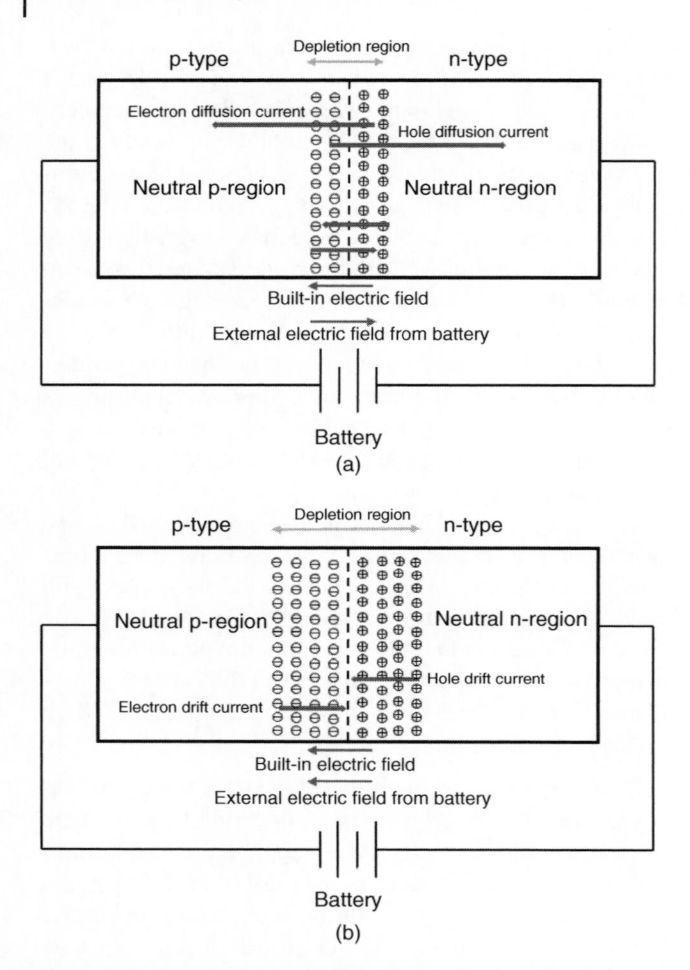

Figure 5.9 The structure of diode. Forward voltage (a) and reverse voltage (b).

electrons are majority charge carriers and holes are minority charge carriers. In a p-semiconductor, holes are majority charge carriers and electrons are minority charge carriers. The battery voltage is assumed to be V. In case (1), the potential energy becomes $q(V_{bi} - V)$, the barrier becomes lower, the diffusion current increases, and the diode turns on. This is called the diode with the forward voltage. The forward voltage is sometimes called forward bias, and the current is called forward current. In case (2), the potential energy becomes $q(V_{bi} + V)$, the potential barrier is increased, there is no diffusion current, and there is only a small drift current. This is called diode with reverse voltage. The reverse voltage is sometimes called reverse bias, and the current is called reverse leakage current.

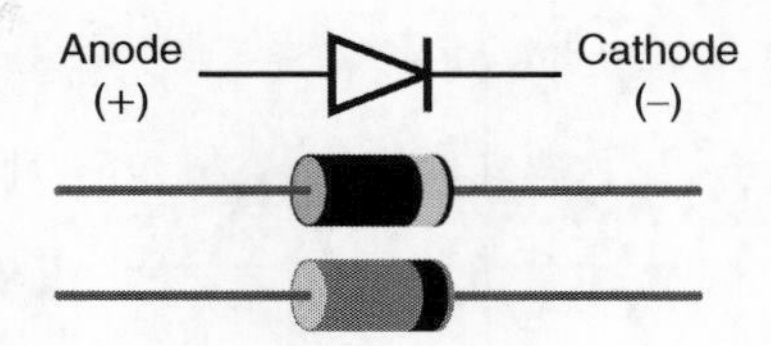

Figure 5.10 Semiconductor diode. Source: Reprinted with permission of Servodroid.ru.

Figure 5.10 is the symbol and photo of the diode. From the discussion, we know that the diode is unidirectional, so the symbol of the diode is in the form of an arrow, indicating that the current is unidirectional in the diode, from the anode to the cathode. This is completely different from resistance, which is bidirectional. Figure 5.11 is the relationship curve between the current and the voltage of the diode and the resistor, the curve is often called the I–V curve. The diode I–V curve was measured from a real Si diode product [3]. When the forward voltage is bigger than around 0.6 V, the forward current begins to increase. This voltage is the threshold voltage (V_{Th}, also called offset voltage in other books [2], 0.3 V for germanium and 0.2 V for Schottky). We will talk about Schottky diodes later. The threshold voltage is expected to be approximately the potential barrier of the junction. Only when the applied voltage (battery) exceeds V_{Th}, the current starts to flow in the diode. At this time, the diode is called conduction. The reverse current is very small compared to the forward current and is called leakage current. When the reverse voltage is bigger than around 115 V, the reverse current is increased significantly. This voltage is the breakdown voltage. Let us look at depletion layer diagram with the reverse bias of the diode in Figure 5.9 and the I–V curve in Figure 5.11. It shows that a diode with the reverse bias is a capacitor. The depletion

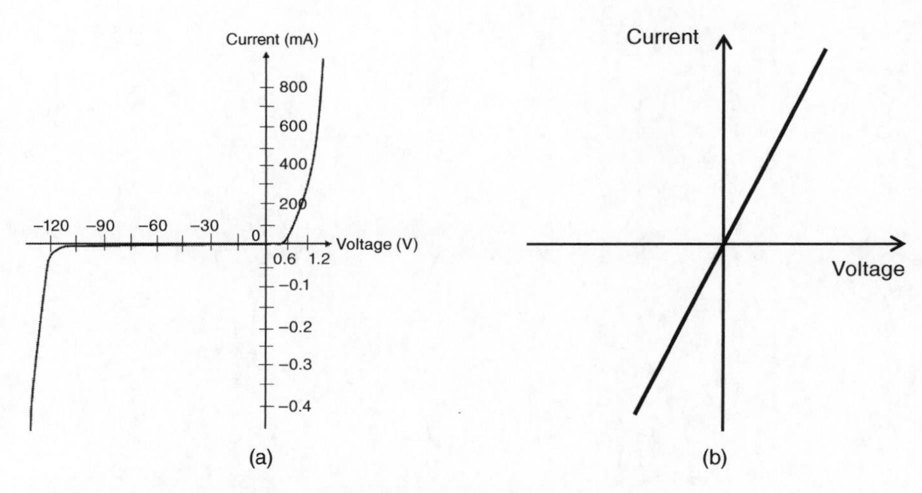

Figure 5.11 The I–V curve of a diode (a) and resistor (b).

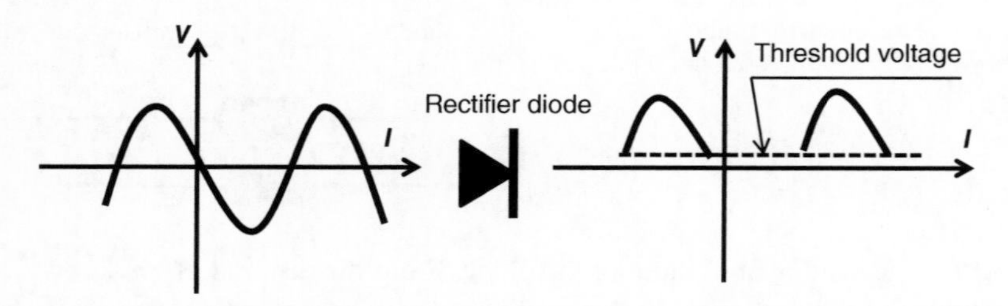

Figure 5.12 Schematic diagram of electric signal before and after a rectifier diode.

layer is equivalent to the dielectric. The regions of p-type and n-type are equivalent to the electrode plates. In fact, the reverse-biased diode capacitor is a widely used structure in the integrated circuit. An important application of diodes is rectifiers. A rectifier diode allows only one-way current conduction, please see Figure 5.12.

References

1 Sze, S.M. *Physics of Semiconductor Devices*, 2e, p. 849, p. 13, pp. 19–25.
2 Streetman, B.G. *Solid State Electronic Devices*, 4e, p. 66, p. 202.
3 模拟电子技术基础(上册), 童诗白主编, 人民教育出版社, 1980, 21页.

6

Transistor and Integrated Circuit

In Chapter 5, we introduced semiconductor diodes. One problem with the diode is that it has no amplification function. Based on the diode, the semiconductor transistor was invented. An important feature of the transistor is that it has an amplification function and can replace the vacuum tube. Based on the transistor, the integrated circuit (IC) was invented. Since then, mankind has entered the chip era.

There are different kinds of transistors, including bipolar, junction field effect, and metal–oxide–semiconductor field effect transistors (MOSFETs). We will discuss these transistors in the following sections.

6.1 Bipolar Transistor

The earliest invented transistor is a bipolar transistor, which is made by combining two p–n junctions. This kind of transistor has two structures, one is n–p–n (N–P–N) and the other is p–n–p (P–N–P). Please see Figure 6.1, which is a schematic diagram of the transistor structures and the corresponding symbols. There are three terminals in a bipolar transistor. They are emitter, collector, and base. I_E is emitter current; I_C is collector current; and I_B is the base current. In this transistor, two types of charge carriers-electrons and holes are involved in the current conduction, so it is called a bipolar transistor, or bipolar junction transistor. Let us take the n–p–n transistor as an example. Figure 6.2 is the basic circuit diagram of a transistor in circuit use. In this circuit, the emitter provides electrons and is the main provider of charge carriers. To make the carriers move, the junction of B and E should be forward biased. During the manufacturing process, the doping concentration of the base region is much lower than that of the emitter. So, the hole current in the base region is much smaller than the electron current in the emitter region. Applying a reverse bias to the junction of B and C to collect electrons flowing through the base region [1]. In this figure,

Semiconductor Microchips and Fabrication: A Practical Guide to Theory and Manufacturing, First Edition. Yaguang Lian.

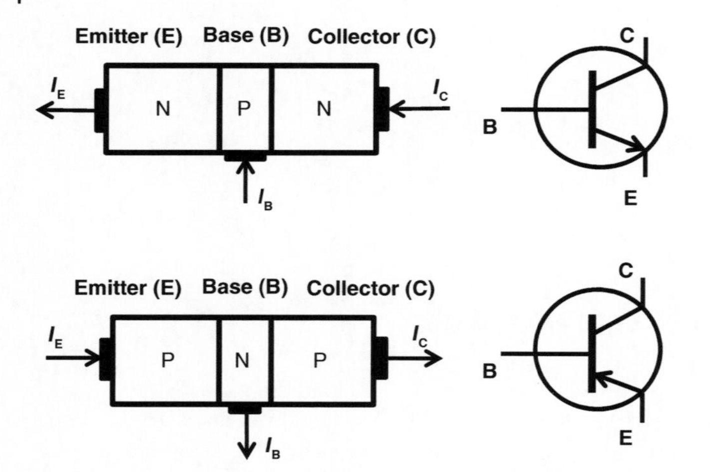

Figure 6.1 The basic structures of bipolar transistor and symbols.

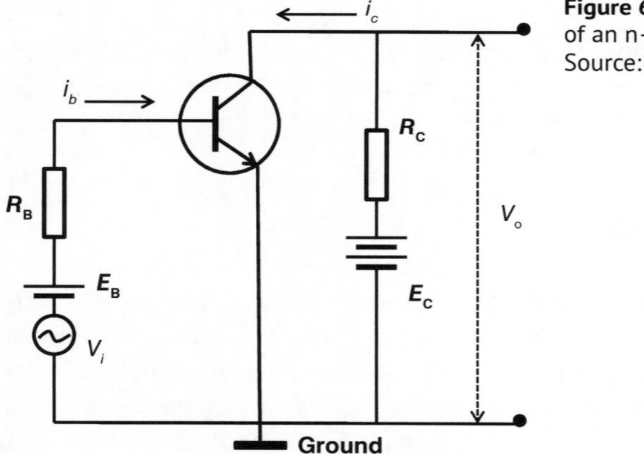

Figure 6.2 The basic schematic of an n–p–n transistor [1]. Source: Chinese Technical Books.

V_i is the input signal voltage, V_o is the output signal voltage, E_B is used to control the junction between the base and the emitter, and E_c is used to collect electrons' current out of the emitter. A small V_i causes a small current i_b to change, which brings about a large change of the current i_c, and finally gives a large output voltage V_o. This is the amplification process of the transistor. Figure 6.3 is the I–V curve of an n–p–n bipolar transistor, which can be used as an amplifier or switch. If it works in amplification mode, it is an amplifier. If it works in saturation mode, the switch is on, because the transistor is in a high current and low-voltage state. If it works in cutoff state, the switch is off, because the transistor is at a high voltage and low-current state.

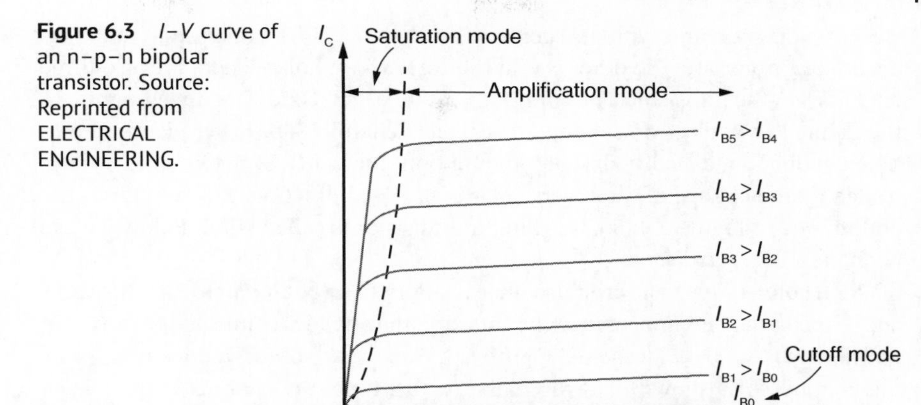

Figure 6.3 I–V curve of an n–p–n bipolar transistor. Source: Reprinted from ELECTRICAL ENGINEERING.

6.2 Junction Field Effect Transistor

Another kind of transistor is a field effect transistor (FET). In this transistor, there are three main types: junction field effect transistor (JFET), metal–semiconductor field effect transistor (MESFET), and MISFET. All these FETs are unipolar transistors because the current is only generated by the flow of majority carriers that can be electrons or holes. In this section, we introduce JFET. Figure 6.4 is the JFET

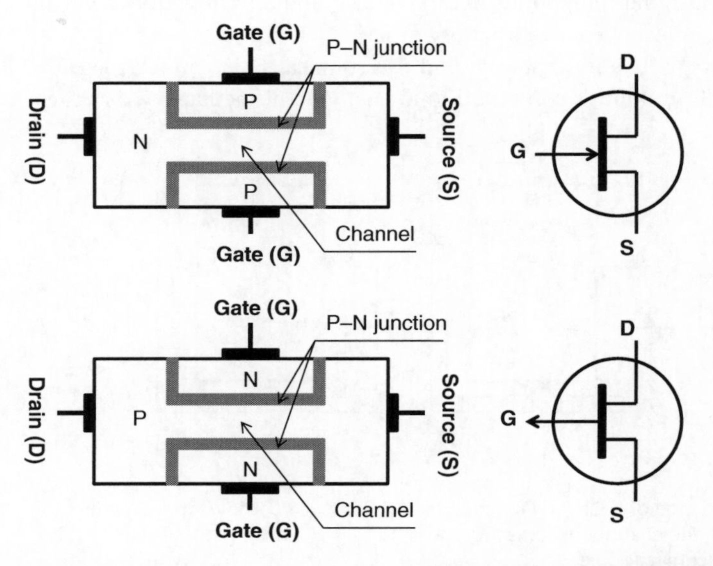

Figure 6.4 JFET structure and symbol. Top is N-channel and bottom is P-channel.

structure diagram and symbol. There are two types of JFETs, one is that the carriers are electrons, and the other is that the carriers are holes. There are also three terminals in the transistor. They are Drain, Source, and Gate. The upper picture in the Figure shows electron conduction, called N-channel. The lower picture shows hole conduction, called P-channel. In addition, the width of the channel is controlled by the change of the reverse bias of the P–N junction. Please refer to the bottom one of Figure 5.9. So, this kind of transistor is called JFET. Below we use N-channel JFET to discuss.

When connecting an adjustable voltage battery to an N-channel JFET, the cathode of the battery is connected to the gate, and the anode is connected to the drain and source ($V_{DS} = 0$). Please see Figure 6.5. If the gate–source voltage is $V_{GS} = 0$, the channel is as shown in Figure 6.5a. At this time, the channel is the widest and can pass the maximum electrical current. Figure 6.5b shows that when the reverse voltage increases, $V_{GS} < 0$, the depletion layer becomes wider and the channel becomes narrower, and the current that can pass becomes smaller. Continue to increase the reverse voltage, and finally the two depletion layers contact each other, see Figure 6.5c. The current channel is cut off and no current will flow through the device. At this time, $V_{GS} = V_P$ is called pinch-off voltage. The battery must be connected in the reverse bias mode of the P–N junction as shown in Figure 6.5. It cannot be connected in the forward bias mode. In this case, the P–N junction is turned on and the current will flow through the P–N junction at both sides. The current channel in the middle cannot be adjusted. In the figure, "No contact" means that the point is not in contact, and an arrow drawn on the battery means an adjustable voltage battery.

Let us connect the battery to the D and S terminals, D is connected to the anode of the battery, and S is connected to the cathode of the battery, as shown

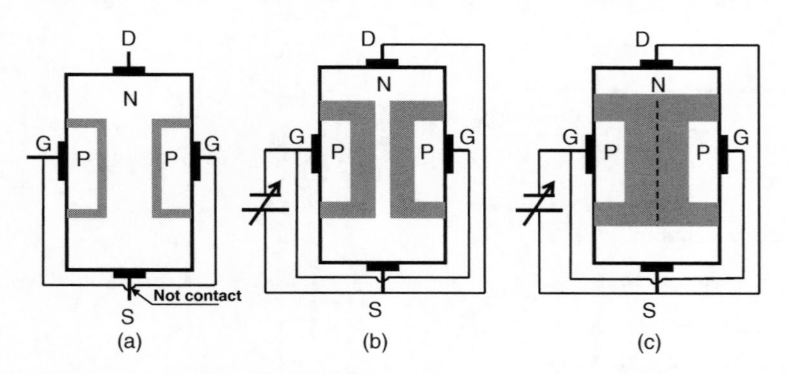

Figure 6.5 The change of JFET P–N junction vs. V_{GS} (a) $V_{GS} = 0$, the channel is the widest, (b) $V_{GS} < 0$, the channel becomes narrow, (c) $V_{GS} = V_P$, the channel is cut off. [1]. Source: Chinese Technical Books.

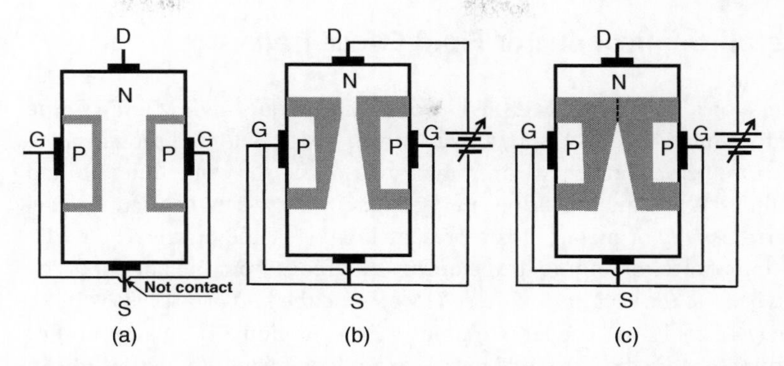

Figure 6.6 The change of JFET P–N junction vs. V_{DS} (a) $V_{DS} = 0$, the channel is the widest, (b) $V_{DS} > 0$, the channel becomes narrow unevenly, (c) $V_{DS} = -V_P$, the channel at *D* is cut off [1]. Source: Chinese Technical Books.

in Figure 6.6. The change of the current channel is similar to the situation in Figure 6.5, but there are some differences. The voltage distribution in Figure 6.5 is uniform, so the channel changes uniformly. But in Figure 6.6, the voltage of D is higher than that of S. As the voltage of V_{DS} increases, the depletion layer at D (the upper part of the figure) meets first. The transformation of the channel is shown from 6.6a to c. The current flowing through the channel initially increases with the increase in voltage. When the depletion layers touch each other, the current will not increase anymore and basically stabilizes at a value, which is called the saturation current. The I–V curve of a typical N-channel JFET is shown in Figure 6.7. In the figure, I_{DSS} is the maximum saturation current when the bias voltage of G–S junction is zero.

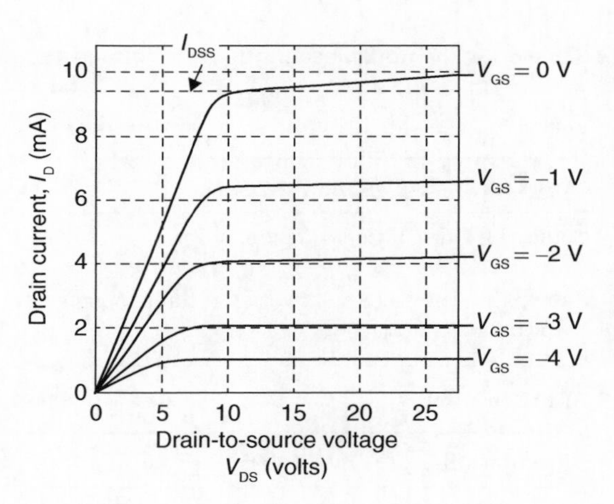

Figure 6.7 N-channel JFET I–V curve. Source: Reprinted with permission of Nuts & Volts magazine.

6.3 Metal–Semiconductor Field Effect Transistor

Now let us look at how a MESFET works. To understand MESFET, the metal–semiconductor contact should be known first. Before discussing the metal–semiconductor contact, we introduce two concepts: work function and electron affinity. Work function is the energy difference between the vacuum level and the Fermi level. For metals, it is represented by $q\Phi_{\mathrm{m}}$, the unit of Φ_{m} is volt. The definition of electron affinity is the energy difference from the bottom of the conduction band to the vacuum, expressed by $q\chi$, and the unit of χ is also volt [2]. Aluminum (Al), gold (Au), copper (Cu), nickel (Ni), titanium (Ti), platinum (Pt), and chromium (Cr) are the most used metals in semiconductor production. Please see their work functions in Table 6.1. Silicon, germanium, and gallium arsenide are the most used semiconductors. Their electron affinities are listed in Table 6.2.

Figure 6.8 is a schematic diagram of the energy band structure before (a) and after (b) of the metal and n-type semiconductor contact. In the figure, qV_{bi} is the potential energy of the depletion region, and $q\Phi_{\mathrm{Bn}}$ is the height of the barrier. “B” means potential barrier and “n” is n-type, W is the width of the depletion layer. When an external electric field E is applied, the barrier height will become lower. This phenomenon is called the Schottky effect, to commemorate the German physicist Walter H. Schottky (July 23, 1886–March 4, 1976). The barrier is called the Schottky barrier, please see Figure 6.9. Metal–semiconductor contact, also called Schottky contact, can be used to make diodes, which are called Schottky diodes. As mentioned in Chapter 5, the threshold voltage of a Schottky diode is approximately 0.2 V.

Metal–semiconductor contact, besides Schottky contact, there is also an important type of contact-ohmic contact. Because the resistance of the semiconductor is

Table 6.1 The work functions of the metals most used in the semiconductor.

Metal	Al	Au	Cu	Ni	Ti	Pt	Cr
Work function (eV)	4.06–4.26	5.1–5.47	4.53–5.1	5.04–5.35	4.33	5.12–5.93	4.5

Source: [3] Lide / Taylor & Francis.

Table 6.2 The electron affinities of three most used semiconductors [2].

Semiconductor	Si	Ge	GaAs
Electron affinity (eV)	4.05	4.0	4.07

Source: [2] Sze / John Wiley & Sons.

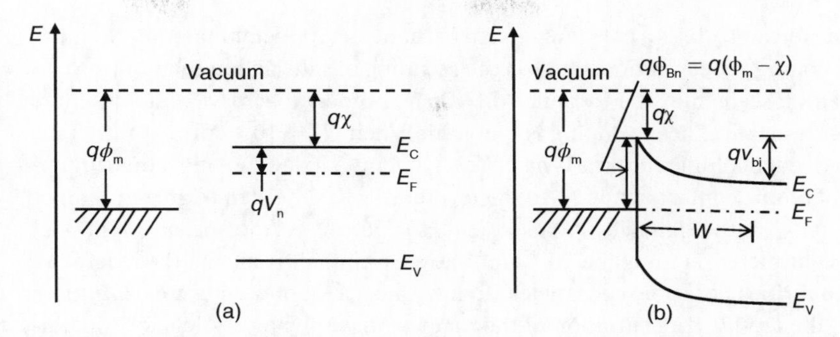

Figure 6.8 The band structure of metal and n-type, (a) before contact and after (b) after contact. Source: [2] Sze / John Wiley & Sons.

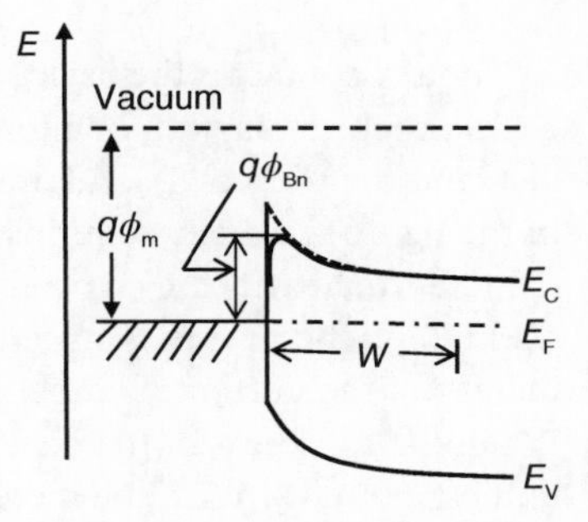

Figure 6.9 Schematic diagram of the energy band of the Schottky effect.

much bigger than the resistance of the metal, we need to make metal electrodes on the surface of the semiconductor device to draw out the electrical signals. In the basic structure diagrams from Figures 6.1–6.6 in this section, the short thick black lines at the electrodes of base, gate, etc., represent metal electrodes. Metal is also needed to connect different devices together. This is the concept of integrated circuits. On a semiconductor substrate, metals are used to connect different devices in different places, we will explain this technology in detail later. In this case, the metal–semiconductor contact must be an ohmic contact. We have said in this section that Schottky contact is a diode, and its I–V characteristics is shown on the left in Figure 5.11, while the I–V curve of ohmic contact should be the same as the one on the right of Figure 5.11. We introduce ohmic contact resistance R_c here. In order to meet the requirements of Figure 5.11 on the right, in the case of donor doping that is the most used ohmic contact doping, R_c needs to follow the relationship below [2]:

$$R_c \sim \exp\left[R_{c0}\left(\frac{\Phi \mathrm{Bn}}{\sqrt{ND}}\right)\right] \tag{6.1}$$

In the formula, R_{c0} is a constant related to the semiconductor material, N_D is the donor doping concentration, and ~is the meaning of satisfaction relation.

It can be seen from the formula (6.1) that in order to achieve a small ohmic contact resistance, heavy doping is required. When $N_D \geq 10^{19}$ cm^{-3}, it meets the requirement of ohmic contact. When $N_D \leq 10^{17}$ cm^{-3}, it does not meet the requirement of ohmic contact at all [2]. In the formula, "≥" is the sign of greater than or equal to, "≤" is the sign of less than or equal to and cm^{-3} is the unit of volume-per cubic centimeter. Let us take 10^{19} cm^{-3} as an example, it means that there are 10^{19} impurities per cubic centimeter. To have a concept of doping concentration, we list the atomic concentration of the main semiconductor crystals for comparison. The unit is the number of atoms per cubic centimeter [2]: Si = 5.0×10^{22}, Ge = 4.42×10^{22}, and GaAs = 4.42×10^{22}. In addition, the formula also shows that in order to have a small ohmic contact resistance, the barrier height should also be small.

Schottky contact is the basic structure used to make MESFET. Figure 6.10 is the basic structure diagram of an n-channel MESFET. The gate (G) is Schottky contact, and the source (S), and drain (D) are ohmic contacts. "Insulating substrate" in the figure is usually semi-insulating GaAs. n$^+$ refers to heavily doped n region (p$^+$ refers to heavily doped p region). The working principle of MESFET is like that of JFET. As the reverse voltage V_{GS} increases, the thickness of the depletion layer increases, the current channel becomes thinner, and electric current decreases. When the thickness of the depletion region occupies the entire channel, the current cannot flow. With the increase of V_D, the thickness of the depletion layer of the D electrode is bigger than that of the S electrode. Until the depletion layer of the D electrode contacts the insulating substrate, the current then enters the saturation region. Figure 6.11 is the I–V curve of a MESFET. GaAs devices and ICs are mainly made by using this structure. GaAs used for devices and ICs are usually n-type material.

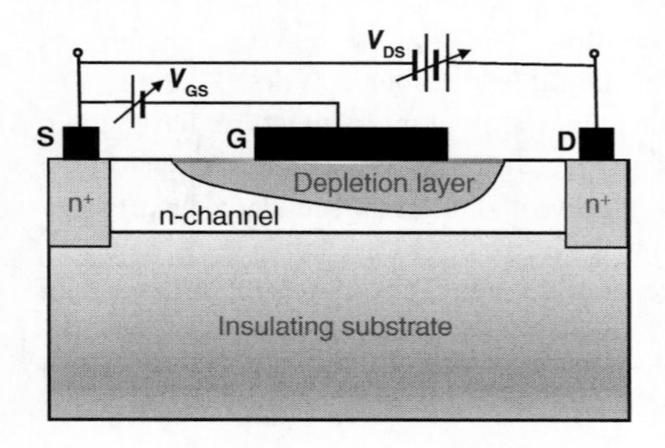

Figure 6.10 The basic structure of a MESFET.

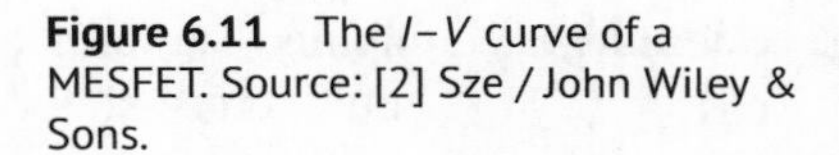

Figure 6.11 The *I*–*V* curve of a MESFET. Source: [2] Sze / John Wiley & Sons.

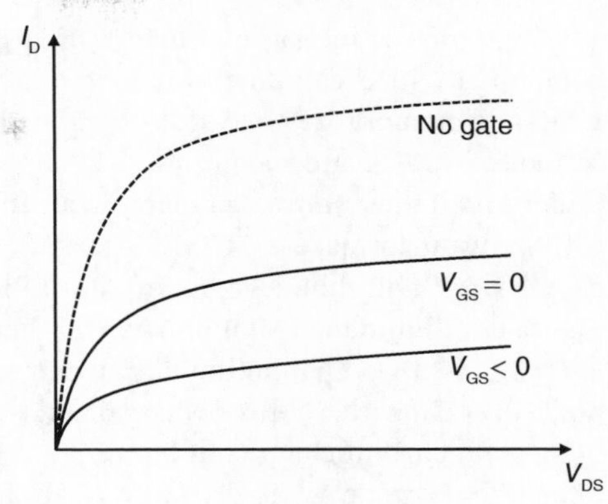

6.4 Metal–Insulator–Semiconductor Field Effect Transistor

Now let us discuss MISFETs. Before discussing MISFET, let us talk about metal–insulator–semiconductor (MIS). Figure 6.12 is a schematic diagram of the basic structure of MIS. The working principle of MIS can be understood through the metal–semiconductor contact, but the existence of the insulator cuts off the current path between the metal and the semiconductor. Therefore, according to the different situations, V_G can be positive voltage or negative voltage. The structures of the current channel are different related to the direction of the bias. V_G is equivalent to the voltage V applied to the metal electrode in Figure 6.12. The electric field generated by this voltage is coupled to the semiconductor through the insulator. The electric charges are created in the semiconductor surface close to the insulator. It is to control the current channel. The device is also called a charge-coupled device. If the insulator is silicon dioxide (SiO_2, oxide), the MISFET becomes a MOSFET. The semiconductor used for MOSFET is mainly

Figure 6.12 The basic structure of MIS. Source: [2] Sze / John Wiley & Sons.

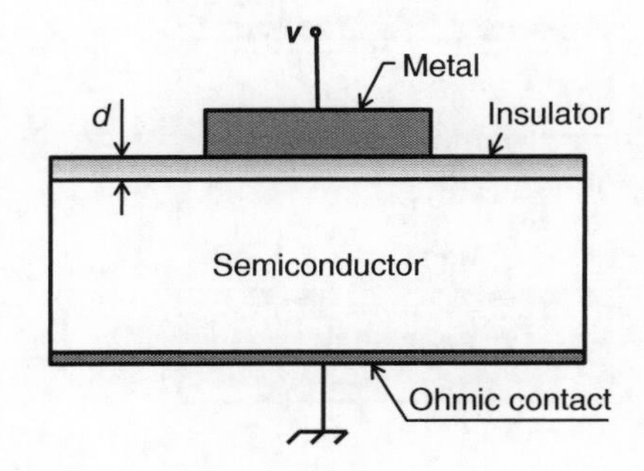

silicon, which is the basic structure of large-scale integrated circuits. The core components in a computer–Microprocessor (μP) or CPU (Central Processing Unit) and memory are large-scale integrated circuits. Since the number of devices in modern CPUs and memories is so large (we will talk about them later), we prefer to call them (ultra) very large-scale integrated circuits.

If we want to make a MOSFET usable, the electrodes should be put on the gate, source, and drain. There are four different kinds of MOSFETs, please see Figure 6.13. From top to bottom, we call them Type 1, Type 2, Type 3, and Type 4.

Type 1 is an n-channel enhancement type, and this kind of MOSFET is a normally off device. In this device, silicon is p-type. When the gate voltage is zero, there is no current channel. If a positive voltage is applied to the gate, the holes under the electrode will be driven away, attracting electrons, and forming an n-type conductive channel. So, it is called n-channel enhancement type. In addition, when the gate voltage is zero, there is no conductive path, thus it is normally close.

Type 2 is an n-channel depletion type, and this kind of MOSFET is a normally on device. In this device, silicon is p-type. When the gate voltage is zero, there is

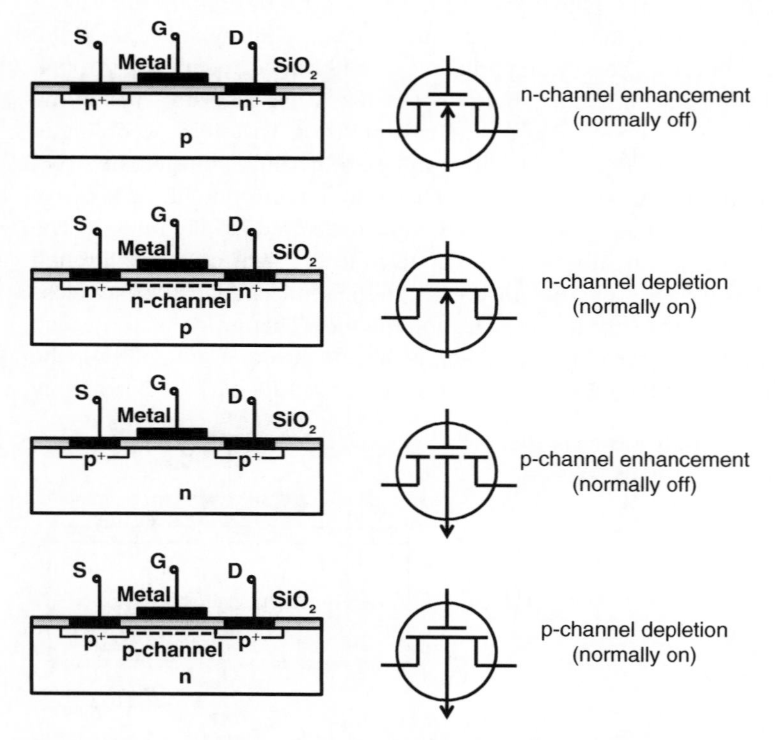

Figure 6.13 The basic structures and symbols of MOSFET. Source: [2] Sze / John Wiley & Sons.

already an n-type channel under the electrode. We need to apply a negative voltage on the gate to drive away electrons to form a depletion layer and reduce the thickness of the channel. Finally, the channel is cut off, so it is called an n-channel depletion type. In addition, when the gate voltage is zero, there is a conductive path, thus it is a normally open.

Similarly, when silicon is n-type, there is a p-channel enhancement and normally off device, which is type 3. The other is a p-channel depletion and normally on device, which is type 4.

The amount of voltage that must be applied to the gate to create an enough conductive channel (i.e. to turn the MOSFET on) is the threshold voltage V_{Th} of MOSFET. For n-channel, the V_{Th} is positive. For p-channel, the V_{Th} is negative. In the Figure 6.13, MOSFET has gate (G), source (S) and drain (D). There is also an electrode connected to the substrate (Figure 6.12). This electrode can provide a reference voltage to affect the operation of the device. The materials of the G, S, and D electrodes can be metal, or heavily doped polysilicon and silicide ($TiSi_2$) on the top which is named as polycide. The basic device parameters are gate length (channel length) L, which refers to the distance between two n^+–p or p^+–n junctions, and the thickness d of SiO_2. The other two parameters are the gate width (channel width) Z and the substrate doping concentration N_A (acceptor) or N_D (donor).

Figure 6.14 is schematic diagrams of the energy bands for different MOS structures. The diagram on the top of the figure shows metal and p-type semiconductor, and the lower diagram shows metal and n-type semiconductor. Take the p-type semiconductor in the above figure as an example. (i) When a negative voltage is applied to the metal electrode, more holes will be attracted to the semiconductor surface, causing the holes to accumulate at the interface between the oxide layer and the semiconductor. (ii) When a zero voltage is applied to the metal electrode, the energy level will not change. The energy band is in a flat-band state. (iii) When a small positive voltage is applied to the metal electrode, holes will be driven away from the interface between the oxide layer and the semiconductor. This is the depletion state (depletion). (iv) When the positive voltage on the metal electrode is large enough, electrons accumulate at the interface between the oxide layer and the semiconductor, causing the type of semiconductor at the surface to be inverse, changing from p-type to n-type, and generating n-type current channel. This is the type 1 FET mentioned above. The diagram on the bottom of the figure shows n-type semiconductor that corresponds to the type 3 FET mentioned above. The band diagrams do not correspond to Type 2 and Type 4 FETs. Figure 6.15 is I–V curves of enhanced MOSFETs.

Based on MOS, complementary MOS is developed, which is abbreviated as CMOS (complementary metal-oxide-semiconductor). CMOS uses two n-type and p-type MOSFETs. Their operating characteristics are complementary to

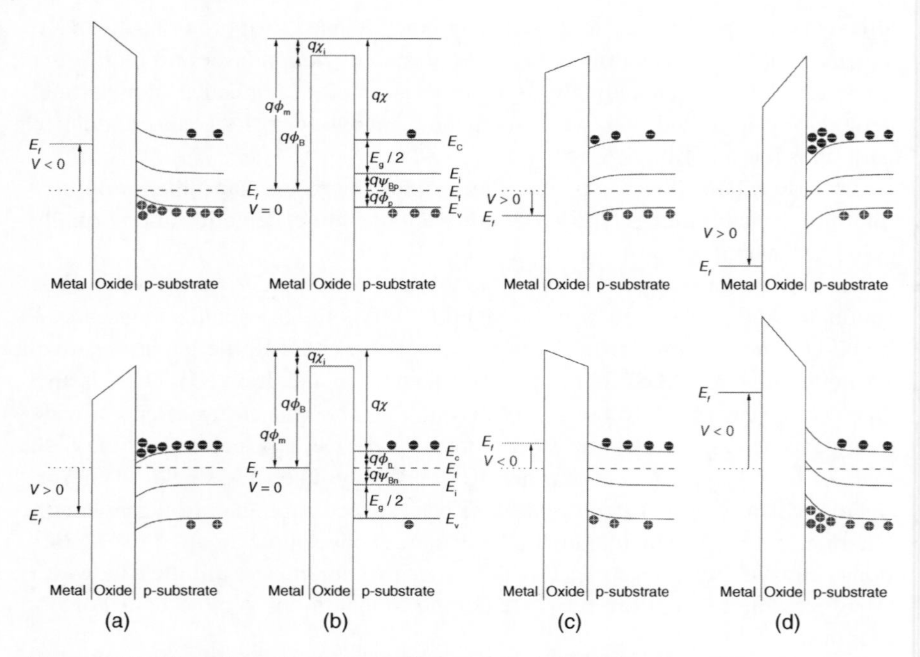

Figure 6.14 The schematic diagram of energy band of metal–oxide–semiconductor (a) $V < 0$, holes accumulate at the interface (image above). $V > 0$, electrons accumulate at the interface (image below), (b) $V = 0$, the energy band is in a flat-band state, (c) $V > 0$, holes are driven away from the interface (image above). $V < 0$, electrons are driven away from the interface (image below), (d) $V > 0$ more, n-type channel is generated (image above). $V < 0$ more, p-type channel is generated (image below). Source: Reprinted with permission of IuE, TU Wien.

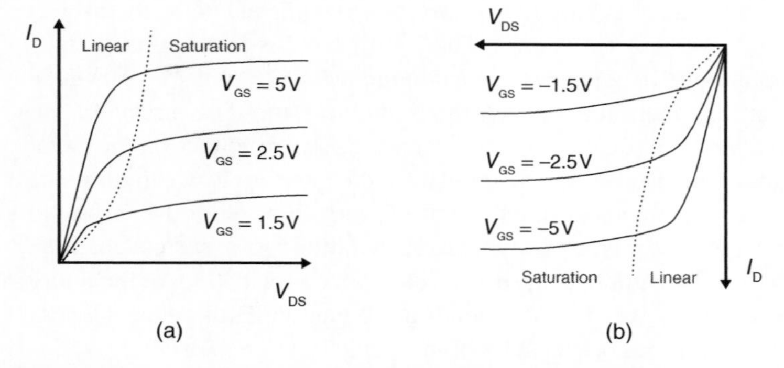

Figure 6.15 I–V curves of enhanced MOSFETs. (a) is n-channel, (b) is p-channel.

each other. For example, if the positive voltage of an electric signal is applied to the gate of an n-type device, the negative voltage is applied to the gate of a p-type device, or vice versa. Connect these two types of devices symmetrically to form a CMOS. CMOS has two important characteristics – low power consumption and easy to make logic gates, so it is widely used in the production of computer chips. The MOSFET mentioned above is the basic structure of CPU and memory, and it appears in the form of CMOS. To understand logic gates, we need to make a brief introduction to binary operations.

Computers are based on binary operation. Our daily life is based on decimal operation and watches use six decimals. The binary operation is used because it is the most convenient way of operation for the computers. Logic gate is composed of CMOS. When the gate is on, the output voltage is low, we assume it is "0." When the gate is off and the output voltage is high, we assume it is "1." Or set "0" and "1" in the opposite state. The numbers have only "0" and "1" in binary operation.

The decimal operation has 10 numbers: 0, 1, 2, 3, 4, 5, 6, 7, 8, and 9. The reason why we use it is that people have 10 fingers, which is a natural choice. Binary operation is called digital operation, or logical operation. The corresponding technology is digital technology. Now the most electronic systems and equipment, whether they are as large as communication systems, giant computers, or as small as home computers, mobile phones, cameras, etc., use digital technology. The calculating rules of binary operation are named Boolean algebra, which was put forward by the British mathematician George Boole (November 12, 1815–December 8, 1864). Using Boolean algebra, we can easily use CMOS to design basic logic gates. They are NOT gate, AND gate, OR gate, NAND gate, NOR gate, Exclusive OR gate "XOR", and Exclusive NOR gate "NXOR." The CPU used in a computer is composed of thousands of such basic logic gates. Since the logic gates composed of CMOS only consider the two states of switching, their anti-noise ability is strong. Easy to implement logic gates, low power consumption, and anti-noise, these three characteristics make CMOS technology to play an irreplaceable role in the IT field. Figure 6.16 is a logic gate-NOT gate formed by CMOS.

The gate is composed of one n-channel enhancement MOSFET (T_1) and one p-channel enhancement MOSFET (T_2). When the input V_{in} is low voltage, T_1 is off and T_2 is on, the output V_{out} is high voltage (around V_{dd}). When the input V_{in} is high voltage, T_1 is on and T_2 is off, the output V_{out} is low voltage (around 0). This realizes the function of a NOT gate. Since the output signal of the NOT gate is opposite to the input signal, a NOT gate is also called an inverter.

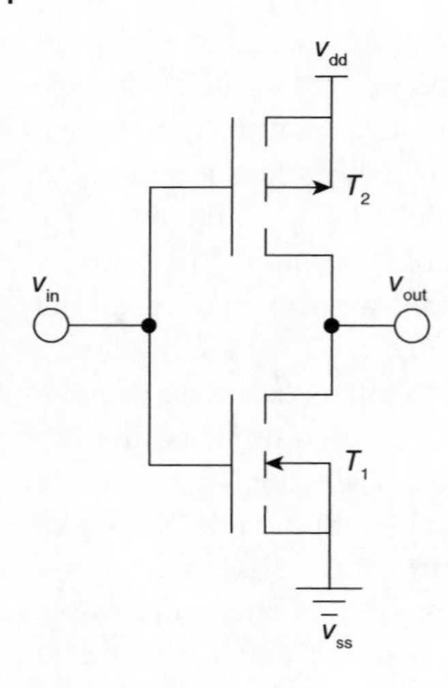

Figure 6.16 CMOS NOT gate.

References

1 童诗白主编. (1980). 模拟电子技术基础(上册), 人民教育出版社, 43 页, 38 页, 64 页, 66 页.

2 Sze, S.M. *Physics of Semiconductor Devices*, 2e, p. 246, p. 850, p. 247, p. 304, p. 305, pp. 336–339, p. 363, p. 455.

3 Lide, D.R. (2008). *CRC Handbook of Chemistry and Physics*, 12–114.

7

The Development History of Semiconductor Industry

In the previous chapter, we briefly introduced different types of semiconductor transistors and described integrated circuits. In this chapter, we introduce semiconductor products, the development history of the industry, and clean rooms.

7.1 The Instruction of Semiconductor Products and Structures

The abovementioned CMOS NOT gate is just a simple integrated circuit. When a transistor or diode is completed, package them to become a product of transistor (Figure 2.6) or diode (Figure 5.10) seen on the market. After opening a package, we can see a die (chip) inside, please see Figure 7.1, and refer to Figure 6.1. The case is used as the collector of the transistor. A single transistor and diode cannot work normally. They need to cooperate with other devices to be used as a circuit. Figure 6.2 is a simple diagram of transistor circuit. The circuit can be built on a circuit board (see Figure 7.4) to realize the circuit function. But the circuit board has a problem-it is bulky. Now we may ask, can transistors, diodes, resistors, and capacitors be implemented on one chip so that a small chip can have the function of a huge circuit board? The answer is yes. Based on this idea, Jack Kilby and Robert Noyce invented the integrated circuit in 1958 (refer to Chapter 3). The IC is to put transistors, diodes, resistors, and capacitors onto a small piece of semiconductor to realize the function of the circuit. The invention of the integrated circuit truly realized the small size and low power consumption of the circuit, and the world has entered the IT era since then.

Semiconductor Microchips and Fabrication: A Practical Guide to Theory and Manufacturing,
First Edition. Yaguang Lian.

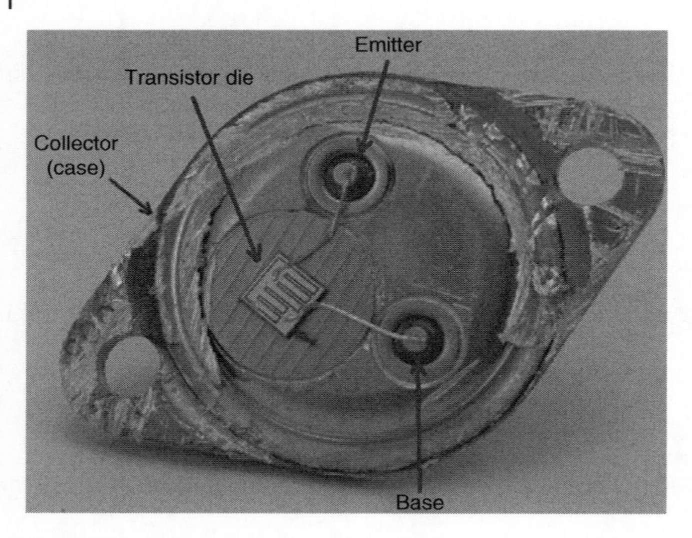

Figure 7.1 An unpackaged transistor. Source: Reproduced with permission of muzique.com.

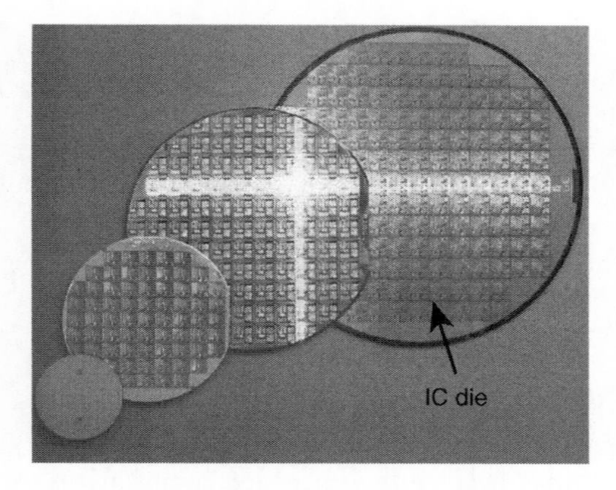

Figure 7.2 Silicon wafers with IC dies on them. Source: Unknown Source / Wikimedia Commons / CC BY-SA 3.0.

Figure 7.2 shows wafers of different sizes (described later) and IC dies on a wafer. Figure 7.3 shows a packaged CPU product and the CPU dies after opening the shell. Figure 7.4 is a computer motherboard. The motherboard is a circuit board, and the CPU is installed on it.

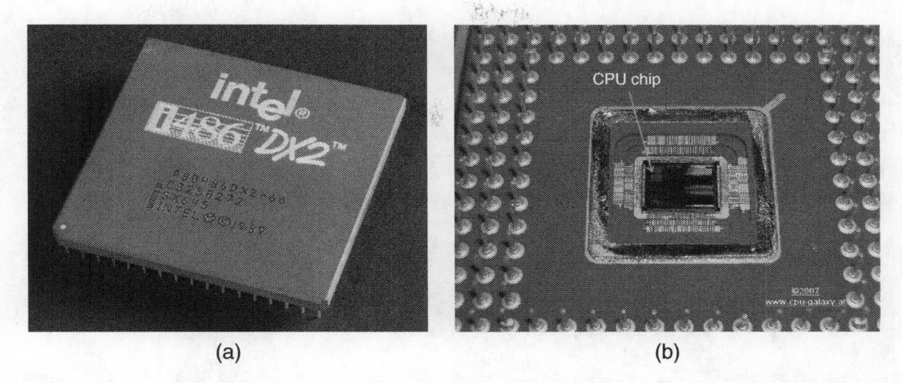

(a) (b)

Figure 7.3 A packaged CPU (a) and an unpackaged CPU (b). Source: (a) Intel Corporation, (b) Reproduced with permission of gmx.at.

Figure 7.4 A mother board of computer and a CPU on it.

7.2 A Brief History of the Semiconductor Industry

The semiconductor industry originated in Silicon Valley. After inventing the transistor, William Shockley left Bell Labs to California in 1953. After three years of working at the California Institute of Technology in Southern California near Los Angeles, he went to Mountain View of Northern California near San Francisco in 1956. Here, he established the Shockley Semiconductor Laboratory. Dissatisfied with Shockley's management method, the eight engineers working in the laboratory left there and founded Fairchild Semiconductor. Shockley called them "traitorous eight." Later, in 1968, two of them, Robert Noyce and Gordon Moore, left Fairchild to form Intel Corporation [1] (refer to Chapter 3).

In 1974, Intel developed the world's first widely used microprocessor-Intel 8080, please see Figure 7.5. In 1975, Bill Gates and Paul Allen designed programs based on the Intel 8080 for MITSs (Micro Instrumentation and Telemetry Systems), which began the course of Microsoft Corporation [2].

Figure 7.5 Microprocessor Intel 8080. Source: Konstantin Lanzet / Wikimedia Commons / CC BY-SA 3.0.

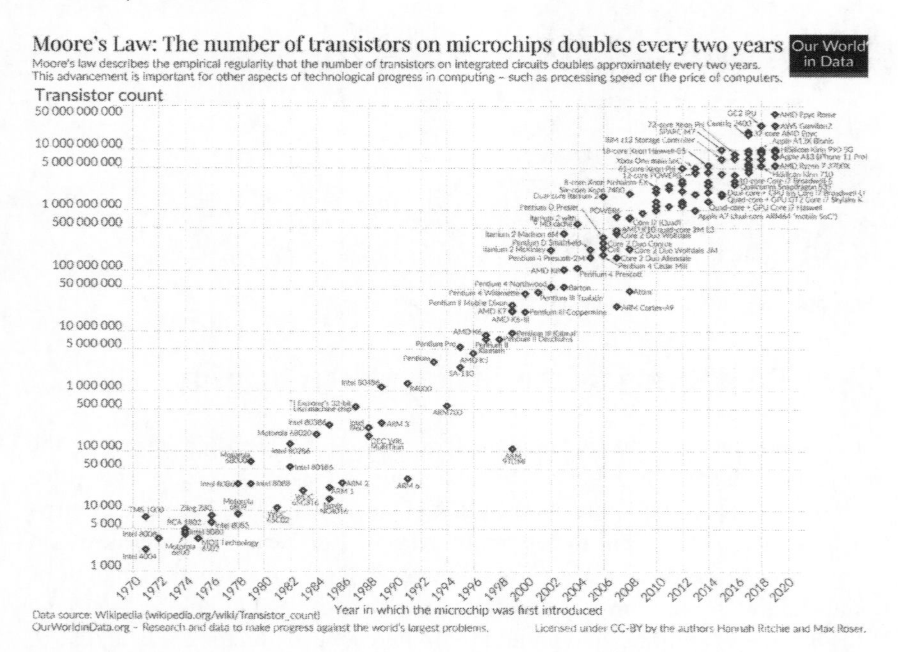

Figure 7.6 The number of transistors on IC chips. Source: Reproduced with permission of OurWorldData.org.

In 1965, Gordon Moore published an article in which he described that the number of devices contained in an integrated circuit would double every year. In 1975, he revised his prediction, turning one year into two years, that is, the number of devices that can be accommodated on an integrated circuit will double about every two years. This is Moore's Law. As the number of devices increases on a die, the performance of chips will be greatly improved. Until now, the semiconductor industry still follows the predictions of Moore's Law. The law also guides long-term planning in this field and the setting of R&R projects.

Figure 7.6 shows the change in the number of transistors on integrated circuit chips from 1971 to 2020. The horizontal coordinate of the figure is year, and the vertical coordinate is the number of transistors. The company and product names in the figure do not need to be concerned. It can be seen from the figure that in 1971, the number of transistors on a chip was about 1000. By 2020, the number of transistors on a chip was nearly 50 billion!

7.3 Changes in the Size of Transistors and Silicon Wafers

As the number of transistors contained on a die increase, the size of the transistors becomes smaller and smaller. We use a word to mark the size of the transistor-feature size. The larger the feature size, the larger the size of the transistor. The smaller the feature size, the smaller the size of the transistor. In addition to the small size of a transistor, there are also two important characteristics: fast speed and low power consumption. At the beginning, the feature size was marked by the gate length (L or L_G) of the field effect transistor, refer to Figure 6.13. Since the gate length is the smallest size in a transistor, the feature size is sometimes called the "minimum feature size," please see Figure 7.7. Due to the manufacturing process, the gate length and the channel length of the gate are sometimes inconsistent. Figure 7.7 illustrates this difference clearly. The "Poly" in the figure

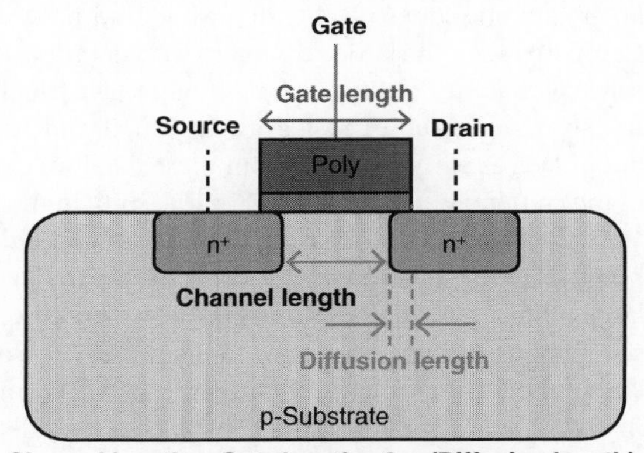

Figure 7.7 The schematic diagram of gate length of a MOSFET Source: Reprinted with permission of http://vlsi-soc.blogspot.com.

refers to the polysilicon gate. Later, the half pitch (hp) of dynamic random-access memory (DRAM) was used as the minimum feature size for many years, please see Figure 7.8. After that, the industry used other device sizes as feature size. In short, the feature size or minimum feature size is used to identify the size of the device.

Figure 7.8 The meaning of a pitch, the hp is half of a pitch. Source: International Technology Roadmap for Semiconductors 2002 Update.

By using of feature size, the International Technology Roadmap for Semiconductors (ITRSs) provides guidance and assistance for the semiconductor manufacturing process and design specifications involved in the Technology Node of the semiconductor industry in different periods. Technology node, sometimes it is called process node, process technology, or simply node. Different node means different generation of the circuits and structures. Generally, small technology node represents small feature size. In recent years, the feature size has lost its original meaning in some semiconductor fabs or foundries. For example, the process nodes in some nm chips only represent a specific generation of chips made in a particular technology. The nm here is nanometer, and the length is one billionth (10^{-9}) of a meter. To give readers an idea of this size, the lattice constant of silicon is written here for reference: it is about 0.54 nm [3]. At the atomic level, we also commonly use a unit of length-Angstrom (Å), to commemorate the Swedish physicist Anders Jonas Ångström (August 13, 1814–June 21, 1874). 1 Å = 0.1 nm, so the lattice constant of silicon is about 5.4 Å. The driving force of the technology node is the Moore's Law. Figure 7.9 is an example of the development blueprint. In this figure, "Transistor Density Mtr/mm^2" refers to the transistor density, in the unit of million/mm^2. "HVM" is the abbreviation of "High-volume manufacturing." It can be seen from this figure that at the 10 nm node, more than 100 million transistors are made on an area of one square millimeter. Now, IBM has introduced 2-nm node chip [5]. Compared with the lattice constant, we can know how small this size is! It is conceivable that the density of 2 nm is higher. The paper says IBM could put 50 billion transistors onto a chip the size of a fingernail, and 2-nm-node chips could be rolling out of fabs as early as 2024. From these data, we can see how tiny the transistors in the contemporary chips are. The difficulty in the manufacture is very, very big.

As the number of transistors contained on a chip increase, the size of a chip becomes larger and larger. At the meantime, the size of the silicon wafer becomes larger and larger. Figure 7.10 shows the wafer sizes from the early 1960s when the semiconductor industry started to the present. We also use inches to indicate the size of the wafer, 1 in. = 25.4 mm. The size of silicon wafers has grown from 1 in. in early 1960 to 12 in. in 2001. 12 in. Si wafer is still used by most companies.

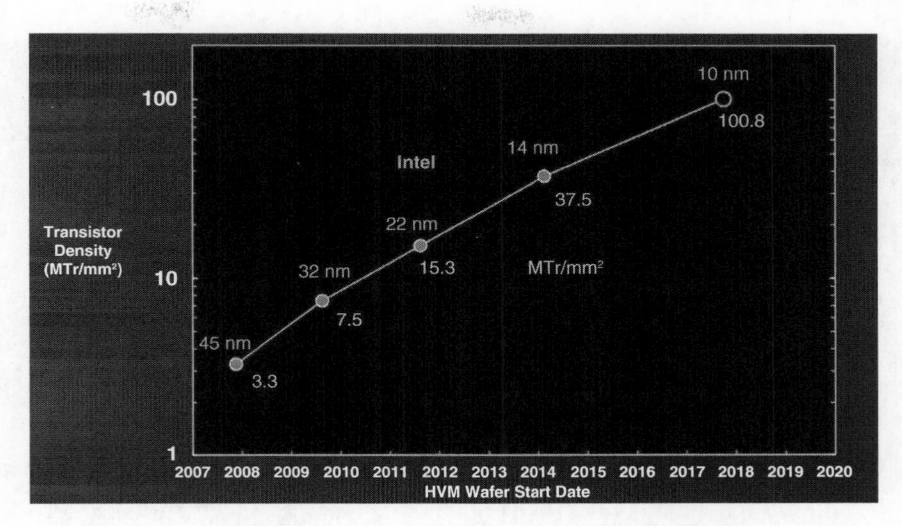

Figure 7.9 Technology development blueprint, MTr is million transistors [4].

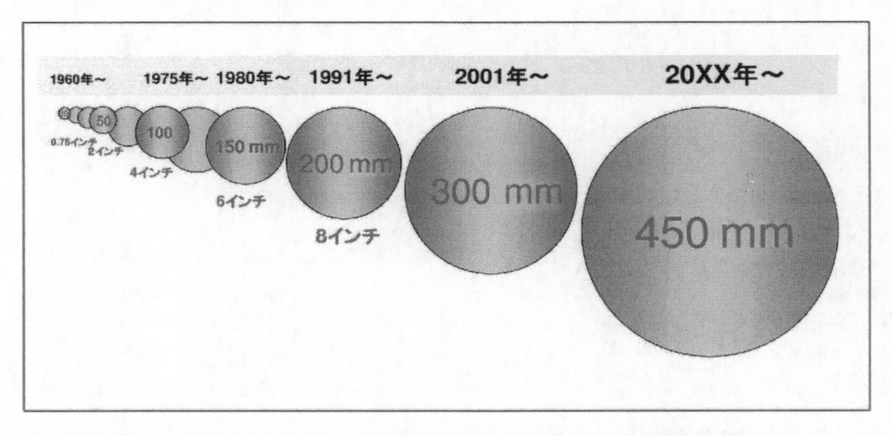

Figure 7.10 Silicon wafer size changes with the year Source: Copyright©2015 Tokyo Electron Limited, All Rights Reserved.

7.4 Clean Room

The transistors on the chip are getting smaller and smaller. Compared with the transistors, human hair is like a big mountain. The diameter of human hair is around 100 μm. 1 μm is one millionth (10^{-6}) of a meter. Figure 7.11 shows the sizes of some natural things. The size classification in the figure is from the perspective of semiconductor technology. The scale of the macroworld is much larger than that, and the size above 1 cm also belongs to the macroworld. So, when

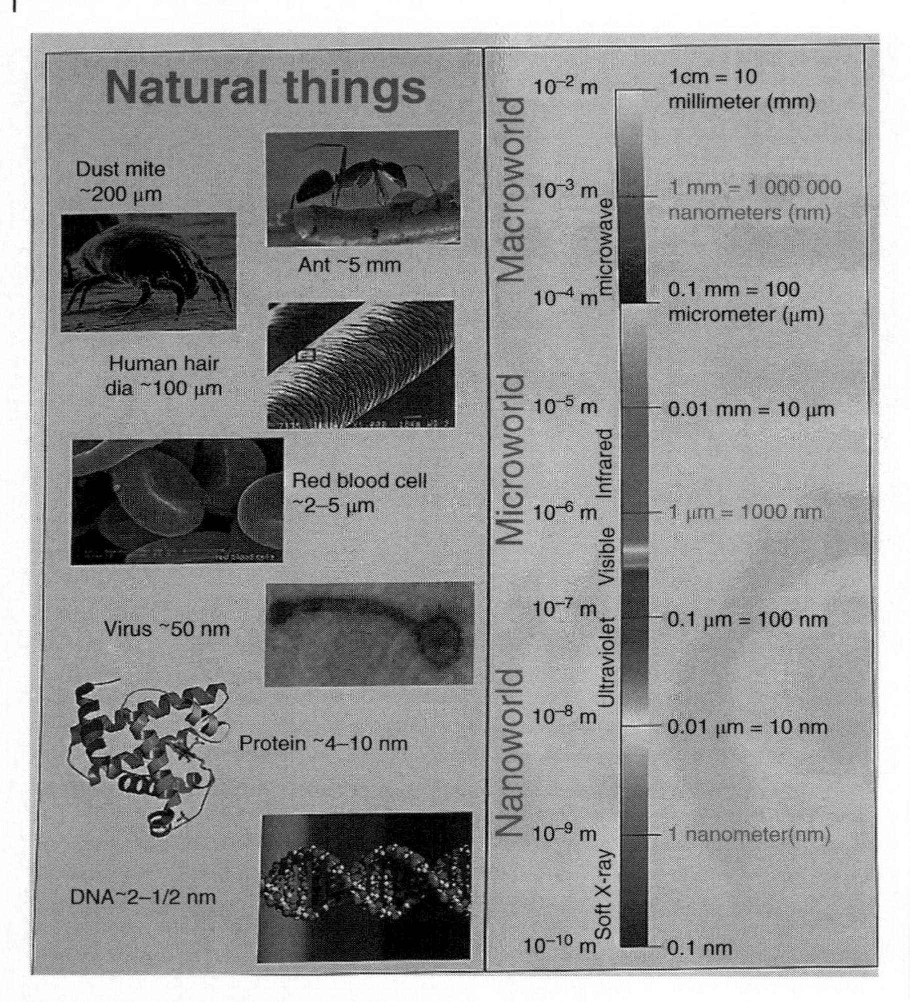

Figure 7.11 The scale of things-Macroworld to nanoworld.

we make semiconductor microchips, most of the work is done in the clean room. Please see Figure 7.12, which shows a clean room and an engineer wearing clean garment. The cleanliness of the air in the clean room follows the standards. One standard is Federal Standard 209E (FED-STD-209E) which was published by the Institute of Environmental Sciences and Technology (IEST) on September 11, 1992. FED-STD-209E had been widely used in the world, and it is still used in many places. In this standard, the cleanliness of air in the clean room is classified by "Class-X." For example, "Class-10" means that in each cubic foot of the air,

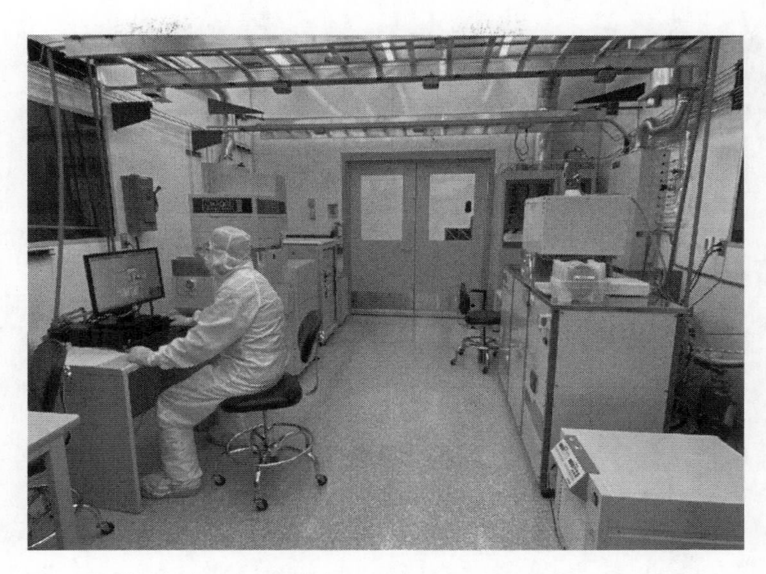

Figure 7.12 A clean room and an engineer wearing clean room garment.

total particles greater than 0.5 μm in size should be less than 10. Terms such as "Class-10" and "Class-100" are indelibly imprinted on our minds and in our speech. However, FED-STD-209E was canceled and superseded on November 29, 2001 [6], and replaced by the standards by International Organization for Standardization (ISO): International Standards for Cleanrooms and associated controlled environments, ISO 14644-1 Part 1: Classification of air cleanliness. Table 7.1 is ISO 14644-1 cleanroom standards [7].

Table 7.1 ISO classes of air cleanliness by particle concentration.

	Maximum allowable concentrations (particles/m^3) for particles ≥ the considered sizes, shown below "×" Particles in low concentrations make classification inappropriate				
ISO class number	**0.1 μm**	**0.2 μm**	**0.3 μm**	**0.5 μm**	**l μm**
1	10	×	×	×	×
2	100	24	10	×	×
3	1000	237	102	35	×
4	10 000	2370	1020	352	83
5	100 000	23 700	10 200	3520	832
6	1 000 000	237 000	102 000	35 200	8320

Figure 7.13 Ultra-high purity oxygen used in the process.

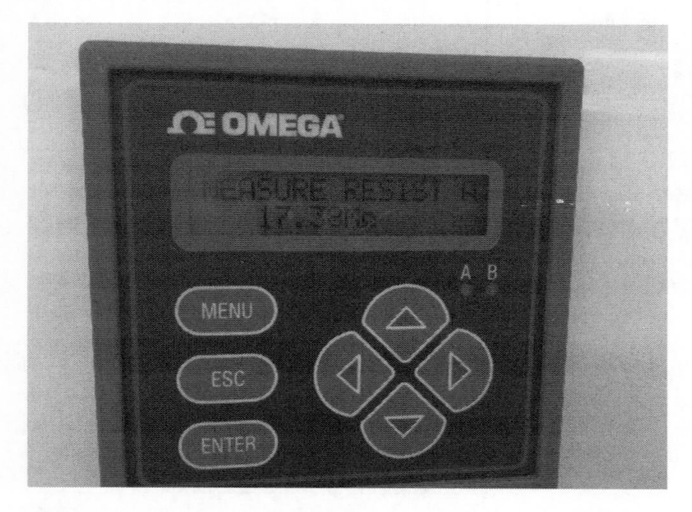

Figure 7.14 The resistivity of DI water with the unit of MΩ·cm.

In addition to the clean room, the entire process uses high-purity gases, chemicals, and other materials. Figure 7.13 is the oxygen used in semiconductor processing. The ultra-high purity (UHP) is 99.999%. The water used is deionized (DI) water. Figure 7.14 shows the resistivity of DI water. For comparison, the resistivity of pure water is about 18.2 MΩ·cm at room temperature [8], the resistivity of distilled water is around 500 kΩ · cm, and the resistivity of tap water is around 1–5 KΩ · cm [9].

7.5 Planar Process

Let us look at the first transistor in Figure 2.6, and the structure diagrams of the Figures 6.1, 6.4–6.6. They are all three-dimensional structures. This kind of structure is complicated to manufacture. Let us look at the real chips from Figures 7.1–7.3, they are all planar two-dimensional structures. It makes the production of transistors and integrated circuits simple. This manufacturing technology is a planar process (technology). In the process of making ICs, planar technology is first used to make individual devices, such as transistors and diodes, on a silicon wafer. After individual devices are completed, they are connected with metals (interconnection process). Please see Figure 7.15 for a schematic diagram of the planar process. This technology was invented by Jean Amédée Hoerni (September 26, 1924–January 12, 1997), one of the "traitorous eight." He applied for a planar process patent while working at Fairchild Semiconductor. In this figure, Hoerni drew the processes that included wafer, dies, oxidation, diffusion, openings by photoresist techniques, etching, and Ohmic contacts. All these processes have established the foundation of modern semiconductor manufacturing technology. We will discuss them later. Figure 7.16 is an early integration circuit made by Fairchild using planar technology. Figure 7.17 is a partial photo of a contemporary computer chip.

In Chapter 4, we mentioned passive components – resistors, capacitors, and inductors, which are used in an electrical circuit. But in an electronic circuit, besides the passive ones, there are active components – transistors and diodes. Among these five kinds of components, transistors and diodes, resistors and capacitors are easily integrated in the circuit through planar technology.

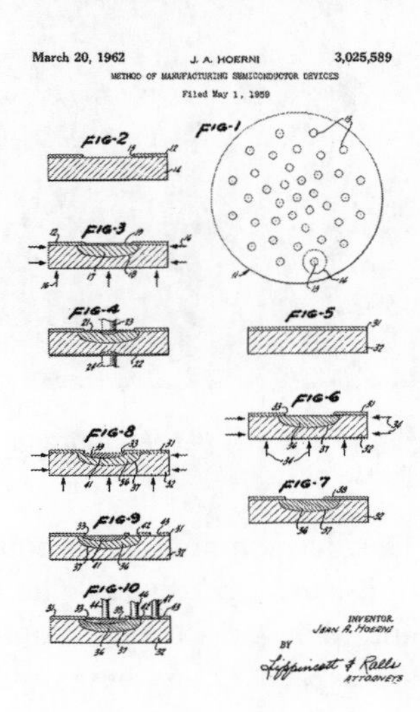

Figure 7.15 Schematic diagram of the planar process [10].

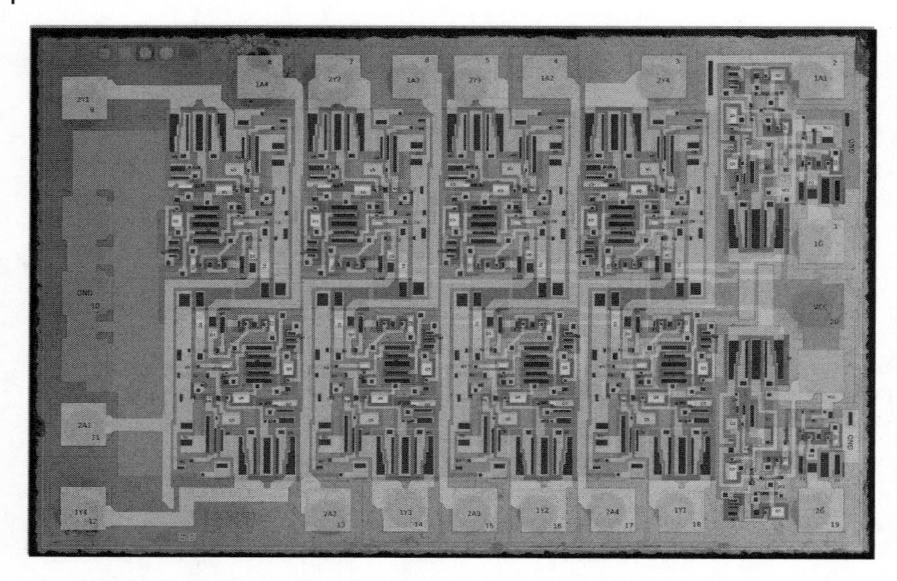

Figure 7.16 An IC chip made by Fairchild Semiconductor. Source: Robert Baruch / Wikimedia Commons / CC BY-SA 4.0.

Figure 7.17 A partial photo of a contemporary computer chip. Source: Reproduced from "ExtremeTech."

The development of various technologies is mainly for the manufacture of transistors and diodes. The resistors can be made by metals, doped polysilicon and alloy, or a certain doping area of silicon. MOS and reverse-biased diodes are capacitors, so they are not difficult to fabricate. However, it is not easy to

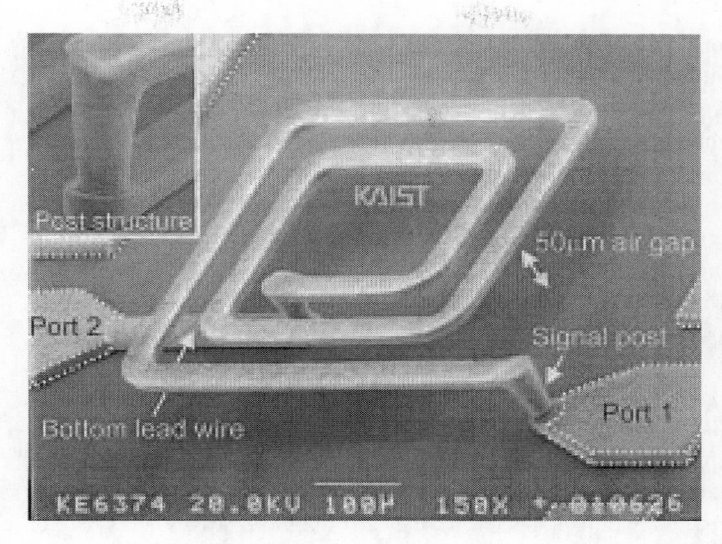

Figure 7.18 A spiral inductor for Si RF ICs. Source: [11] Jun Bo et al. 2002 / with permission of IEEE.

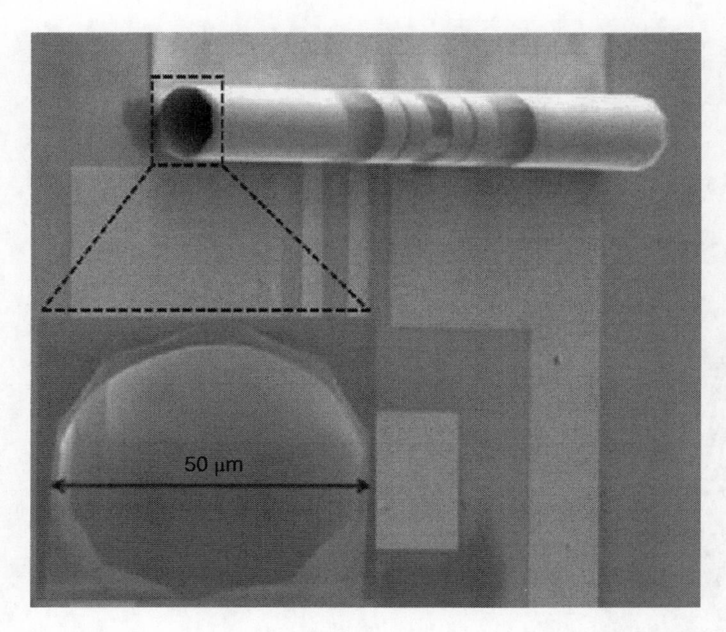

Figure 7.19 A transformer with μm size. Source: [12] Huang et al. 2018 / with permission of Springer Nature.

make an inductor because it is a coil structure and three-dimensional. Although there are many attempts to make an inductor, it is still difficult to integrate with the mainstream CMOS process. Figure 7.18 is a spiral inductor used in Si RF ICs. RF is the abbreviation of radio frequency. This inductor is made of metal microstructures and the manufacturing process is CMOS-compatible. In 2018, Xiuling Li and Huang Wen et al. used the self-curling characteristic of a stressed silicon nitride film, a μm tube or even nm one was made. A thin metal film was deposited on the inner wall of the tube, thus forming a μm or even smaller inductor coil. By making metal coils of different turns on a silicon nitride tube, a tiny transformer could be formed, see Figure 7.19. Although the transformer is currently used in RF circuits, it is hoped that in the near future, this transformer can be used for chip voltage conversion of small household appliances, so that these small appliances can be directly connected to the power socket on the wall instead of requiring a voltage converter (Figure 1.1).

The concept of RF was introduced above, which is a part of the electromagnetic spectrum. The frequency range (spectrum) of electromagnetic waves is very large.

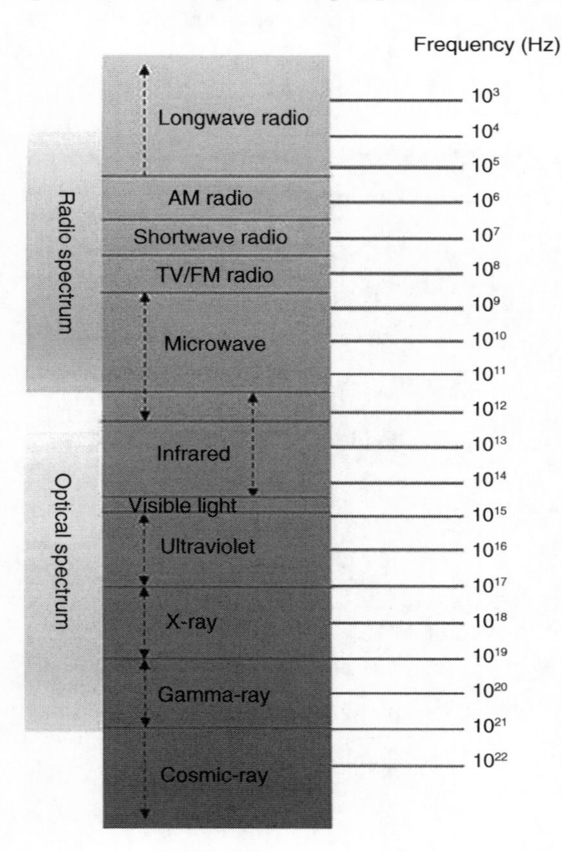

Figure 7.20 Electromagnetic spectrum.

We use different names to indicate the different ranges. Please see the electromagnetic spectrum in Figure 7.20. The frequency range from TV to amplitude modulation (AM) at the bottom of the figure is classified as a RF range. The radio we use is divided into three frequency ranges: medium wave (MW), that is AM broadcast. AM is followed by short wave (SW) and then TV/FM radio. FM is the abbreviation of frequency modulation. The wavelength of SW is shorter than that of AM. FM is shorter than SW. In the spectrum, electromagnetic waves in the RF and microwave ranges are widely used in radio communication and detection (radar) technology. The concept of electromagnetic spectrum is very important. We will continue to use it in Chapters 13–15.

References

1 Williams, J.B. (2017). *The Electronics Revolution Inventing the Future*, 104. Springer Praxis Books.

2 Microsoft Visitor Center Information for Students. Key Events in Microsoft History.

3 Sze, S.M. (1985). *Physics of Semiconductor Devices*, 2e, 850. Wiley.

4 Mistry, K. (2017). *10 nm Technology Leadership, Technology and Manufacturing Day*. Intel.

5 Johnson, D. (2021). Big blue gets small > IBM's 2-nanometer chip is a world's first. *IEEE Spectrum*, (August 2021), p. 7.

6 FED-STD-209 (2001). *Notice of Cancellation FED-STD-209 Notice 1*. The Institute of Environmental Sciences and Technology (IEST).

7 ISO 14644-1 (2015). *International Standard*, 2e. International Organization for Standardization (ISO).

8 Light, T.S., Licht, S., Bevilacqua, A.C., and Morash, K.R. (2005). The fundamental conductivity and resistivity of water. *Electrochemical and Solid-State Letters* 8 (1): E16–E19.

9 Wiater, J. (2012). Electric shock hazard limitation in water during lightning strike. *Electrical Review* 52–53.

10 Hoerni, J.A. (1959). Method of manufacturing semiconductor devices, US Patent 3 025,589, filed May 1, 1959.

11 Jun-Bo Yoon, Yun-Seok Choi, Byeong-Il Kim, Yun Seong Eo, Euisik Yoon, "CMOS-compatible surface-micromachined suspended-spiral inductors for multi-GHz silicon RF ICs", *IEEE Electron Device Letters*, Vol. 23, No. 10 October 2002, P. 591–593

12 Huang, W., Zhou, J., Froeter, P.J. et al. (2018). Three-dimensional radio-frequency transformers based on a self-rolled-up membrane platform. *Nature Electronics* 1: 305–313.

8

Semiconductor Photonic Devices

In the previous chapters, we discussed semiconductor microchips that deal with electrical signals. The characteristic of these microchips is that the input is an electrical signal, and the output is also an electrical signal. The protagonist of photonic devices is photons. Their main feature is that the input is electrical signals (energy) and the output is optical signals (radiation). Or on the contrary, the input is optical signals, and the output is electrical signals. Photonic devices can be divided into three categories: (i) convert electrical energy into optical radiation, (ii) detect optical signals through electronic technology, and (iii) convert optical radiation into electrical energy [1]. To understand this kind of device, we need to understand the basic characteristics of light and the principles of luminescence. In this chapter, we introduce these basic characteristics and principles, the difference between spontaneous emission and stimulated emission, and related devices.

8.1 Light-Emitting Devices and Light-Emitting Principles

Photonic devices, sometimes they are called optoelectronic devices. In this type of devices, photons play a key role. It can be seen from Figure 7.20 that although light is also a part of electromagnetic waves, its frequency is much higher than those of RF and microwave used for radio communication and radar. Because photons have short wavelengths and high energy, they will produce some special phenomena in semiconductors.

The use of light by humans can be traced back to ancient times. Torches, candles, and oil lamps have accompanied humans for thousands of years as lighting tools. But the true generation of optoelectronic technology is the invention of electric light. Thomas Alva Edison (February 11, 1847–October 18, 1931) invented the incandescent lamp (electric light bulb) in 1879, which opened the era of converting electrical energy into light radiation. Figure 8.1 shows picture of

Semiconductor Microchips and Fabrication: A Practical Guide to Theory and Manufacturing,
First Edition. Yaguang Lian.

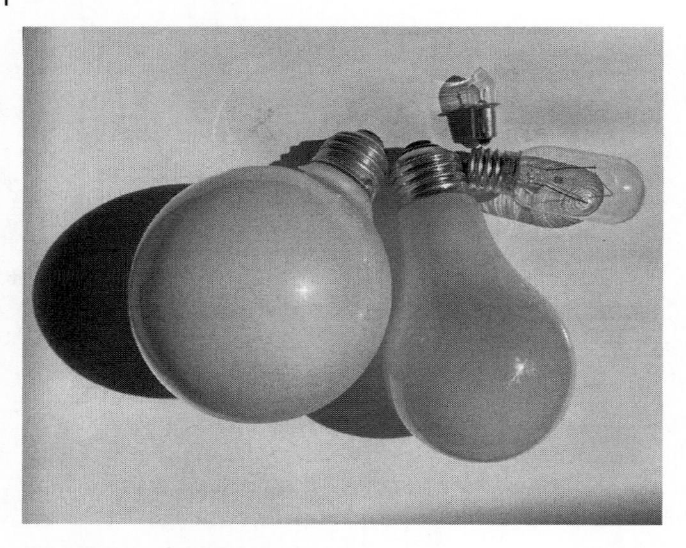

Figure 8.1 Incandescent lamps.

some of our commonly used incandescent lamps. In 1960, Theodore H. Maiman (July 11, 1927–May 5, 2007) and Gordon Gould (July 17, 1920–September 16, 2005) invented the world's first laser – ruby laser, please see Figure 8.2. Although both incandescent lamp and laser convert electrical energy into optical radiation, their light-emitting mechanisms are different. Incandescent lamp is spontaneous radiation and laser is stimulated radiation. From the perspective of energy level or energy band, luminescence is an electron jumping from the high-energy level of the excited state to the low-energy level, emitting a photon with energy equal to the difference between the two energy levels, see Figure 2.2. An electron is excited to transit from a low-energy level to a high-energy level, which can be done by the thermal shock of the crystal lattice, or by absorbing the energy of a photon. When an electron jumps to a high-energy level, if it jumps back to a low-energy level spontaneously and emits a photon, this way of emitting light is called spontaneous emission. When an electron is triggered by a photon with the energy that is the same as the energy level difference and jumps back to a low-energy level, two photons will be emitted at this time. This way of emitting light is called stimulated emission, and the light emitted is laser light. From the point of view of the laser emission process, the laser is a photon amplifier. In fact, the word "laser" is the abbreviation of "light amplification generated by stimulated radiation." The photons emitted by the laser have the same or similar physical properties. Such kind of light is called good coherence. Correspondingly, the coherence of light emitted by spontaneous emission lamps is not good. Figure 8.3 is a schematic diagram showing the difference between spontaneous emission

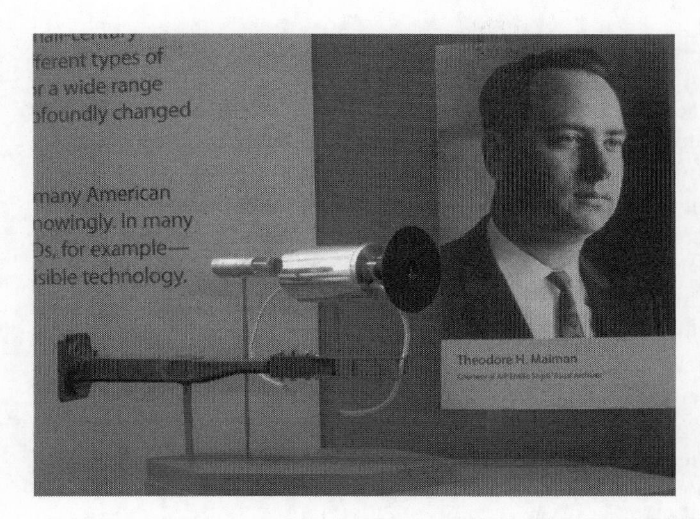

Figure 8.2 Ruby laser. Source: National Museum of American History / Heise Media.

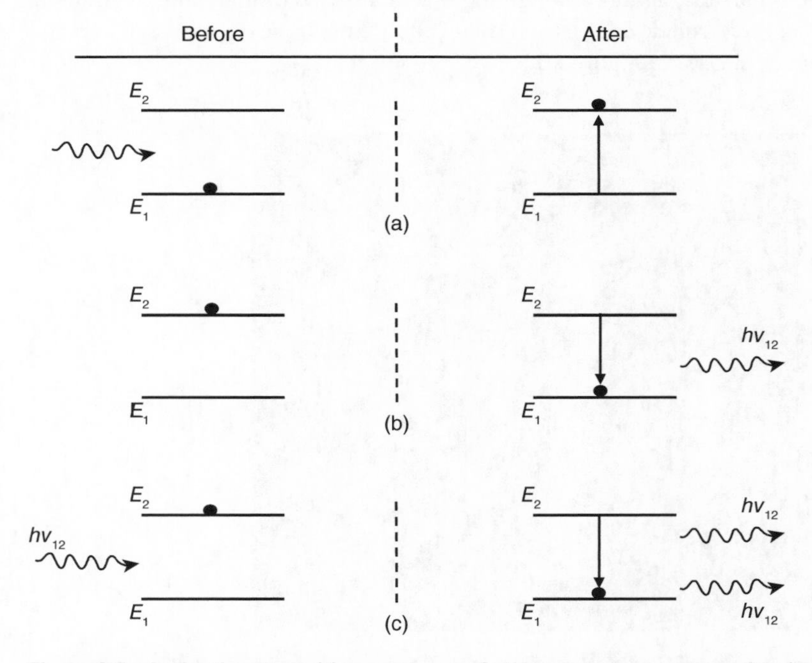

Figure 8.3 In the three transition processes of electrons between energy levels E_1 and E_2. The black dot represents an electron. The one marked "Before" on the left is the initial state, and the one marked "After" on the right is the transition state: (a) Absorb a photon; (b) spontaneous emission; and (c) stimulated emission. Source: [1] Sze / John Wiley & Sons.

and stimulated emission. The theory of stimulated emission was proposed by Einstein in 1917.

As we mentioned earlier, the original intention of the invention of transistor was to replace the vacuum (glass) tube with a solid-state device. After the semiconductor transistor was invented, people naturally asked: Can we use semiconductors to make light-emitting devices? Can this device be used to replace an incandescent lamp and make a smaller laser? The answer is yes. Later, many types of semiconductor photonic devices were invented. At the beginning of this section, we divided such devices into three categories. The first type is electro-to-light; such devices include light-emitting diode (LED), diode, and transistor lasers. The second type is detection of optical signals; such devices have photodetectors. The third type is optical-to-electricity; such devices include photovoltaic devices and solar cells. LEDs are basic devices used to replace incandescent lamps, to make display panels and LED TVs. Please see Figure 8.4. Diode lasers are basic devices used in fiber optic communication, laser printers, CD (Compact Disc), DVD (Digital Versatile Disc), and laser pointers, please see Figure 8.5. Optical detectors are basic devices used in CD players, digital cameras, and optical fiber communication. Please see Figure 8.6. Photovoltaic devices and solar cells are the basic devices used to make solar panels, please see Figure 8.7.

Figure 8.4 LED lamps.

Figure 8.5 Optical fibers.

Figure 8.6 Digital cameras.

Figure 8.7 A solar panel.

8.2 Light-Emitting Diode (LED)

Please refer to Figures 5.2 and 5.3. There are two types of energy band structures for semiconductors, one is direct band gap, and the other is indirect band gap. An electron excited to the conduction band is unstable. It jumps back to the valence band, combines with a hole, and emits a photon at the same time. This process is called electron–hole recombination. Figure 8.8 is a schematic diagram of the electron–hole recombination and photon radiation process in direct band gap and indirect band gap semiconductors. In the figure, E is energy, k is wave

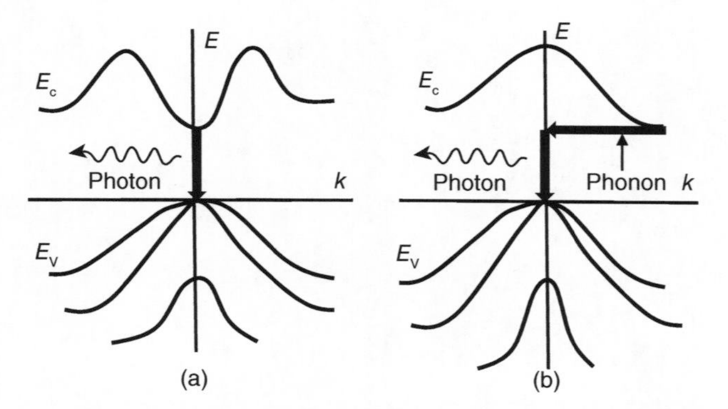

Figure 8.8 Schematic diagram of electron–hole combination and photon radiation in direct band gap (a) and indirect band gap (b). (a) An electron jumps directly from the conduction band to the valence band, combines with a hole, and a photon is emitted, (b) an electron collides with the crystal lattice, jumps indirectly from the conduction band to the valence band, combines with a hole, and a photon is emitted. Source: [2] Kevin / Cambridge University Press.

number. The space formed by k is called k-space, which is a space unit commonly used in solid-state physics:

$$k = \frac{2\pi}{\lambda} \tag{8.1}$$

To further understand the recombination and radiation processes in the figure, it is necessary to introduce the concept of momentum. Momentum is expressed by P. In Newtonian mechanics, P is the product of mass m and velocity v:

$$P = m \cdot v \tag{8.2}$$

In a closed system, there is no exchange of matter and force between the system and the outside world, and its total momentum is a constant, which is the law of conservation of momentum. Figure 8.9 is a schematic diagram of the conservation of momentum of two spheres. In the figure, the velocity is represented by u after collision. If a ball hits a wall, within the range of 0–90° of impact angle (θ), there will be three situations, as shown in Figure 8.10. Scenario a is when the ball hits the wall at a vertical 90° angle, the momentum generated at this time is the largest. Scenario b is when the ball hits the wall at a certain angle θ, the momentum generated at this time becomes smaller. The smaller the θ, the smaller the momentum. In case c, the impact angle is zero, the momentum in this case

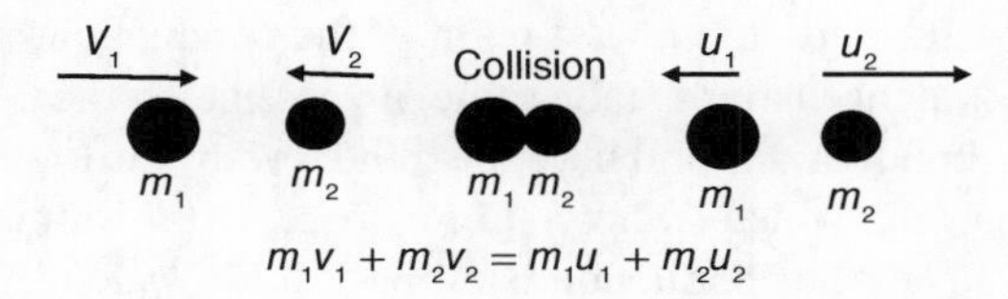

Figure 8.9 The law of conservation of momentum.

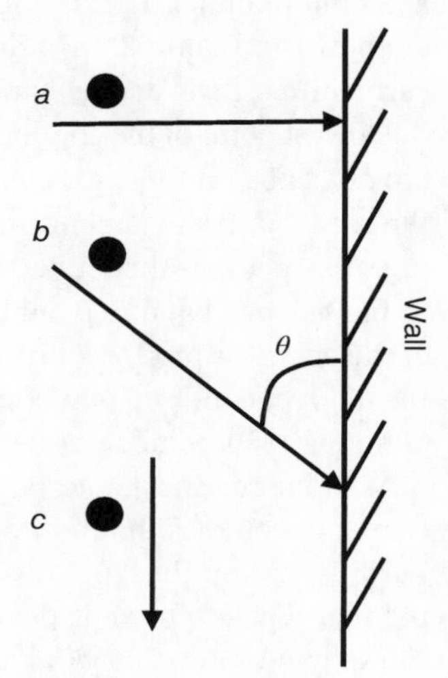

Figure 8.10 Three situations when a ball hits a wall.

is also zero. Generally speaking, the larger the momentum, the larger the impact. The smaller the momentum, the smaller the impact. The concept of momentum is very important, and we will use it later in Sections 15.6.2 and 16.4.

In quantum mechanics, microscopic particles also have momentum. French physicist Louis Victor Pierre Raymond de Broglie (August 15, 1892–March 19, 1987) pointed out that all matter has a wave-particle duality. In the microscopic world, the momentum of a particle is expressed by the de Broglie relation:

$$p = \frac{h}{\lambda} \tag{8.3}$$

The wavelength of λ is called de Broglie wavelength of the particle. By using relation (8.1), de Broglie relation can be rewritten as follows:

$$p = \hbar k \tag{8.4}$$

$\hbar$ is called reduced Planck constant. Its expression is as follows, and h is the Planck constant:

$$\hbar = \frac{h}{2\pi} \tag{8.5}$$

In quantum mechanics, the momentum of microscopic particles is also conserved. Let us look at Figure 8.8 again. Direct band gap semiconductors, such as GaAs, InP, etc., whose energy band structure is a direct band gap. In this structure, the lowest point of the conduction band and the highest point of the valence band are the same, no change on the k-axis. The electrons excited to the conduction band have the trend toward the lowest energy point of the conduction band, which is shown in Figure 5.2. At this time, the electrons will jump directly from the conduction band back to the valence band, recombine with the holes, and emit photons. The luminous efficiency is high of this energy band structure, as shown in Figure 8.8a. Indirect band gap semiconductors, such as silicon and germanium, their energy band structure is a indirect band gap. In this structure, the lowest point of the conduction band and the highest point of the valence band are not same on the k-axis. An electron in the conduction band first collide with the crystal lattice. Momentum is conserved, and some energy is lost. Then it enters the same place as the highest point of the valence band on the k-axis, jumps back to the valence band, recombines with a hole, and emits a photon. Because part of the energy is lost by collision with the crystal lattice, the luminous efficiency of the indirect band gap semiconductor is low, as shown in Figure 8.8b. In the figure, "Phonon" is used to describe the collision of an electron with the crystal lattice.

As discussed in Chapter 5, an important feature of commonly used compound semiconductors is that most of their energy bands are direct bandgap structures. For this reason, this kind of semiconductor is widely used in the manufacture of the first type of photonic deviceslight emitting devices, such as LEDs and lasers. These two kinds of devices are now widely used in the society because of their

Figure 8.11 Nick Holonyak and the invention of LED.

low cost, high efficiency, wide frequency spectrum, relatively simple drive circuit, high reliability, and long working life. LED was invented in the year 1962. At that year, Nick Holonyak and S.F. Bevacqua developed an LED that emitted visible red light [3]. Please see Figure 8.11.

The light emission of LED is a spontaneous emission. Its basic structure is a diode. When a forward voltage is applied, electrons diffuse into the P region and holes into the N region. The electrons that reach P region combine with the holes and emit photons. The holes that reach N region combine with the electrons and emit photons. This type of device converts electrical energy into light radiation. Comparing with diode, there are some special considerations in the design of LED. Its light emission mode is divided into surface emission and edge emission, please see Figure 8.12. "n^+, p^+" are N-type and P-type heavily doped (see Chapter 6). The LED chip is packaged in a dome-shaped plastic package, please see Figure 8.13.

To further understand LED, we need to introduce the three primary colors in television. They are red, green, and blue. Mixing these three colors can get white color, as shown in Figure 8.14. That means if we want to use LEDs to replace incandescent lamps, or use LEDs to make televisions, we need red, green, and blue LEDs. Let us analyze the required semiconductor band gap from the wavelength of light. The wavelength of red light is 620–750 nm, green light is 495–570 nm, and blue light is 450–495 nm. According to formula (2.1), and the following relationship between wave velocity V, frequency f, and wavelength λ:

$$V = \lambda \cdot f \tag{8.6}$$

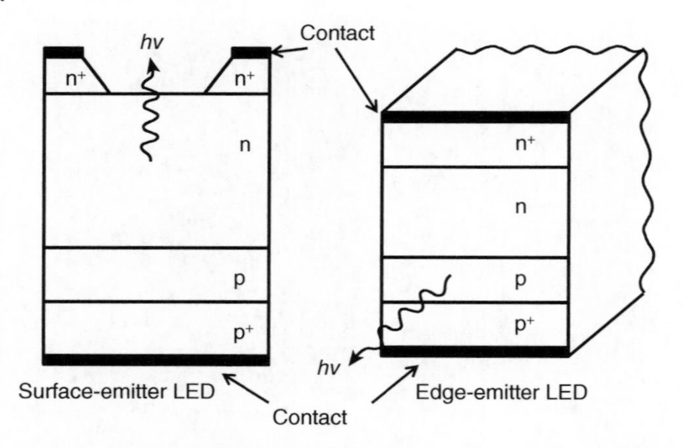

Figure 8.12 The basic structure of LED. Source: [2] Kevin / Cambridge University Press.

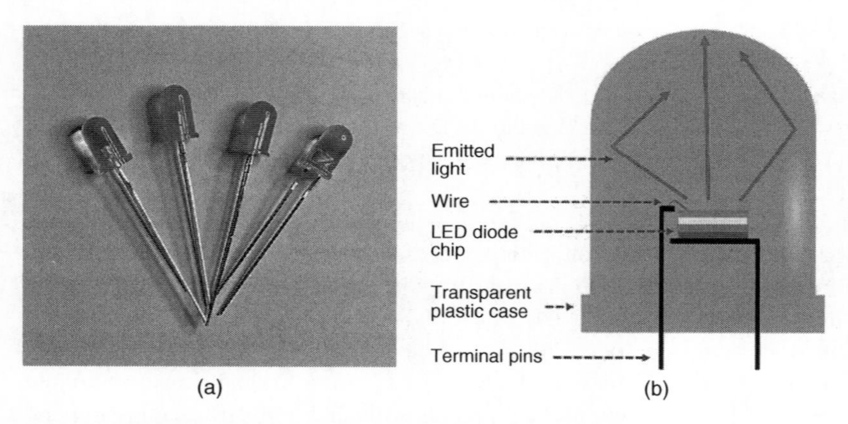

Figure 8.13 The products (a) and structure (b) of LED.

As mentioned in Chapter 2, the frequency in quantum mechanics is denoted by ν, and the speed of light is denoted by C. C is the first letter of "constant" because light moves at a fixed (constant) speed $C = 3 \times 10^8$ m/s to propagate in vacuum. Combined with Planck's constant $h = 4.134 \times 10^{-15}$ eV/s, we can have: for red light, $E = 2.00$–1.65 eV; for green light, $E = 2.51$–2.17 eV; and for blue light, $E = 2.76$–2.51 eV.

Pure III–V materials are not easy to make LEDs. Usually, the ternary materials are grown on an III–V substrate used as the light-emitting region. For red light and green light, the materials used to make LEDs are GaAs and GaP and their corresponding ternary materials $GaAs_{1-x}P_x$. GaAs is a direct band gap, $E_g = 1.42$ eV; GaP is an indirect band gap, $E_g = 2.26$ eV. As x changes from 0 to 1, the E_g of $GaAs_{1-x}P_x$ changes from 1.42 to 2.26 eV. When $x = 0.45$, the energy

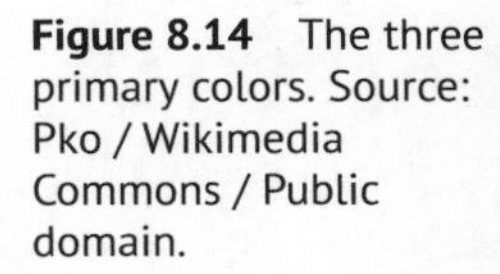

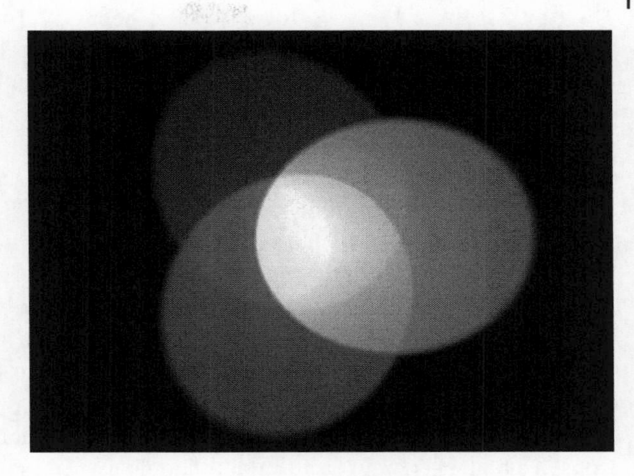

Figure 8.14 The three primary colors. Source: Pko / Wikimedia Commons / Public domain.

band changes from a direct band gap to an indirect band gap [1]. In 1962, Nick Holonyak used $GaAs_{1-x}P_x$ to create a red LED. In 1967, George Craford, a student of Holonyak, used GaAsP grown on GaAs substrates to make orange, yellow, and green LEDs [4]. To improve the luminous efficiency of the GaP indirect band gap, some special impurities should be added to this material to improve the recombination rate of electrons and holes.

After the invention of red and green LEDs, blue LED research progressed slowly. $GaAs_{1-x}P_x$ could not meet the needs of blue light because its maximum band gap is less than the energy corresponding to the blue wavelength. The ternary materials InGaN ($In_xGa_{1-x}N$) based on gallium nitride (GaN) substrate are used to make the light-emitting region of blue LEDs, which are made of a mixture of GaN and InN. GaN and InN are direct band gaps. The band gap of InGaN can vary from 3.4 eV for GaN to 0.69 eV for InN, which spans the energy of blue light and can meet the requirements of manufacturing blue LEDs. However, the preparation of GaN material encountered many difficulties. Until the end of 1980 and the beginning of 1990, three Japanese scientists, Shuji Nakamura, Hiroshi Amano, and Isamu Akasaki, broke through the difficulties in the preparation of GaN substrate and invented the GaN-based blue LED [5].

The invention of blue LED makes blue light, the last light source of the primary colors to come out. It is possible to replace the incandescent lamp with solid-state light source LED and develop a new type of TV-LED TV. The LED TV uses the three-color ratio to adjust the color, while the LED bulb uses the light emitted by blue LED to excite the fluorescent material in the bulb to emit light and convert blue light of the LED into white light. The conversion efficiency of contemporary white LED bulbs from electrical energy to light radiation can exceed 50%, while the conversion rate of ordinary incandescent lamps is only 4%. The service life of LED bulbs can reach 100 000 hours. Correspondingly, the life of fluorescent lamps is 10 000 hours, while that of incandescent lamps is 1000 hours [6].

8.3 Semiconductor Diode Laser

We discussed LED in the previous section. In this section, we will discuss the diode laser(or laser diode). As mentioned earlier, the laser is a photon amplifier. It has good coherence. To understand the coherence, it is necessary to make a brief introduction to the coherence of light.

In physics, the good coherence for two beams of waves means that their frequency and waveform are identical, and their phase difference is constant. Figure 8.15 is a schematic diagram of the phase difference between two waves. Figure 8.16 is a schematic diagram of coherent light and incoherent light. In the figure, monochromatic means that light waves have same frequency. Laser is monochromatic light.

As can be seen from Figure 8.3, to achieve laser emission, the population of electrons in a higher-energy (excited) state should exceed that in a lower-energy state, which is the population inversion. In addition, to achieve the coherence of the emitted light, the device must have a resonant cavity. To ensure laser to occur, the gain exceeds the total losses. This condition can turn on the current density injected into the diode junction. There is a minimum current density necessary for a semiconductor device to lase, which is threshold current density. These are three conditions – population inversion, threshold current density, and resonance – that need to be met to achieve semiconductor laser emission [2]. Resonance refers to the phenomenon that the amplitude increases sharply when the frequency of the external force is the same or close to the natural frequency of the system. Now let us explain the three conditions one by one:

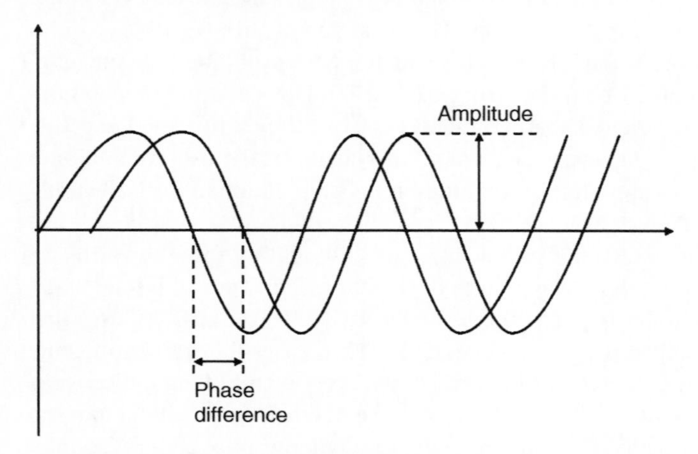

Figure 8.15 The phase difference between two waves.

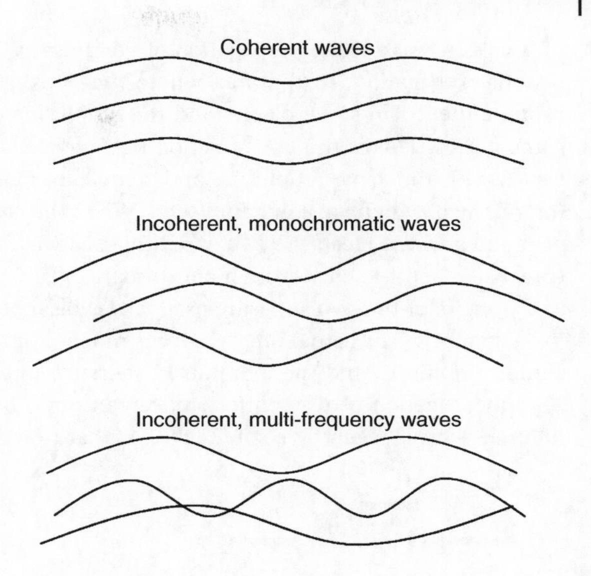

Figure 8.16 Schematic diagram of coherent light and incoherent light.

8.3.1 Resonant Cavity

The resonant cavity in the laser can produce laser oscillation. The most used structure is the Fabry–Perot resonator, also known as the Fabry–Perot interferometer. It was developed by French physicists Charles Fabry (June 11, 1867–December 11, 1945) and Alfred Perot (November 3, 1863–November 28, 1925) in 1899. The resonator is usually composed of two parallel transparent plates plus reflective surfaces, or two parallel mirrors. Light forms a standing wave through multiple reflections on two surfaces. The standing wave is a wave confined to two parallel reflecting surfaces. It oscillates with time but does not propagate in space. Correspondingly, the waves propagating in space are traveling waves. Figure 8.17 is a schematic diagram of the Fabry–Perot resonator structure. To further understand how it works, we should know the interference

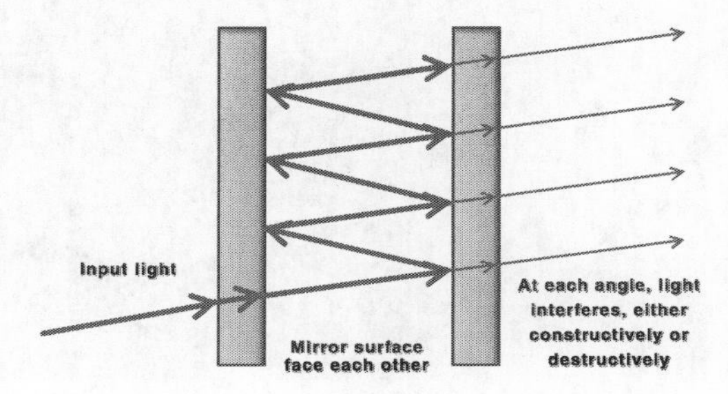

Figure 8.17 The structure of Fabry–Perot resonator. Source: Reprinted with permission of http://starkeffects.com.

of waves. A wave is the propagation of vibration. When two waves meet on the way of propagation, in the area where they meet, the amplitude of the wave is strengthened in some places, and the amplitude of the wave is weakened in some places. This is the phenomenon of interference of waves. Only when two waves with the same amplitude and frequency meet, they can produce stable interference, which are coherent waves. When the crest–crest and trough–trough of the two waves meet, the two superimposed waves will have larger crests and troughs. When the crest-trough and trough–crest of two waves meet, the crests and troughs of the two superimposed waves disappear. Figure 8.18 shows these two situations. The actual interference is much more complicated than these two simple situations, and there are also coherence phenomena of multiple waves. The phenomenon of wave interference can often be encountered in daily life. Figure 8.19 was taken by a small lake. At that time, there were a few Canadian

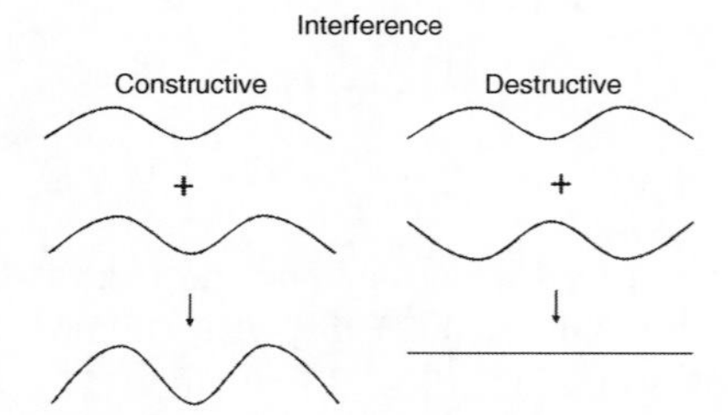

Figure 8.18 Schematic diagram of the interference of two waves.

Figure 8.19 The interference of water waves.

geese in the lake. The water waves caused by them had interference. It can be seen clearly from the figure that the waves are strengthened in some areas and are weakened in some areas. The Fabry–Perot resonator in Figure 8.17 is the interference of coherent waves, but the second wave is the reflected wave. In other words, the resonator is the interference of the incident wave and the reflected wave. So, the Fabry–Perot resonator is also called the Fabry–Perot interferometer.

Here, we have talked about the reflection of light beams. In addition to reflection, another important characteristic of light is refraction. For reflection, what we frequently encounter is mirror reflection. At the interface of different propagation media, light beams can be reflected, and refracted as well. Let us discuss further.

8.3.2 Reflection and Refraction of Light

When light beams impinge on a surface, if the surface is an ideal mirror surface, the light can be completely reflected from the surface; this phenomenon is called total reflection of light. If the surface is smooth, but not an ideal reflected mirror surface, such as a stationary water surface, some light beams are reflected by the surface, while some light beams pass through the surface and enter the water. Please see Figure 8.20. The light enters the water from the air. In physics, it is called the light from the first transparent optical medium (air) to the second transparent optical medium (water). Of course, the light can continue to pass through the third transparent optical medium… Here, we use two transparent optical media of air and water as examples to discuss the reflection and refraction of light.

When a beam of light hits the surface of water at an angle θ_1 with the normal to the surface, θ_1 is called the incident angle. Part of the light is reflected by the water surface and reflects to the other side of the normal to the surface with same angle,

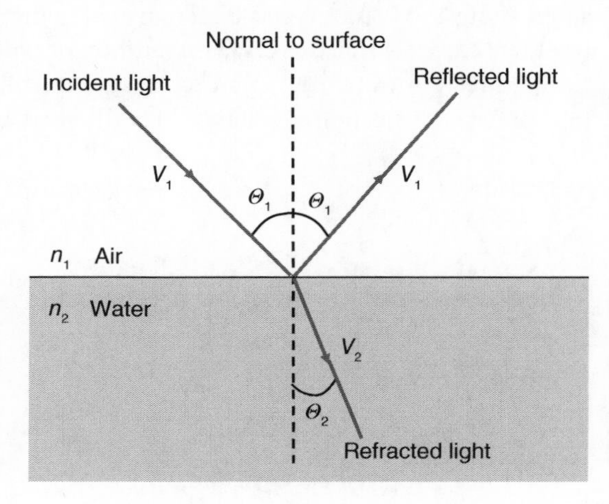

Figure 8.20 The reflection and refraction of light.

which is the law of light reflection. Part of the light enters the water, the angle in the water is θ_2, $\theta_1 \neq \theta_2$, θ_2 is called the refraction angle, which is the refraction of light. The speed of light in different media is different. If V_1 represents the speed in air and V_2 represents the speed in water, then the propagation of light between two different media satisfies the following formula:

$$\frac{\sin\theta_2}{\sin\theta_1} = \frac{V_2}{V_1} = \frac{n_1}{n_2} \tag{8.7}$$

The n_1 and n_2 in the formula are the refractive indices of light in air and water, respectively. In general, the refractive index (index) is expressed by n. The commonly used indices are: 1 for vacuum, around 1.33 for water, and around 1.45 for quartz [7]. The index of air is almost 1. This formula is called Snell's law, in memory of Dutch astronomer Willebrord Snellius (June 13, 1580–October 30, 1626). The $\sin\theta_1$ and $\sin\theta_2$ in the formula are sine functions of the incident angle and refraction angle, respectively.

The index of water is bigger than that of air. So, water is called denser medium, and air is called rarer medium. According to formula (8.6), $\theta_1 > \theta_2$. We can imagine that when light is shot from water to air (denser medium to rarer medium), if the angle of incidence θ changes from small to large, the light refracted to the air tilts toward the water surface until θ exceeds a certain value, the beam of light will be completely refracted into the water. In this way, although the water surface is not an ideal mirror surface, total reflection can still occur. It is called total internal reflection. The angle of incidence at which refracted light produces total internal reflection is called the critical angle. Please see Figure 8.21.

In semiconductors, two methods are commonly used to make resonant cavities: (i) Mirror resonance: The method of making mirrors is to cleave along the direction perpendicular to the p–n junction. Cleavage means that the crystal splits along a certain crystal direction under the action of external force. We use the cleavage surface as a mirror surface; or polish the surface perpendicular to the direction of p–n junction to make a mirror surface. (ii) Heterojunction total internal reflection resonance: The method is to use a heterojunction to

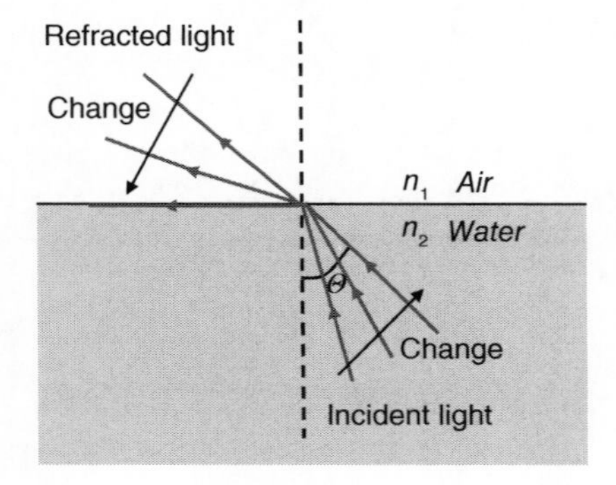

Figure 8.21 The critical angle of refracted light.

make a resonant cavity with total internal reflection in the interface between two materials. These two materials have different refractive indices which are used as denser medium and rarer medium to confine the light. In semiconductor lasers, these two resonant cavities are used in combination. Now let us discuss the heterojunction.

8.3.3 Heterojunction Materials

All devices we discussed above are made of homojunction materials, that is, the devices use one material, such as silicon, gallium arsenide, etc. In fact, we also use another technology to make devices, which is heterojunction technology. This technology is widely used in III–V semiconductors. Heterojunction technology is used to make two materials with different band gap widths together to form a p–n junction, that is different from homojunction material. Please see Figure 8.22. Heterojunction technology was first proposed by Herbert Kroemer. One of its important advantages is the faster operation speed of the devices. Nowadays, terahertz (THz, 10^{12} Hz) is frequently mentioned. The device for THz uses heterojunction technology. Another advantage is that the refractive indices of the two materials are different, so that a Fabry–Perot resonant cavity can be made. Please see Figure 8.23. The homojunction cannot make a resonant cavity due to the same refractive index. The Fabry–Perot cavity in the figure has two heterojunctions, so this structure is called a double-heterojunction (DH) laser. In this structure, the light is restricted to the active region (denser medium). This structure is also called a waveguide. The optical fiber (Figure 8.5) uses a waveguide structure. Please see Figure 8.24. The core of the optical fiber is made of glass or plastic. Figure 8.25 is a schematic diagram of a p–n double heterojunction laser, in which the refractive index of AlGaAs is lower than that of GaAs.

We have discussed the simple structure of a semiconductor laser. Actual lasers are much more complex than this. There are many structures, such as quantum well lasers, quantum dot lasers, and vertical cavity surface emitting lasers. We will not discuss them here.

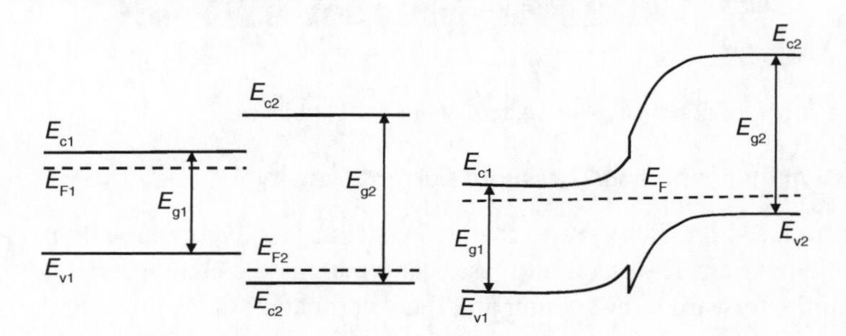

Figure 8.22 An example of p–n heterogeneous junction.

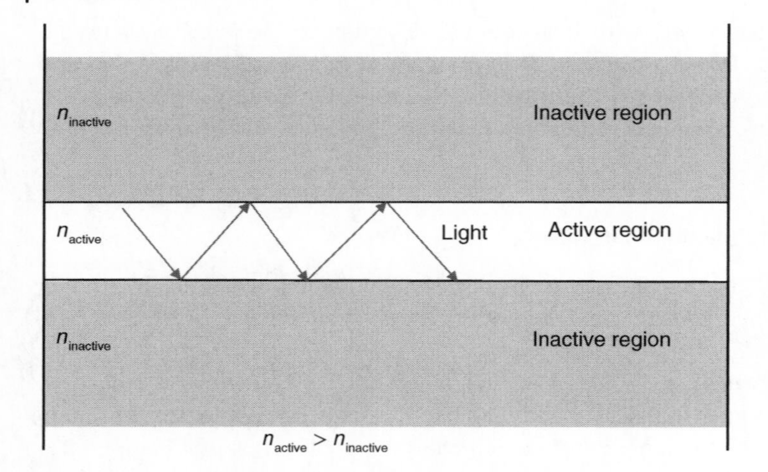

Figure 8.23 Schematic diagram of the basic structure of the Fabry–Perot cavity made by heterojunction.

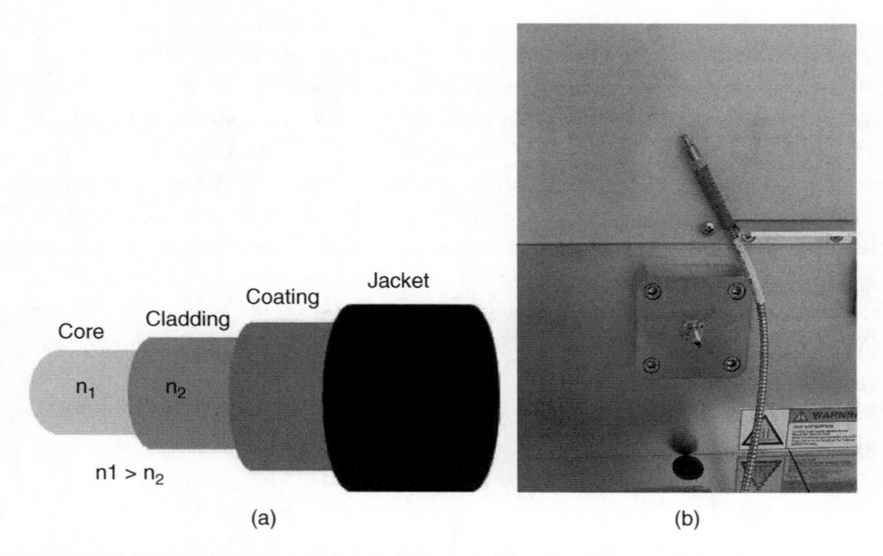

Figure 8.24 The structure of optical fiber (a) and a fiber optic cable (b).

8.3.4 Population Inversion and Threshold Current Density

We have just introduced the Fabry–Perot cavity, now let us talk about population inversion. For diode lasers, the essence of laser generation is the charge carriers flowing through the forward p–n junction and the combination of electrons and holes. In thermal equilibrium, the number of the carriers (mostly electrons) in

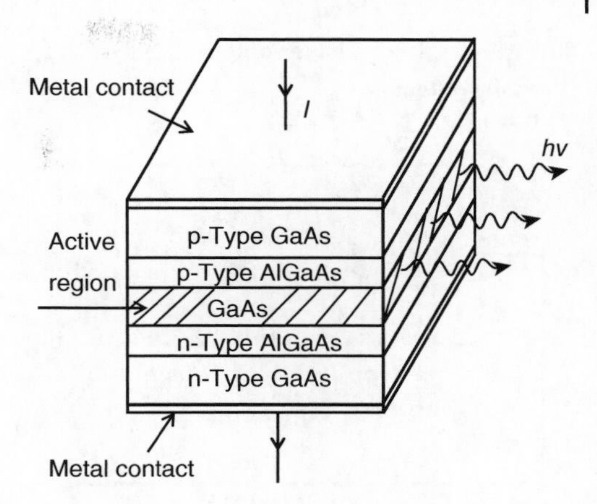

Figure 8.25 Schematic diagram of DH laser [2].

the excited state is always less than the number of the carriers in the ground state. The population inversion means that the number of carriers in the excited state is more than the number of carriers in the ground state, please refer to Figure 8.3. In this case, when the population is reversed, there are more electrons in the excited state E_2 than in the ground state E_1. At this time, photons with energy $h\nu_{12}$ are incident into the p–n junction, which will trigger these electrons to jump back to the ground state and recombine with holes, thereby generate more photons with energy $h\nu_{12}$. The number of photons of stimulated emission is more than that of absorbed photons. This phenomenon is quantum amplification [1]. In diode lasers, this quantum amplification is performed in a Fabry–Perot cavity.

The lifetime of an electron in the excited state is very short, that is, the electron can only stay in the excited state for a short time. To ensure population inversion and quantum amplification, the current injected in the forward direction must be large enough (see Figures 5.8 and 8.22), and the current must exceed a threshold so that the photon gain is greater than the photon loss. We often use the threshold current density J_{th} to express this threshold, where "th" means "threshold". Usually, the current density is expressed by J, which refers to the current passing through a unit area. Please see Figure 1.3. The expression is shown in formula (8.8). For the use of lasers, the smaller the J_{th}, the better the laser.

$$J = \frac{I}{A} \tag{8.8}$$

Now with the technologies of mirror reflection, heterojunction resonance, population inversion, and threshold current density, a diode laser is made, as shown in Figure 8.26. In this figure, the dotted line is the heterojunction interface, which is parallel to the p–n junction. Vertical to these interfaces, one side is made of a fully reflecting surface and the other is made of a partially reflecting surface, and the laser is emitted from the partially reflecting surface (laser output). Compared to Figures 8.17 and 8.26 has a similar structure.

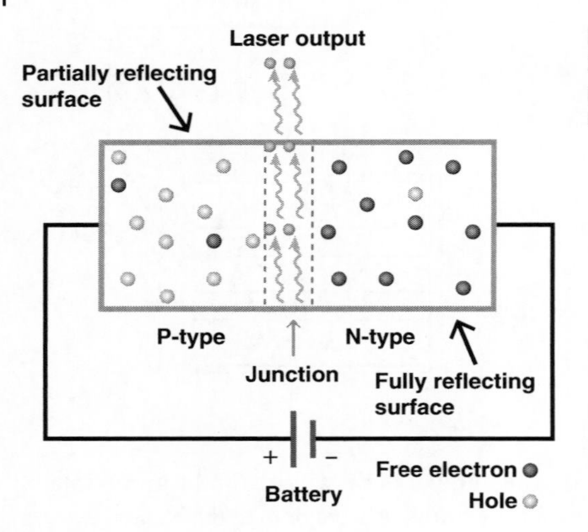

Figure 8.26 A diode laser. Source: Reprinted with permission of http://physics-and-radio-electronics.com.

References

1 Sze, S.M. (1985). *Physics of Semiconductor Devices*, 2e. Wiley, p. 681, p. 707, p. 690, p. 720.

2 Kevin F.B. (1999). *The Physics of Semiconductors with Applications to Optoelectronic Devices*. Cambridge University Press, p. 492–493, p. 676, p. 680–686, p. 689.

3 Holonyak, N. and Bevacqua, S.F. (1962). Coherent (visible) light emission from ga ($As_{1-x}P_x$) junctions. *Applied Physics Letters* 1: 82.

4 Bush, S. (2010). 50 year history of the LED. *Electronics Weekly*. (22 September).

5 Nakamura, S., Mukai, T., and Senoh, M. (1994). Candela-class high-brightness InGaN/AlGaN double-heterostructure blue-light emitting diodes. *Applied Physics Letters* 64 (13): 1687–1689.

6 Diep, F. (2014). Why a blue LED is worth a Noble Prize. *Popular Science* (7 October).

7 饭田修一, 大野和郎, 神前熙等. (1979). 物理学常用数表, [日], 科学出版社, 115–116 页。

9

Semiconductor Light Detection and Photocell

As mentioned at the beginning of Chapter 8, semiconductor photonic devices are divided into three categories: (i) convert electrical energy into optical radiation; (ii) detect optical signals through electronic technology; (iii) convert optical radiation into electrical energy. The semiconductor LEDs and lasers in Chapter 8 are devices that convert electrical energy into light radiation. This chapter introduces semiconductor photodetectors and photovoltaic cells (solar cells). What these two types of devices have in common is the conversion of optical signals (power) into electrical signals (power). The main application of light detecting devices is digital cameras and fiber optic communications.

9.1 Digital Camera and CCD

Digital cameras do not use film, they use photon sensors and detectors to convert light into electrical charges. In this chapter, we use "detector" to describe this type of device. There are two main technologies that are used to make detector (image sensor) for a digital camera, i.e. charge-coupled device (CCD) and CMOS. Here, we use CCD as an example to discuss the technology of image sensor. With the MOS structure, the granularity and spatial resolution of the camera depend on the density of independent detector units. Each detector unit is called a pixel. The pixel density is a key factor in determining the spatial resolution [1]. The granularity we mentioned here is a concept taken from the past film camera. The coarser the granularity of the film, the worse the effect will be after magnification. The higher the spatial resolution, the smaller the details that can be seen. With finer granularity and higher resolution, that is, more pixels, the quality of the photo is higher, but the memory and disk space occupied will be larger.

A CCD is basically a MOS diode array very close in space. When using it, the information is expressed in terms of the amount of charge (called a charge packet), which is different from the other devices, because these devices use the magnitude

Semiconductor Microchips and Fabrication: A Practical Guide to Theory and Manufacturing,
First Edition. Yaguang Lian.

of current and voltage to express information. A deeply depleted MOS diode is the basic unit of a CCD. There are two types of CCDs – surface channel (SCCD) and buried channel (BCCD) [2]. We are going to use SCCD to explain the structure and operation of a CCD in this section. In the SCCD, charges are concentrated and transferred at the semiconductor surface.

Now let us take p-type silicon as an example to see how CCD works. As we have said before, if the device is illuminated by light and the energy of the photon is equal to or bigger than the forbidden band width of the semiconductor, the electrons will transfer from the ground state to the excited state, thus forming electron–hole pairs. Please see Figure 9.1a, which is a schematic diagram of a MOS diode. When a higher-positive voltage is applied to the gate metal, a deep depletion region will be generated at the interface of SiO_2 and Si, as shown in Figure 9.1b. At this time, when photons irradiate the device, electron–hole pairs will be generated. The electrons will be attracted to the interface between SiO_2 and Si by the applied positive voltage, and the holes will be repelled away from the interface, as shown in Figure 9.1c. The Q_{sig} in the figure is signal charge. The stronger the light

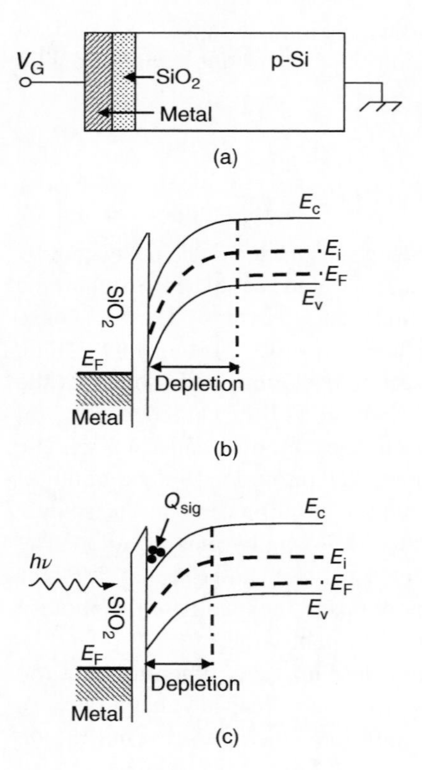

Figure 9.1 p-Type silicon CCD structure and energy band diagram (a) a MOS diode, (b) a higher positive voltage is applied to the gate to create a deep depletion region, (c) after illumination, electrons accumulate at the $Si-SiO_2$ interface. Source: [2] Sze / John Wiley & Sons.

is, the more the charges accumulated, the stronger the buildup of negative voltage is, and the thinner the depletion region will be. Please compare Figure 9.1b,c.

As we mentioned earlier, a depletion layer is a capacitor, so we can use a MOS capacitor array to represent a MOS diode array. When the light of an image illuminates the capacitor array through the lens, each capacitor on the array will generate a charge (in our case, electrons) accumulation. The accumulation of charge is proportional to the intensity of the light irradiated here. The charge corresponds to a part of the object being illuminated. Once the capacitor array is illuminated by the image light, a control circuit will transfer the charge stored in a capacitor to its adjacent capacitor. The last capacitor in the array will put the charge into a charge amplifier, which converts the charge into a voltage. Repeating this process, the control circuit will convert the entire content of the image stored in the semiconductor capacitor array into a sequence voltage. In digital devices, these voltages are sorted, digitized, and stored in memory. Figure 9.2 is a cross-sectional view of a typical three-phase CCD. The basic device structure is to grow a dielectric layer of SiO_2 on a semiconductor substrate. On this continuous medium, MOS diodes (capacitors) are arranged very close together. Figure 9.2a shows the middle gate, which is connected to a higher-gate voltage and is used as charges

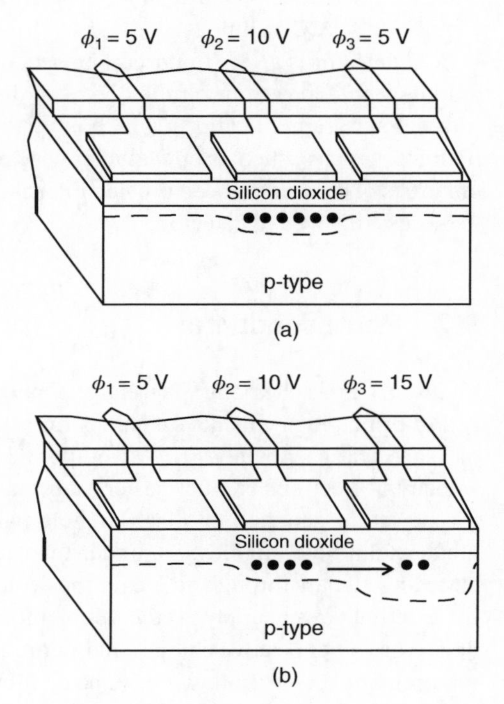

Figure 9.2 Cross-sectional view of a three-phase charge coupled device. (a) The high voltage at ϕ_2 is used for charges storage. (b) Apply higher-pulse voltage to ϕ_3 terminal for charges transfer. Source: [2] Sze / John Wiley & Sons.

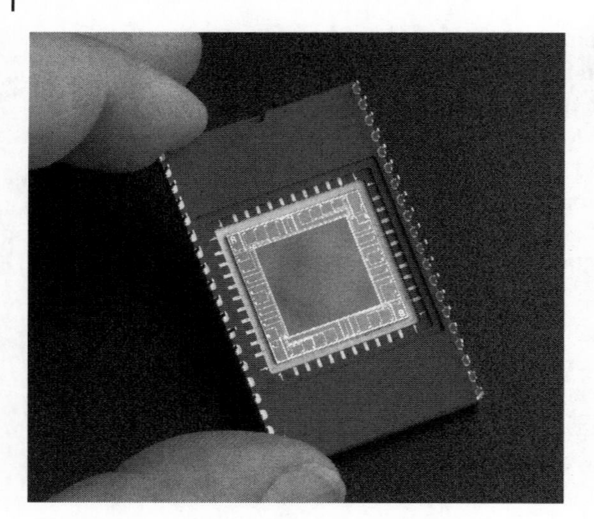

Figure 9.3 One CCD chip. Source: NASA / Wikimedia Commons / Public domain.

storage unit. When a higher-pulse voltage is applied to the gate on the right, the charges begin to transfer to this capacitor (Figure 9.2b). CCD was invented by American scientists Willard Boyle and George E. Smith at Bell Labs in 1969. Figure 9.3 is a CCD chip.

CCD detector is a very important device. It completes the conversion of photographic technology from film to digital. A defect of this device is that there is no internal gain (amplification), which is very important to detect small signals at high frequencies, such as optical fiber communication systems. There are different kinds of devices that can provide internal gain. Below we use photoconductor to discuss this type of detector.

9.2 Photoconductor

Figure 9.4 is a schematic diagram of a photoconductor structure. The amplification principle of this device comes from the electrical neutrality requirement of space charge. In most III–V compound semiconductor materials, the drift velocity of electrons is much faster than that of holes. When a photon irradiates the device to generate an electron–hole pair. Because the electron mobility is greater than hole mobility. So, under the action of an external electric field, the electron will pass through and leave the device in a short time and be absorbed by the external power supply. At the same time, a hole remains inside the device, the device has a net positive charge. In order to maintain the space charge neutrality requirement, an electron will be injected into the device from an external circuit,

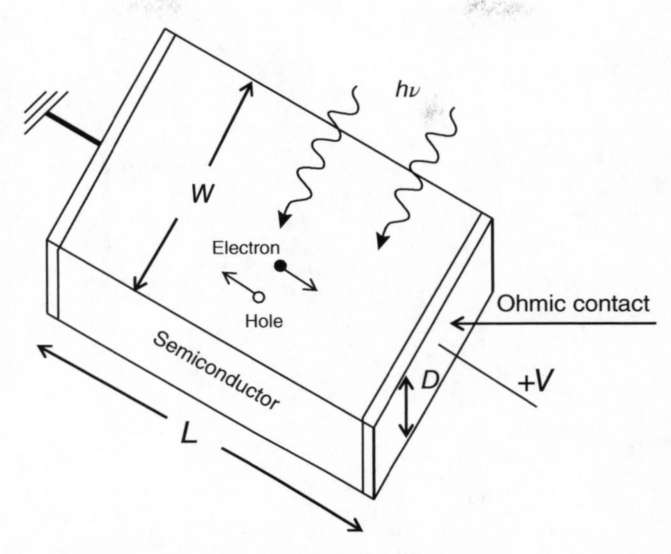

Figure 9.4 Schematic diagram of a photoconductor. Source: [1] Brennan / Cambridge University Press.

and this new electron will also pass through the device. If the second electron passes through the device before recombining with the hole, the device still has a net positive charge. Once again, an electron is injected into the device to maintain the space charge neutrality requirement. This process will continue until the hole generated by the light recombines with the injected electron. This procedure shows that there are many electrons passing through the device before recombining with holes. For example, in a device with a length of 1 μm, the transit time of an electron is on the order of 10 ps, where ps is picoseconds, and $1\ \text{ps} = 10^{-12}$ seconds. The typical lifetime of a hole (referring to the time from being excited to recombination with an electron) is 10 ns, where ns is nanoseconds, and $1\ \text{ns} = 10^{-9}$ seconds. As a result, on average, one photon generates one electron, and 1000 electrons will pass through the device. This leads to a gain of 1000, which is called photoconductive gain [1]. Usually, the actual gain of the device is less than 1000.

9.3 Transistor Laser

We have just introduced LEDs, diode lasers, and photodetectors. These devices are widely used in lighting, television, camera, and fiber optic communications. In fiber optic systems, we need photonic devices and electronic devices to work

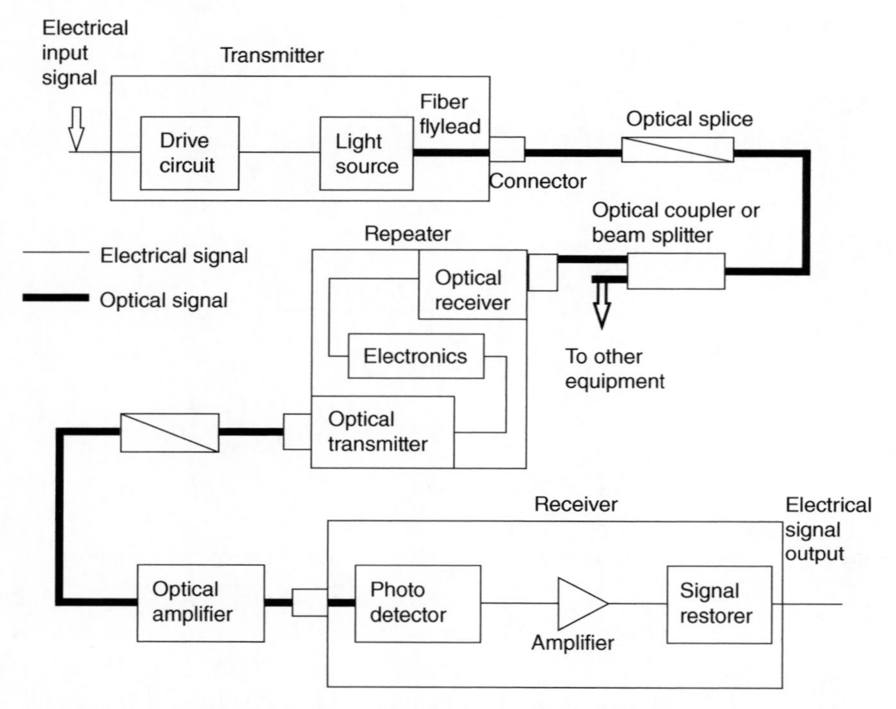

Figure 9.5 Schematic diagram of the basic overall structure of the optical fiber communication system.

together to run the system, which is why this subject is also called optoelectronics. Figure 9.5 is an overall schematic diagram of an optical fiber communication system. In this figure, the thin lines represent electrical signals, and the thick lines represent optical signals. So far, the devices we have introduced, they deal with electricity–electricity, electricity–light, or light–electricity.

To connect optical devices and electrical devices, optical fibers and waveguides are needed. These technologies are difficult to be compatible with the existing mainstream semiconductor processes, and it is difficult to integrate them into a single chip.

In the early of 2004, Milton Feng and others invented the light-emitting transistor (LET) [3]. Figure 9.6 shows the material structure, device structure, and luminescence photos of the LET. Figure 9.6a is the structure diagram of the material, which also uses heterojunctions, but its structure is carefully designed and complicated. Figure 9.6b is a typical bipolar transistor structure which

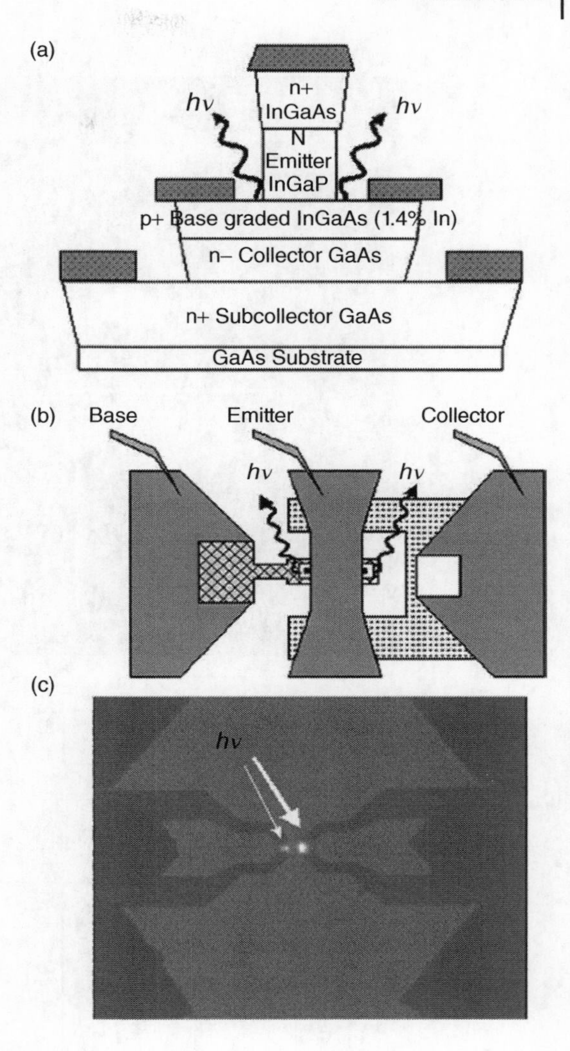

Figure 9.6 Light-emitting transistor (a) the heterojunction structure of the transistor, (b) the layout of the transistor, (c) the photo of light-emitting.

includes the emitter, collector, and base. Figure 9.6c is a photo of light emission. In late 2004, Feng, G. Walter, and others produced the world's first transistor laser [4], please see Figure 9.7. The LET and the transistor laser help realize the simultaneous processing of electrical and optical signals on a chip. They give us the opportunity to realize the complete integration of optoelectronics on a chip.

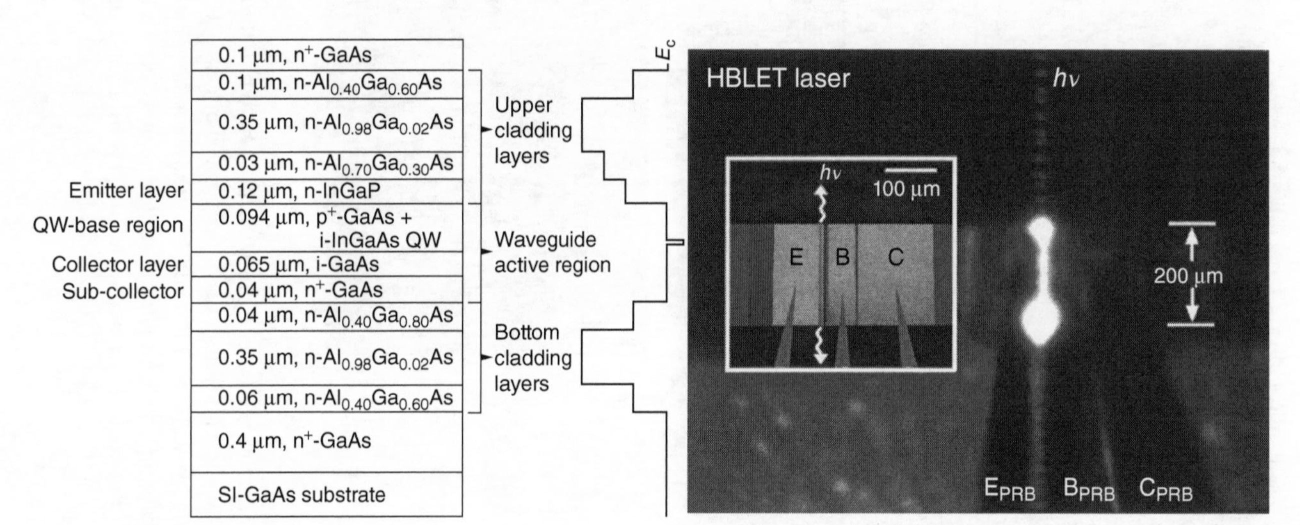

Figure 9.7 Transistor laser. HBLET, heterojunction bipolar light-emitting transistor. Source: Walter et al. 2004 / with permission of AIP Publishing LLC.

9.4 Solar Cell

Now let us discuss the working principle and basic structure of solar cells (Figure 8.7). A solar cell is a kind of electric device, which uses photovoltaic effect to directly convert light into electricity. The photovoltaic effect is that under the irradiation of light, a material can generate voltage and current. Photovoltaics is achieved by using semiconductor materials, mostly silicon. We now use silicon as an example to discuss photovoltaic cells (solar cells).

As shown in Figures 5.8 and 5.9, a p–n junction contains a depletion layer. In the depletion region on both sides of the p–n junction, there are negative ions in the p-type Si and positive ions in the n-type Si. A built-in electric field is established in the depletion layer, from the n-region to the p-region. When the p–n junction is irradiated by sunlight, the part of photons whose energy is greater than the band gap of Si is absorbed, and electron–hole pairs can be generated. Under the attraction of the built-in electric field, electrons drift to the n-region and holes drift to the p-region. The width of the depletion layer becomes narrower, and a voltage is built up in the p–n junction. Since there is no circuit and load at this time, this voltage is called the open circuit voltage (OCV), as shown in Figure 9.8a. The V_{OC} in the figure is the OCV. When the p–n junction is connected to a resistor (load) through a wire, current will flow through the resistor, as shown in Figure 9.8b. This is a solar cell. When many solar cells are arranged together, it becomes a solar panel.

The silicon wafers used in the production of devices are mirror polished. In order to reduce light reflection and increase absorption on the surface, usually antireflective coatings are put on the Si surface. We also use dry etching (discuss later) to modify silicon surface to form a rough structure. This technology is called "Black Silicon-BSi," because the surface of silicon treated in this way is black. There are many structures of BSi. Figure 9.9 shows the structure of silicon nanocones [5]. One of the reasons why the light absorption can be increased is that on such

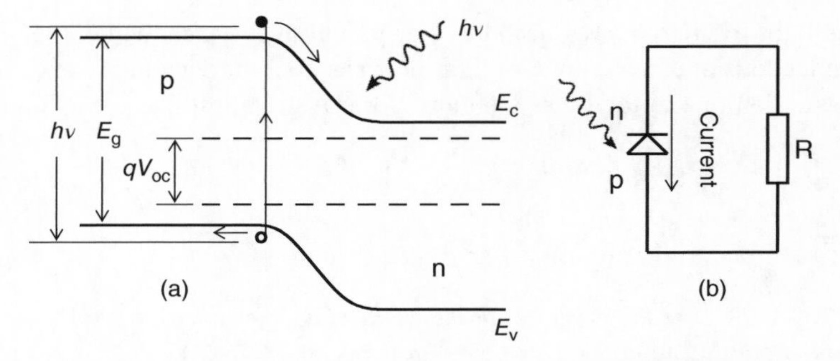

Figure 9.8 (a) Schematic diagram of p–n junction energy band of solar cell. (b) A simple solar cell circuit. Source: [2] Sze / John Wiley & Sons.

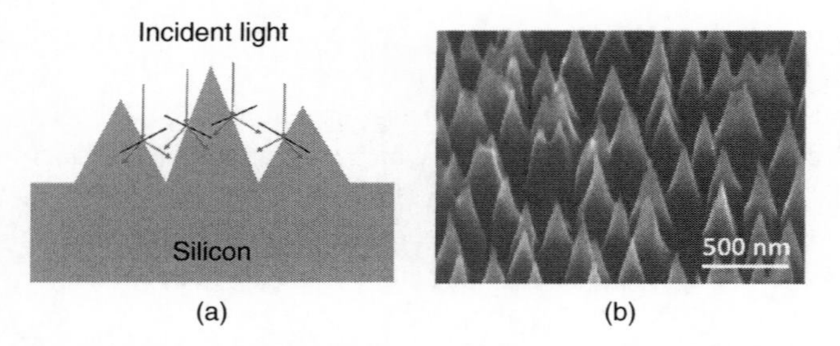

Figure 9.9 Nanocone structures of Si surface. (a) Schematic diagram of light refraction and reflection in the surface of Si nanocone. (b) SEM picture of the nanocone structure.

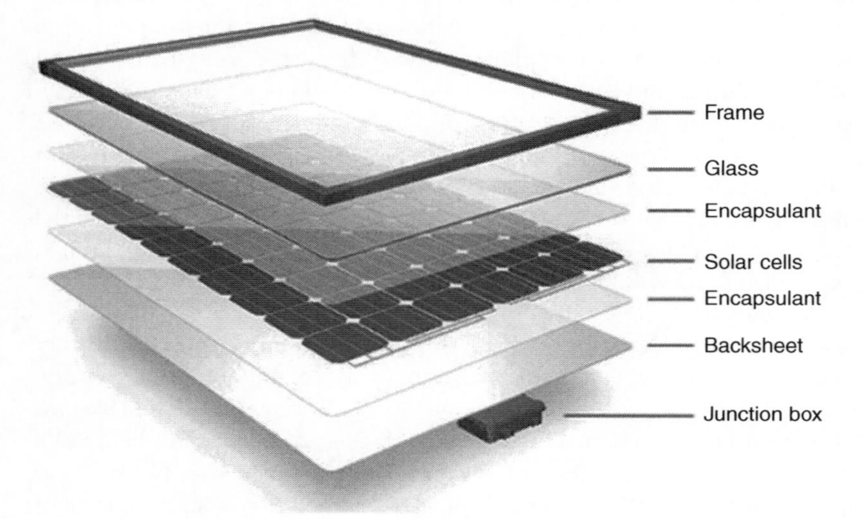

Figure 9.10 The structure of a solar panel. Source: Moulin [6].

a surface, the light irradiating area becomes bigger, and in the meanwhile, the number of reflections and refractions of light between the interface of air and silicon increases (left one in the figure). Figure 9.10 is a structural diagram of a solar panel.

References

1 Brennan, K.F. (1999). *The Physics of Semiconductors with Applications to Optoelectronic Devices*. Cambridge University Press pp. 609, 630, 631.

2 Sze, S.M. (1985). *Physics of Semiconductor Devices*, 2e. Wiley pp. 407, 408, 412, 794.

3 Feng, M., Holonyak, N. Jr.,, and Hafez, W. (2004). Light-emitting transistor: light emission from InGaP/GaAs heterojunction bipolar transistors. *Applied Physics Letters* 84 (1): 151–152.
4 Walter, G., Holonyak, N., Feng, M., and Chan, R. (2004). Laser operation of a heterojunction bipolar light-emitting transistor. *Applied Physics Letters* 84 (20): 4768–4770.
5 Chen, Y. et al. (2011). Ultrahigh throughput silicon nanomanufacturing by simultaneous reactive ion synthesis and etching. *ACS Nano* 5 (10): 8002–8012.
6 Moulin. (2018). Important of materials in PV modules: recommended bes t practices to select PV modules. *Energetica India* (09 October).

10

Manufacture of Silicon Wafer

Silicon plays a pivotal role in the semiconductor industry. It ranks second in the earth's storage capacity after oxygen. Approximately 25% of the earth's crust is silicon. The melting point of silicon is very high, reaching 1420 °C; in comparison, the melting point of germanium used to make the first transistor is much lower than that of silicon, only 937 °C. The band gap of silicon is 1.14 eV, and its allowable operating temperature can reach 150 °C; the band gap of germanium is 0.67 eV, and its allowable operating temperature is lower than 100 °C, please refer Section 5.2. In addition, silicon is the only material that can grow insulating dielectric by thermal oxidation among all semiconductors. The grown dielectric film of silicon dioxide (SiO_2) offers many wondrous properties: First, it is compatible with Si in many respects. That is, it adheres tightly to Si, does not dissolve in water, and is chemically stable up to the melting point of Si. Second, it is an excellent dielectric with a high-electric field breakdown strength, and it exhibits a stable interface with Si. Third, SiO_2 passivates the silicon surface, preventing surface leakage. Finally, it serves as an effective diffusion mask (discuss later) [1]. The role of Si–SiO_2 structures in semiconductor manufacturing is so important that they are like King and Queen in the kingdom of semiconductor.

Because of the abovementioned reasons, although the transistor was invented with germanium, silicon later replaced germanium and became the protagonist of the semiconductor industry until today. Here we mainly use silicon and planar technology as examples to introduce the manufacturing process of semiconductor microchips. The design, manufacturing, testing, and packaging of integrated circuits are undoubtedly the most complex in the world today. A CMOS-IC process flow contains more than 350 processing steps, which takes six to eight weeks to complete. The process flow refers to a series of chemical and physical operations performed on the wafer of silicon [1]. Most of the operations are completed in the clean room, please see Figure 7.12. The purity of the chemicals, solvents, and gases used in the semiconductor process is very high. The purity of the gas reaches ultra-high purity, as shown in Figure 7.13. The purity of the liquid chemicals and

Semiconductor Microchips and Fabrication: A Practical Guide to Theory and Manufacturing,
First Edition. Yaguang Lian.

Figure 10.1 The liquids and solvents used in semiconductor process.

solvents can reach semiconductor grade (≥99%) and CMOS grade (99.5%), at least ACS (American Chemical Society) level (≥95%), please see Figure 10.1. The vacuum required in the process can be as low as 10^{-11} Torr, where Torr is a pressure unit. 1 atm = 760 Torr. The temperature range is from liquid helium (−269 °C) to 1200 °C.

Silicon used to make the chip must have a crystalline structure (see Figure 2.4) and its purity must also be high. This chapter will introduce the production process of silicon wafer.

10.1 From Quartzite Ore to Polysilicon

Although the storage capacity of silicon in the earth's crust is very high, it does not exist in the form of pure silicon in nature, but in the form of quartzite ore. Please see Figure 10.2. The main content of quartzite is silica (SiO_2). Figure 10.3 is the process flow chart from quartzite to silicon wafer.

In the first step, the raw material is reduced to metallurgical-grade silicon (MGS), which has a purity of about 98%. The method is to put the quartzite together with coal or wood chips in an electric arc furnace (metallurgical furnace). Figure 10.4 is a schematic diagram of the arc furnace. At high temperatures, the carbon in coal or wood will chemically react with the SiO_2 in silica to obtain MGS. The chemical reaction equation is as follows:

$$SiO_2\,(\text{solid}) + 2C\,(\text{solid}) \rightarrow Si\,(\text{solid}) + 2CO\,(\text{gas}) \tag{10.1}$$

Figure 10.2 Quartzite.

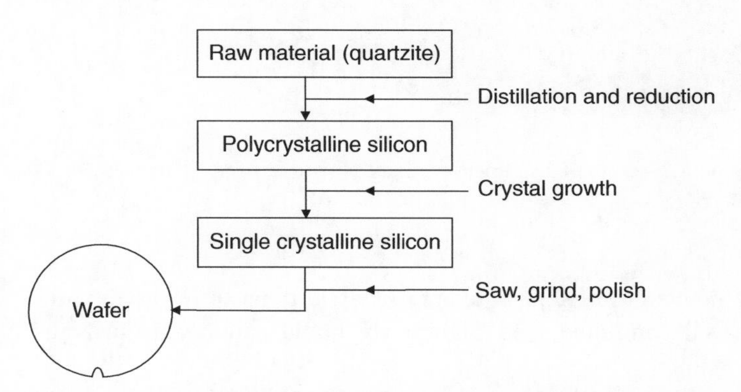

Figure 10.3 The process flow from quartzite to silicon wafer. Source: [2] Wolf and Tauber.

In the second step, the second chemical reaction is carried out. The MGS is ground into powder, and anhydrous hydrogen chloride and MGS are used for the chemical reaction to form trichlorosilane ($SiHCl_3$) by using fluidized bed reactor (FBR):

$$3HCl\ (gas) + Si\ (solid) \rightarrow SiHCl_3\ (liquid) + H_2\ (gas) \tag{10.2}$$

$SiHCl_3$ is a volatile liquid and its boiling point is 31.8 °C. Due to the low-boiling point of trichlorosilane, it is easy to use distillation for purification.

The third step is to obtain high-purity $SiHCl_3$ after distillation and separation. The working principle of distillation is to heat the mixed liquids and use the

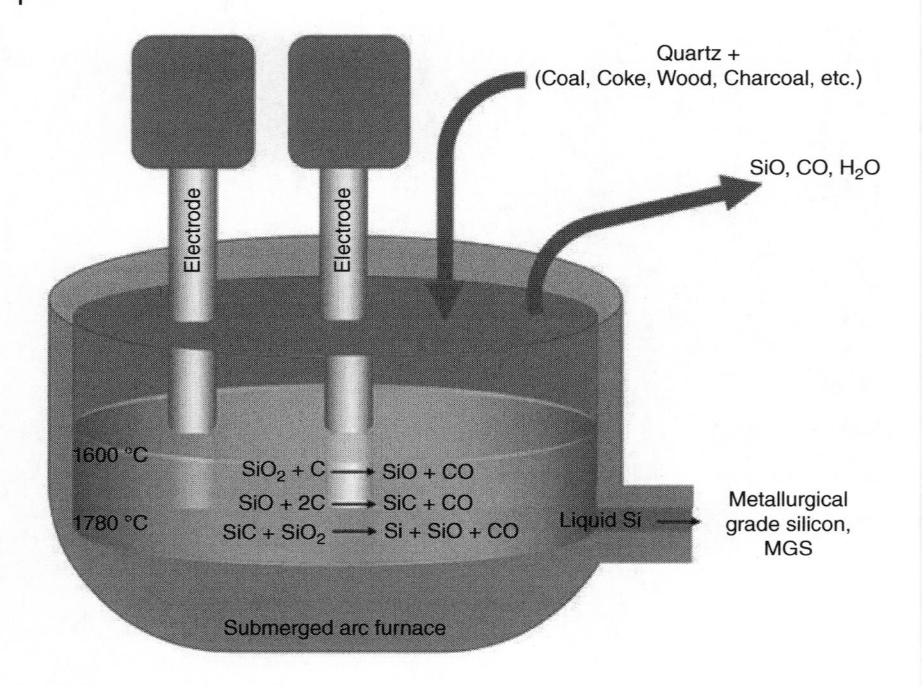

Figure 10.4 Schematic diagram of a submerged electrode arc furnace. Source: [3] MKS Instruments Handbook.

different boiling points of different liquids to separate them from the mixture. The process is well controlled, and high-purity liquid can be distilled and separated.

The fourth step is to add high-purity trichlorosilane and hydrogen to the chemical vapor deposition equipment to obtain high-purity polysilicon. Chemical vapor deposition is represented by CVD. We will have more discussion later. In the CVD process, the reactor (Siemens-type reactor) is heated, and the following chemical reaction occurs with hydrogen:

$$SiHCl_3\,(\text{gas}) + H_2\,(\text{gas}) \rightarrow Si\,(\text{solid}) + 3HCl\,(\text{gas}) \tag{10.3}$$

So far, we have obtained solar cell grade (SoG-Si-99.9999%) and even higher-electronic grade (EGS-99.999999999%, eleven 9 s) purity polysilicon [4]. The abovementioned silicon purification process was first developed by Siemens, so this process is called the Siemens process. Figure 10.5 is a schematic diagram of the Siemens process flow.

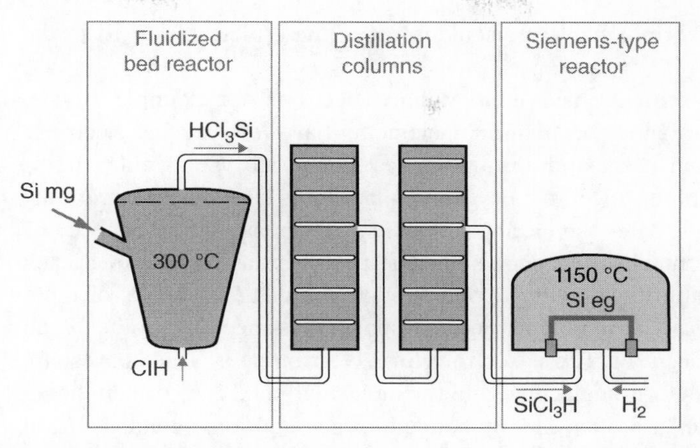

Figure 10.5 Using Siemens technology to produce electronic grade silicon.
Source: [5] Jfermin.

10.2 Chemical Reaction

In the previous section, some equations of chemical reaction were used. The reaction equations represent new substances produced by different substances through chemical reactions. A chemical reaction is a common phenomenon in nature, and it is also widely used in semiconductor technology. In this section, we will further discuss chemical reactions.

For an atom, when the number of outermost valence electrons is 8, the structure of the atom is stable. In the periodic table (Figure 5.1), such elements are listed in the rightmost column, group 18, which are six elements such as helium, neon, and argon. Helium in this group is a special case. It has two electrons in total and is also a stable structure. These elements exist in the form of monoatomic gases. Because of their stable structure, we call them noble gases (inert gases). Usually, noble gases cannot react chemically with atoms of other elements. Under extreme conditions, inert gases can also undergo chemical reactions, but their products are very unstable and split easily. We will give examples in later chapters. From the periodic table, we can see that most of the elements do not meet the requirements of stable structure. When atoms with unstable structures meet, they exchange or share valence electrons in the outer layer, making the number of valence electrons to reach 8. The atoms' structure become stable and form a molecule. In general, chemical reactions usually occur between oxygen-similar and metal-similar elements in the periodic table. Oxygen similar elements include oxygen, nitrogen, fluorine, chlorine, etc. Metal-similar elements include silicon, gallium, hydrogen, aluminum, copper, etc. The products formed by the reaction between them are called compounds. In addition, the atoms of oxygen-similar elements can also share electrons to form a stable molecule. Most of these molecules exist in the form of gas.

Oxygen-similar elements have more valence electrons, for example, oxygen has 6 and fluorine has 7. Metal-similar elements have less valence electrons, for example, sodium has 1 and aluminum has 3. When oxygen-similar atoms and metal-similar atoms meet, the oxygen-similar atoms take electrons from the metal similar atoms. After the electrons transfer, oxygen-similar atoms become negative ions and metal-similar atoms become positive ions. They attract each other by electrostatic force to form compound molecules. The bonds of these molecules are ionic bonds. When oxygen-similar atoms meet each other, each atom contributes electrons. These electrons form electron pairs. The atoms share the electron pairs to form molecules, and the molecular bond is a covalent bond. Figure 10.6 is molecules of covalently bonded oxygen and ionic-bonded silicon dioxide (SiO_2). Since silicon has four valence electrons, it must combine with two oxygen atoms to form a stable molecule SiO_2. Ionic bond and covalent bond are referred to as chemical bonds. Besides these two bonds, there is another type of bond-metallic bond. In a metal, protons have weak attraction to electrons. Some electrons can get rid of the attraction of the nucleus and become free electrons (discuss later). In a metal structure, these free electrons are shared by positively charged ions to build metallic bond.

Generally, when substances meet at room temperature, the chemical reaction starts difficultly, and the reaction rate is slow. One important reason for this is that most substances exist in stable molecular forms, such as O_2, SiO_2, and so on. We learned the experiment of hydrogen and oxygen reacting to produce water in middle school. When hydrogen and oxygen meet, they will explode and react to produce water under the condition of an open flame which is fire:

Figure 10.6 The molecules of oxygen and SiO_2.

$$2H_2 + O_2 \rightarrow 2H_2O \tag{10.4}$$

We cannot help but ask: why is open flame needed? To understand this issue, we must first understand what fire is. The ordinary fire is a chemical reaction between oxygen and carbon. When we are close to the fire, we feel hot, and the temperature is high. This is because the air molecules move fast near the fire and the collisions between the molecules are violent. This collision is so strong that the chemical bonds of the reactant molecules are broken, and the outermost electrons of the atoms are excited or even ionized (refer to Chapter 2). The excited electrons return from the high-energy level to the low-energy level, emitting photons, which is the first reason for the fire to glow. In addition, if the electrons are ionized, the electrons get rid of the bondage of the atomic nucleus and become free electrons. The atoms after the loss of the electrons become positively charged (positive ions).

The charges of electrons and ions are opposite and equal in number. This state is the plasma state. A plasma is defined as a partially ionized gas containing an equal number of positive and negative charges as well as some number of neutral particles [2]. The positive and negative charges in the plasma can also recombine and emit photons, which is the second reason why the fire glows. No matter the chemical bonds of the molecules are broken, the electrons are excited or ionized, all of these make the stable molecular structure unstable, which leads chemical reactions to start easily. From here, we understand why hydrogen and oxygen can explode (explosive reaction) under an open flame. Explosive reaction refers to the extremely fast reaction rate. Correspondingly, even under an open flame, many reaction rates are slow and nonexplosive. Here, the plasma state is introduced. It was mentioned in Section 2.2 that there are usually three states of matter in nature: solid, liquid, and gaseous. In fact, there is a fourth state of matter – the plasma state.

From the above discussion, we can see that the purpose of an open flame is to increase the temperature, strengthen the collision between molecules, and change them from stable structures to unstable structures for chemical reactions. In the semiconductor process, an open flame is not used for chemical reactions. The equipment is heated to achieve chemical reactions, as shown in Figure 10.5. In addition to increasing the temperature, another commonly used method is to use plasma technique to achieve chemical reactions at low temperature. This issue will be discussed in later chapters. There are other ways of chemical reactions, such as the use of catalysts to achieve chemical reactions. We do not cover this topic in this book.

10.3 Pull Single Crystal

Through the Siemens process, we have obtained high-purity polysilicon. The next step is to use the Czochralski – CZ method to make monocrystalline (single-crystal) silicon, which is currently the most used process for growing monocrystalline silicon.

Put the polysilicon obtained from the Siemens process into the crucible and control the temperature to just exceed the melting point of silicon (1414 °C). According to need, boron or phosphorus is added to the molten silicon to obtain p-type or n-type silicon. The operation of this process is as follows: fix a small monocrystalline silicon seed crystal on the lower end of a vertical rod, immerse the seed crystal on the surface of the molten silicon, and slowly rotate and lift it. Silicon will form a single crystal according to the seed crystal structure. Under the action of gravity, a cylindrical silicon single-crystal ingot will be drawn out. The length of ingot for 300 mm wafer is several meters. The entire process is completed in a quartz furnace under the protection of inert gas argon. Please see Figure 10.7.

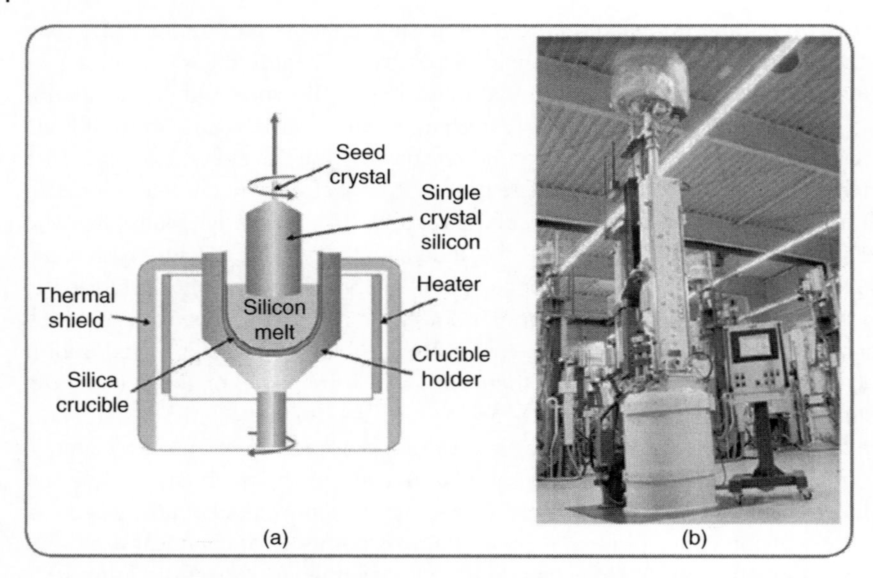

Figure 10.7 (a) Czochralski silicon crystal pulling furnace, (b) Process equipment. Source: Ref. [3], MKS Instruments, Inc.

The Czochralski method was invented by Polish chemist Jan Czochralski (October 23, 1885–April 22, 1953) in 1915.

10.4 Polishing and Slicing

The pulled silicon ingot is not a perfect circle, and its diameter distribution is not uniform. It must be processed to achieve the designed shape and size. In general, the size of the pulled silicon ingot is intentionally larger. After grinding to remove the excess, the ingot is finally reaching the desired cylindrical shape and diameter, as shown in Figure 10.8.

The size of silicon wafer ranges from 25 mm (1 in.) in the 1960s to 300 mm (12 in.) in 2001, see Figure 7.10. From 1 in. to 6 in. (150 mm), the flats are used to indicate the crystal orientation, and they can also be used for the alignment process in lithography (more on this in the following chapter). The flat is divided into primary flat and secondary flat. When the size reaches 8 in. (200 mm) and above, a small notch replaces the flats, see Figure 10.9. There are also wafers that do not have flat/notch.

In Figure 10.9, the words {100} and {111} appear, which refer to the lattice planes and orientations of silicon crystal. Refer to the schematic diagram of the silicon lattice in Figure 2.4. In a coordinate system, silicon crystal is divided into three

Figure 10.8 The grinding process of silicon ingot. Source: siltronic.com.

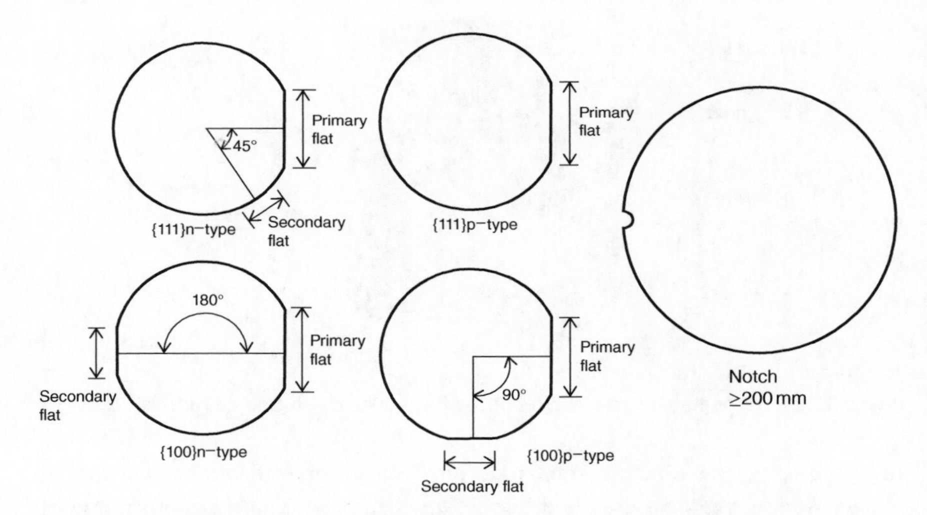

Figure 10.9 The flats and notch in silicon wafer. Source: [2] Wolf and Tauber.

lattice planes (100), (110), and (111). (100) refers to the plane that only intersects the X-axis. (110) is the plane that intersects the X- and Y-axes. (111) is the plane that intersects the X-, Y-, and Z-axes. They are shown in Figure 10.10. The integers 100, 110, and 111 representing the lattice planes are called Miller indices, to commemorate British scientist William Hallowes Miller (April 6, 1801–May 29, 1880.5.20), who established the foundation of modern crystallography. The symbol {100} comprises the planes of (100), (-100), (010), (0-10), (001), and (00-1). The other two symbols {110} and {111} have the same meaning. Like the planes, [100], [110], and [111] are used to indicate the crystal axes. <100> is represented by six directions [100], [-100], [010], [0-10], [001], and [00-1]. <110> and <111>

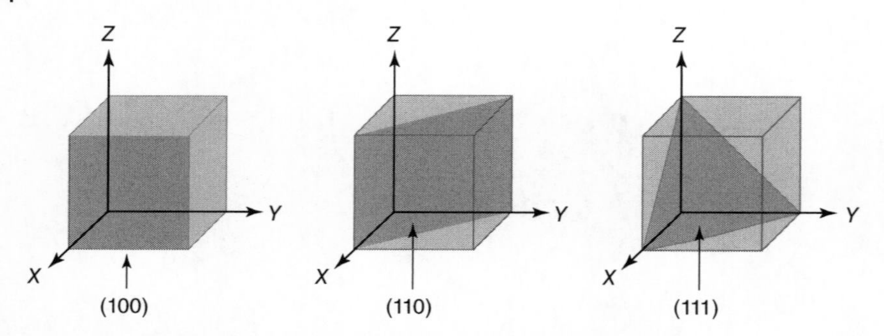

Figure 10.10 Lattice planes of silicon. Source: Reprinted with permission of http://crystal-scientific.com.

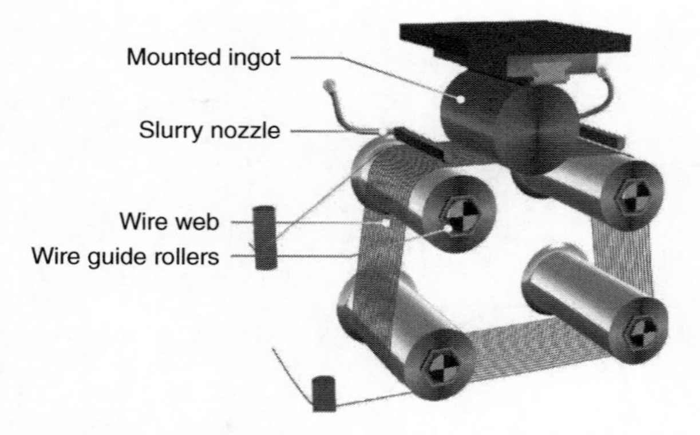

Figure 10.11 Schematic diagram of a multi-wire saw machine. Source: siltronic.com.

have the same meaning. In solid-state physics, the crystal lattice structure of silicon is classified as a diamond tetrahedral structure. There are other crystal lattice structures in semiconductor materials. We do not discuss them here.

After the silicon ingot grinding, it is sliced. The slicing tools include wire saw and internal diameter (ID) saw. Figure 10.11 is a multi-wire saw. This technology makes it possible to slice complete silicon ingot into hundreds of silicon wafers in just one step. After a silicon ingot is sliced into wafers, the next steps include edge rounding, laser marking, lapping, polishing, and cleaning of the wafers.

The edge of sliced wafer is not very smooth, which may cause unnecessary mechanical damages, such as the incident of chipping during wafer handling. In photoresist coating step of lithography process (discuss later), the rough edge can cause the problem of photoresist buildup on the edge, which is called edge bead and makes nonuniform film. The purpose of edge rounding is to avoid these problems. Please see Figure 10.12.

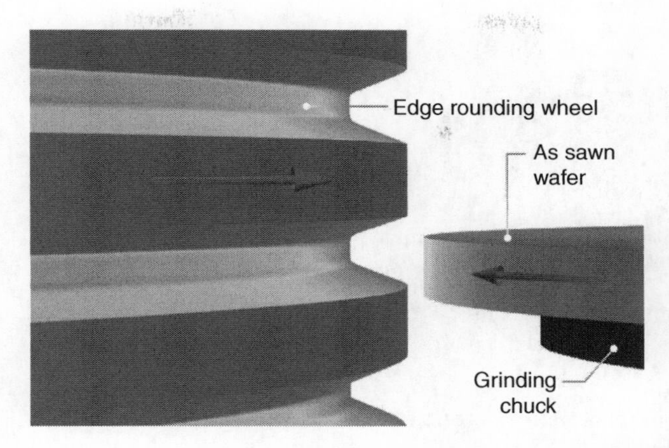

Figure 10.12 Schematic diagram of edge rounding machine. Source: siltronic.com.

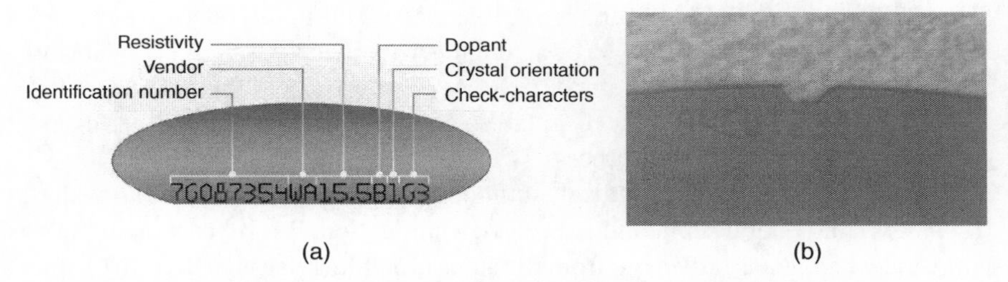

Figure 10.13 The meaning of markings on the edge of wafer (a) The markings show the information of vendor, resistivity, dopant, etc. Source: siltronic.com. (b) The markings on a silicon wafer.

Laser marking is used to create an alphanumeric, or barcode, identification mark on the front of the wafer near the primary flat or notch. According to SEMI (Semiconductor Equipment and Materials Institute) Std. M1.8, laser marking specifies an 18-character field to identify the wafer manufacturer, conductivity type, resistivity, flatness, wafer number, and device type [2]. It allows to know individual wafer or wafer batch in order to have manufacturing traceability. Please see Figure 10.13.

The lapping and polishing process includes chemical etching and mechanical polishing, so this process is called chemical mechanical polishing (CMP). The basic structure of the equipment is shown in Figure 10.14. After the CMP is completed, the wafer is wet-etched and cleaned to remove the mechanical damage and polishing slurry on the surface. Etching mainly uses chemical reaction to remove some thin layers on the surface of the wafer. If the etching is carried out in a liquid etchant, it is wet etching; if the etching is carried out in a gaseous chemical, it is

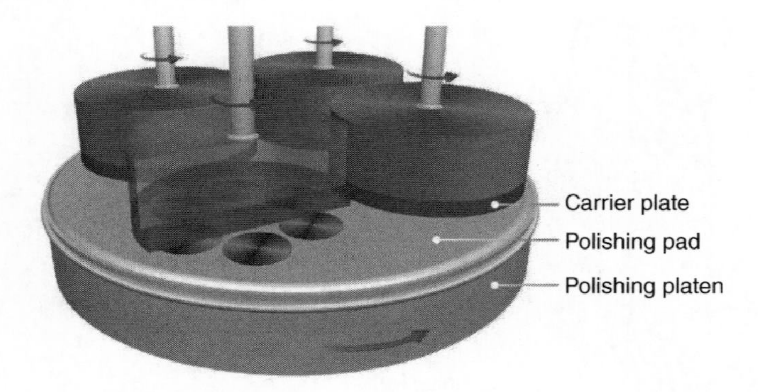

Figure 10.14 CMP machine of Si wafer. Source: siltronic.com.

dry etching. We will further discuss the etching process in later chapters. This step is to use wet etching.

For silicon wafers, RCA cleaning is a set of standard clean (SC) steps. The cleaning method was developed in 1965 by an engineer Werner Kern, who was working at the Radio Corporation of America – RCA. Figure 10.15 is the RCA cleaning flow chart. The solution H_2SO_4/H_2O_2 (sulfuric acid/hydrogen peroxide) is called piranha in the semiconductor process.

CZ growth is the most economical method to produce silicon wafers (known as CZ wafers) for general semiconductor device fabrication. But it has a main problem. Since the ingot is pulled out from the silica crucible (refer Figure 10.7a), some oxygen contamination is always present in the silicon. Graphite crucible has been used to avoid this contamination. However, it produces carbon impurities in the silicon. So, in CZ method, it is difficult to obtain wafers with resistivities greater than 100 Ω cm. Higher-purity silicon can be produced by a method know as Float Zone (FZ). A cylindrical polycrystalline silicon ingot is mounted over an induction coil. An RF electromagnetic field helps melt the silicon from the lower part of the rod. The field regulates the silicon flow through a small hole in the induction coil and onto the monocrystal that lies below. The ingot is not contact with any of the chamber components except for the ambient gas and a seed crystal of known orientation. So FZ silicon wafers have resistivities as high as 1×10^4 Ω·cm [3]. The silicon wafers manufactured by using FZ method are ideal for the use in power semiconductor devices. Please see Figure 10.16.

To meet the tough requirements of some devices such as MOSFET manufacturing, a thin, defect-free crystal layer is additionally deposited onto the polished surface of silicon wafer by using epitaxy technology. Epitaxial materials are inherently free of oxygen and carbon. They also have some advantages and selections, such as controlled and abrupt changes in dopant profiles, n-type silicon over p-type silicon and p-type silicon over n-type silicon [3]. Please see Figure 10.17. To apply the epitaxial layer, the silicon wafer is fastened to a susceptor and heated to a high temperature with the help of infrared lamps. The process gas flow and

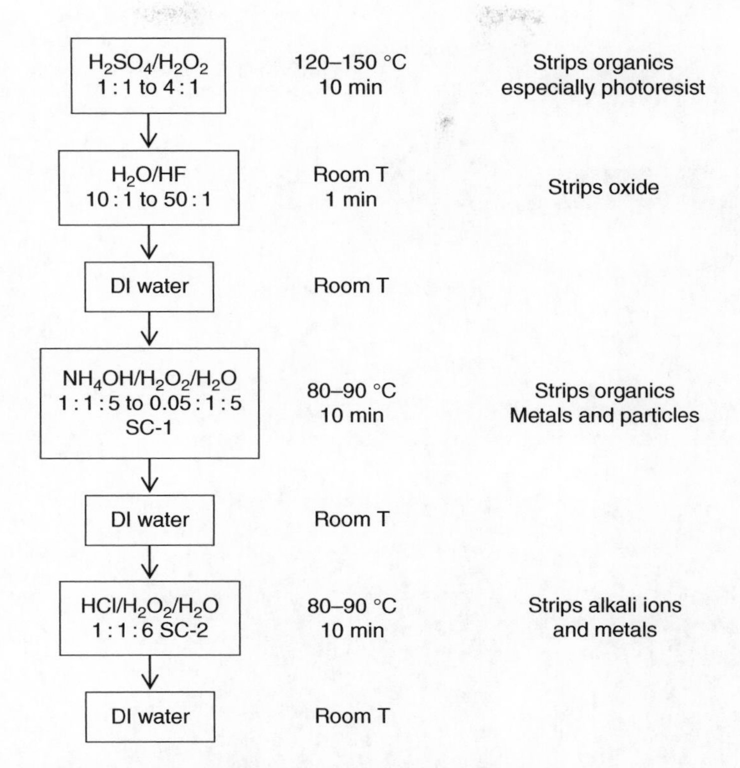

Figure 10.15 The RCA cleaning flow. Source: Wolf [1].

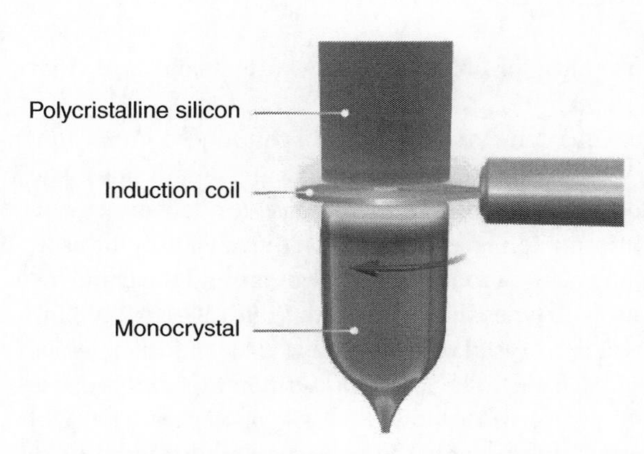

Figure 10.16 Schematic diagram of Float Zone method. Source: siltronic.com.

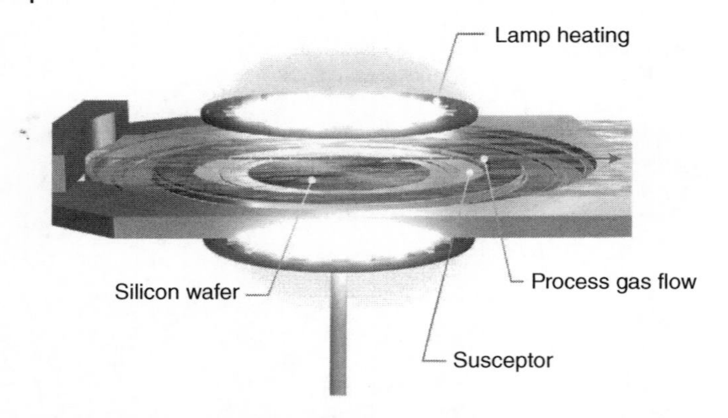

Figure 10.17 Schematic diagram of epitaxy machine. Source: siltronic.com.

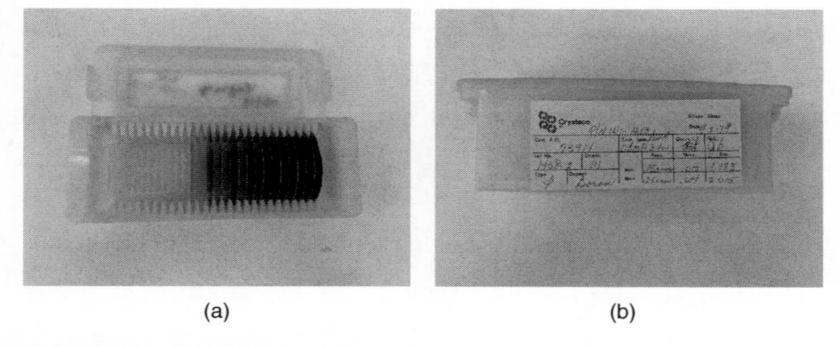

Figure 10.18 2-in. silicon wafers delivery pack and information label.

temperature are carefully controlled in order to create an epitaxial layer that meets the specifications.

So far, the whole process from quartzite to silicon wafer is introduced. According to the different needs, silicon wafers are polished on one side or on both sides. The wafers are put into a cassette and packaged in a carrier case for delivery. A cassette can hold 25 wafers. An information label is posted on the outside of the case. Figure 10.18 shows 2-in. silicon wafers. From the label, we can find the company name, production date, dopant is B, type is P, diameter is from 1.985 to 2.015 in., resistivity is from 0.12 to 0.25 Ω·cm, crystal orientation is (111), and thickness is from 0.012 to 0.014 in. (304.8–355.6 μm). The label was written by hand because the wafers were made in 1979. Figure 10.19 shows 8-in. silicon wafers. The label was printed. We can find information similar to the 2-in. wafer label. Figure 10.20 is the picture of 2-in. and 8-in. silicon wafers. 2-in. is p-type (111) wafer and has a primary flat. 8-in. has a notch.

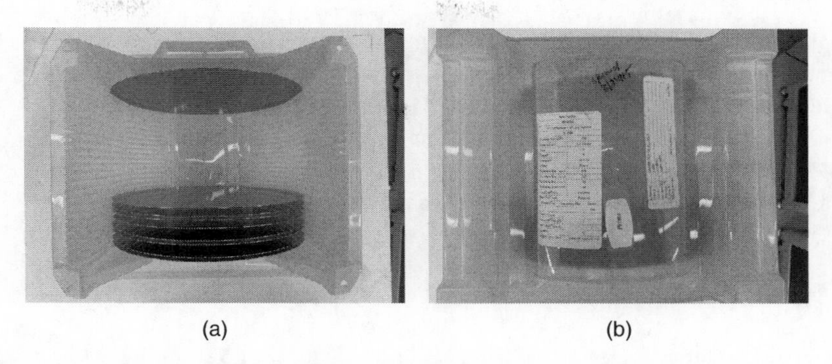

(a) (b)

Figure 10.19 8-in. silicon wafers delivery pack and information label.

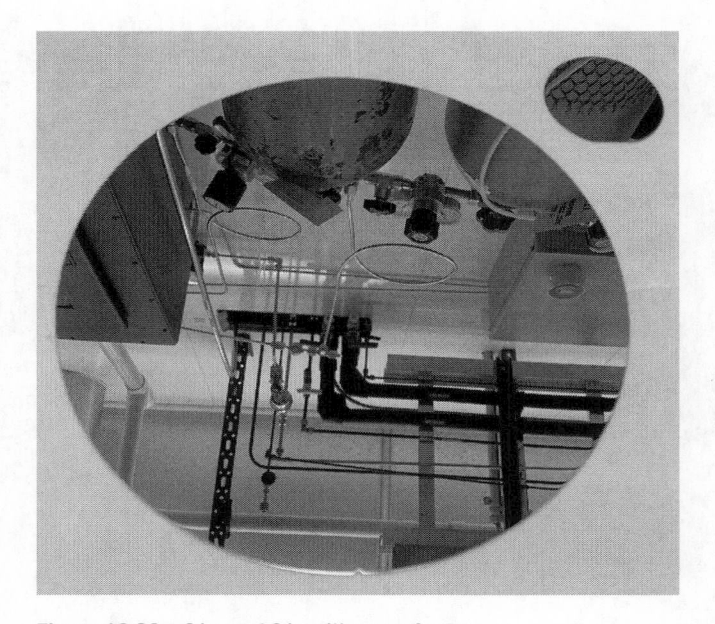

Figure 10.20 8 in. and 2 in. silicon wafers.

References

1 Wolf, S. (2004). *Microchip Manufacturing*. Lattice Press, p. 214, p. 55, p. 133.

2 Wolf, S. and Tauber, R.N. (2000). *Silicon Processing for the VLSI Era*, Process Technology, Lattice Press, 2e, vol. 1, p. 6, p. 438, p. 23, p. 24.

3 MKS Instruments Handbook (2017). Semiconductor Devices and Process Technology by the Office of the CTO, p. 19, p. 21, p. 22, p. 32.
4 Hashim, U., Ehsan, A.A., and Ahmad, I. (2007). High purity polycrystalline silicon growth and characterization. *Chiang Mai Journal of Science* 34 (1): 47–53.
5 Jfermin70, "From sand to the silicon wafer", Steemit.

11

Basic Knowledges of Process

The fabrication of integrated circuits is mainly divided into three parts: design, manufacture, and test. The design is completed through various design softwares, the manufacture is achieved through various processes, and test is performed by using different tools and methods. As mentioned in Chapter 10, a modern manufacturing process contains more than 350 technological steps, many of which are repeated. This book includes very little contents of design and test. In the following chapters, we mainly discuss the technologies commonly used in semiconductor processing. Most of the processes use planar technology, see Figure 7.15. As integrated circuits become more and more complex, the number of interconnected layers also increases. Figure 11.1 is a partial cross-sectional view of a CPU with 45 nm, showing the nine-layer metal interconnect (M9). Regarding metal interconnects, we will discuss them in later chapters.

Semiconductor microchip manufacturing processes mainly include photolithography, etching, deposition, doping, and packaging. Photolithography, etching and deposition techniques are used repeatedly. This chapter gives a brief introduction to the overall structures of integrated circuits, optical knowledge that needs to be understood in lithography, and why plasma is used in the process.

11.1 The Structure of Integrated Circuit (IC)

In this section, we use MOSFET as an example to describe the structure of an IC. Figure 6.13 is a side view of this device. The top view is shown in Figure 11.2. This figure is an n-channel MOSFET. A p-channel has a similar top view to that of the n-channel. The patterns of n^+, source, drain, and gate in the device are completed by photolithography. n^+ refers to n-type heavy doping, usually completed by phosphorus doping, which can be implanted or diffused. The dielectric layer SiO_2 on the device can be completed by CVD (chemical vapor deposition) or thermal oxidation. SiO_2 in the source and drain regions is removed by etching. The metal of

Semiconductor Microchips and Fabrication: A Practical Guide to Theory and Manufacturing,
First Edition. Yaguang Lian.

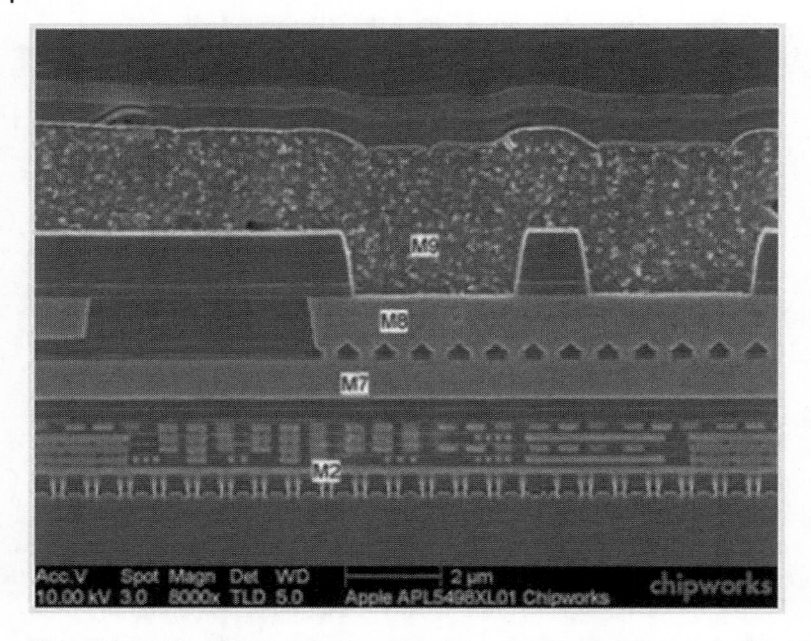

Figure 11.1 A partial cross-sectional view of a CPU. Source: [1] Shimpi et al., 2012 / AnandTech.

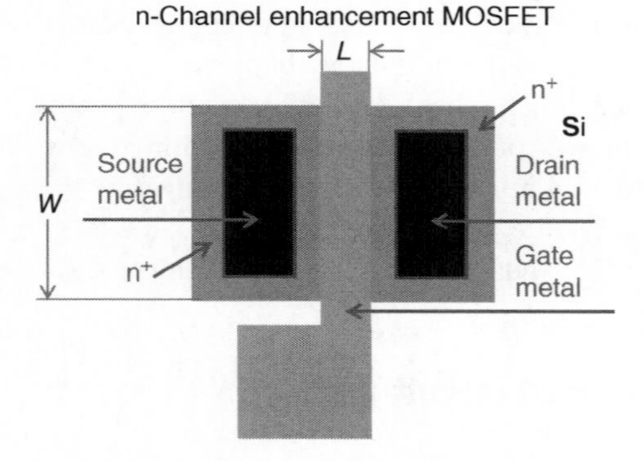

Figure 11.2 Top view of a MOSFET.

the source, drain, and gate is completed by deposition. This kind of deposition is PVD (physical vapor deposition). It mainly includes evaporation and sputtering techniques.

A MESFET has a top view similar to a MOSFET. The devices are made on a semiconductor wafer, and they must have electrically isolated each other. Silicon devices often use SiO_2 isolation, and III–V devices often use proton implantation.

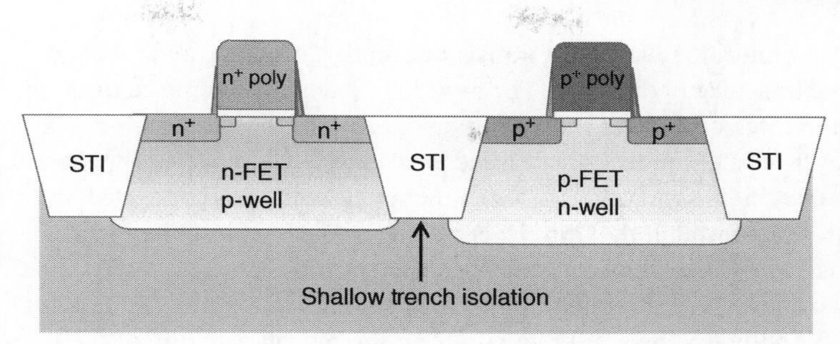

Figure 11.3 Schematic diagram of CMOS structure made with STI technology. Source: [2] Wolf.

For silicon MOSFET, when the feature size is greater than 0.25 μm, the LOCOS (LOCal oxidation of silicon) technology is adopted; When the feature size is less than or equal to 0.25 μm, the STI (shallow trench isolation) technology is adopted. As shown in Figure 11.3. The STI area in the figure is SiO_2, and Poly is polysilicon. In the early technology, the gate metal was usually made of Al, but in modern times it is replaced by heavily doped polysilicon.

After the basic devices-transistors, diodes, resistors, and capacitors are made on the wafer and isolated from each other, SiO_2 film is deposited on these devices as an ILD (interlevel dielectric) by CVD. Vias (connecting holes) are opened on the ILD, and the independent devices are connected by metal through these vias to become an IC. This technology of connecting devices together is called interconnect technology, and this part of an IC is called interconnect layer. Figure 11.4 is a schematic diagram of the interconnect layer between P-channel MOS (PMOS) and N-channel MOS (NMOS). This schematic diagram is only one-layer Interconnect. The nine-layer interconnect shown in Figure 11.1 cannot meet the current IC complexity requirements. For example, Intel's 14 nm chip has

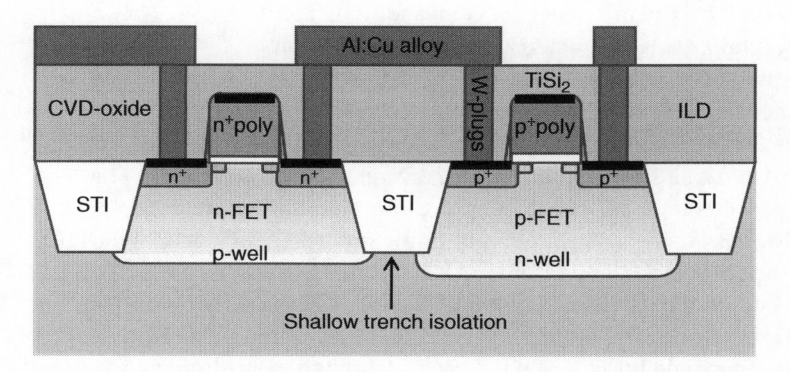

Figure 11.4 Schematic diagram of the interconnection of PMOS and NMOS.

a 13-layer interconnect [3]. Normally, transistors, diodes, resistors, and capacitors are built on the first layer of the wafer. The basic logic gates are completed through the first layer of interconnect (as shown in Figure 6.16). In the next step, through the second layer of interconnect, these basic logic gates are connected into basic functional units. In the next step, these functional units are connected into more complex units through the third layer of interconnect; and so on. Finally, a complete IC is formed, and it meets the design specifications.

As discussed above, no matter how complex an IC is, the main processing steps include: (i) photolithography is used to complete the planar structure of the IC, (ii) deposition is used to grow dielectric films or metal films, (iii) etching is used to remove some dielectric or metal films in some specific areas, (iv) diffusion and implantation are used to complete the n-type and p-type doping of the IC, and (v) testing and packaging are used to complete the final product of the IC. All semiconductor processes follow the basic principles of chemistry and physics. The design of process and equipment is also based on these principles.

11.2 Resolution of Optical System

Among all the processes, photolithography (lithography) is the most important because it determines the technology node of the IC (refer to Section 7.3). The lithography process is completed by an aligner. The ultraviolet light emitted by the machine is used to transfer the designed patterns to the surface of the wafer through a photomask (mask). Most lithography tools are a set of complex optics and control system. The resolution of the lithography machine determines the technology node. We introduced the reflection and refraction of light in Section 8.3. In addition, light has another important characteristic feature-diffraction. Diffraction is a characteristic of waves. The diffraction of light is related to the resolution of the optical system.

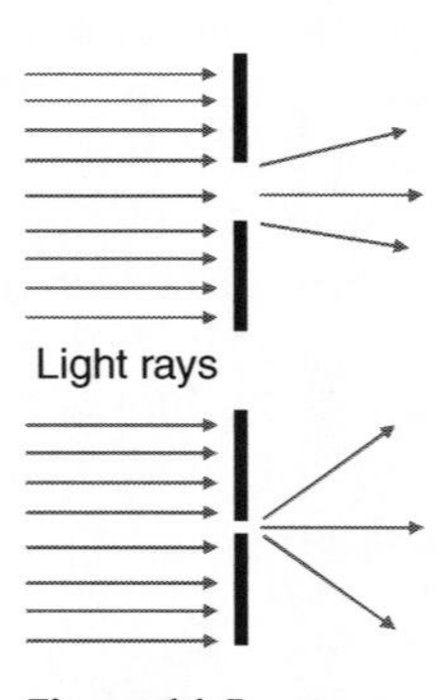

Figure 11.5 Diffraction of light.

When a beam of parallel light passes through a gap or hole, the light will not continue to travel along a straight line, but will diverge, as shown in Figure 11.5. The wider the gap or the larger the diameter of the hole, the smaller the angle of the light divergence. The narrower the gap or the smaller the diameter of the hole, the bigger the angle of the light divergence. This is the diffraction of light. The feature size of contemporary ICs is very small, that is to say, the size of some patterns on the mask used in the lithography is very small. These patterns are composed of a series of small gaps and holes. When light passes through them, diffraction occurs. We will continue to discuss lithography machine and mask in later chapters.

Figure 11.6 Light diffraction intensity profile and Airy disk. Source: [4] Ghoushchi et al., 2015 / Koç University.

Let us take a small hole as an example. Due to diffraction, the intensity distribution of the light passing through the small hole is uneven, with a bright large spot in the middle and dim rings around it. This is the diffraction pattern, as shown in Figure 11.6. The area with the highest intensity in the middle is called Airy disk (or Airy disk), in honor of British scientist Sir George Biddell Airy (July 27, 1801–January 2, 1892). The radius of Airy disk refers to the point from the center of the spot to the point where the first light intensity of the diffraction pattern is zero. Figure 11.6 shows the diameter of Airy disk.

Through the diffraction of light, we can deduce the resolution of the optical system. Please see Figure 11.7. When two light sources (Object 1 and Object 2) illuminate a screen through a small hole, two diffraction images appear on the screen. The diffraction image is shown in Figure 11.8. In the figure, "Just resolved" means that the center peak of one Airy disk just falls on the first light intensity at zero of another Airy disk. If we use θ to represent the angle between the two rays entering the small hole, and r to represent the distance between the two light spots, then when the following formula is satisfied:

$$\theta = 1.22\frac{\lambda}{D} \tag{11.1}$$

$$r = \frac{1.22\lambda}{2n\sin\theta} = \frac{0.61\lambda}{\text{NA}} \tag{11.2}$$

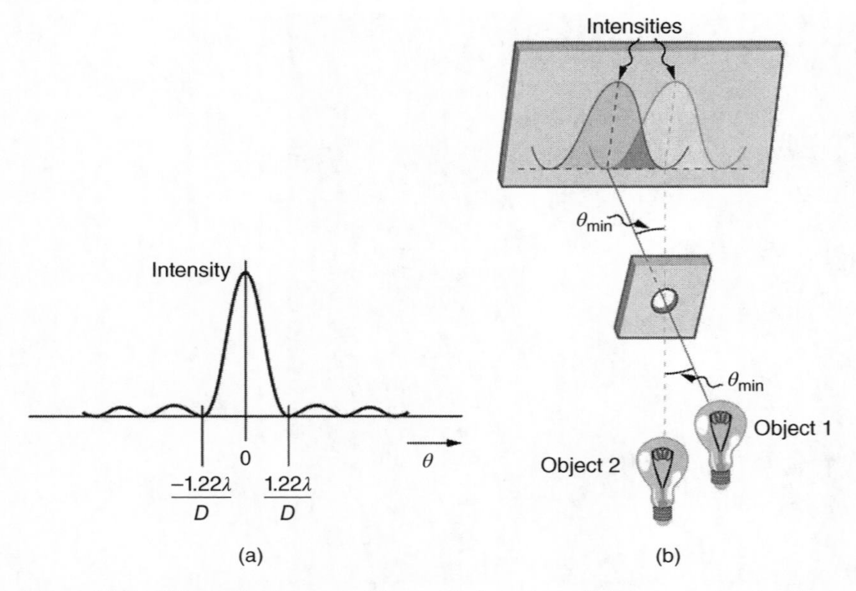

Figure 11.7 Schematic diagram of the resolution of two light sources passing through a small hole. Source: Reprinted with permission of BCcampus.

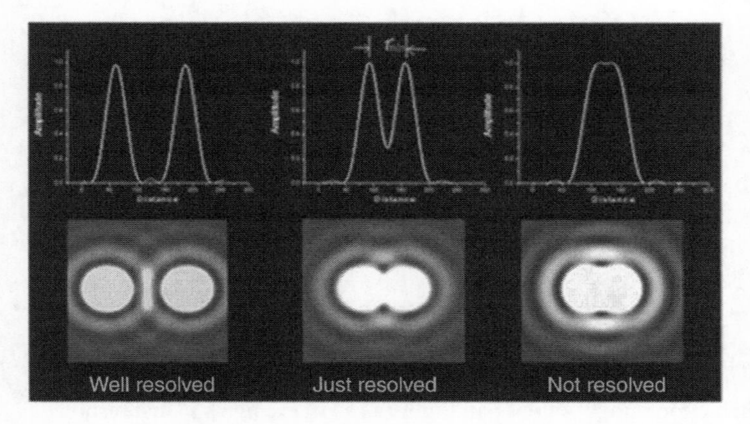

Figure 11.8 Schematic diagram of Rayleigh criterion. Source: Reproduced with permission of Boston Biophysics.

θ is the smallest angle when the two light spots are just resolved, which is half the angle from the center of the small hole to Airy disk. It is represented by θ_{min} in Figure 11.7. r is the minimum distance between the centers of the two Airy disks when the two light spots are just resolved. This value is the radius of Airy disk, which is represented by r_{min} in Figure 11.8.

Formulas (11.1) and (11.2) are Rayleigh criterions, which were developed by British scientist John William Strutt, 3rd Baron Rayleigh (November 12, 1842–June 30, 1919). λ in the formulas is the wavelength of the incident light, D is the diameter of the small hole, n is the refractive index of the surrounding optical medium, and NA is called the numerical aperture:

$$\mathrm{NA} = n \sin\theta \tag{11.3}$$

If formulas (11.1)–(11.3) are applied to the design of a lithography system, it is seen that in order to obtain a high-resolution, a short-wavelength light source and a large-size optical lens are required. Usually the medium around the lens is air. If we immerse the lens in a medium with a higher-refractive index than air, the numerical aperture will increase, and the resolution of the system will be higher. We will continue the discussion in later chapters.

If optical medium (medium) is air, the maximum NA is 1 according to formula (11.3). For a microscope, NA = 0.87 is nearing the limit when air is utilized as the imaging medium [5]. We will continue the discussion in the later chapters.

11.3 Why Plasma Used in the Process

As we mentioned before, etching and deposition are also the main parts of semiconductor processes. In these processes, plasma technique is commonly used, such as RIE (reactive-ion etching), PECVD (plasma-enhanced chemical vapor deposition), and sputtering. In theory, there is no essential difference between RIE and PECVD. Both require plasma and mainly use chemical reactions. In sputtering system, the plasma is required to perform PVD.

The biggest difference between RIE and PECVD is the chemical reaction product. If the product is volatile, that is, the product is in a gaseous state, it is RIE. If the product is nonvolatile, that is, the product is in a solid state, it is PECVD. The chemical reaction product is volatile, which usually means its boiling point is very low. The chemical reaction product is nonvolatile, which usually means its boiling point is very high. For example, in RIE, the most used etching chemicals for Si, SiO_2, and Si_3N_4 are fluorine-containing gases CF_4, CHF_3, and SF_6. After fluorine and silicon react chemically, the boiling point of the product silicon tetrafluoride (SiF_4) is −86 °C:

$$\mathrm{Si\,(solid)} + 4\mathrm{F\,(gas)} \rightarrow \mathrm{SiF_4\,(gas)} \tag{11.4}$$

SiF_4 can be removed from the process chamber by a pump, which is etching. However, the boiling point of silica produced by the reaction of oxygen and silicon is 2950 °C:

$$\mathrm{Si\,(solid)} + 2\mathrm{O\,(gas)} \rightarrow \mathrm{SiO_2\,(solid)} \tag{11.5}$$

SiO_2 is not pumped away, but remains on the surface of the sample, which is deposition.

As mentioned in the previous chapter, to make the chemical reaction to proceed smoothly between stable molecules, we should encourage the molecules to collide with each other, break the chemical bonds, and excite the electrons in the outermost shell. One of the ways to make molecules collide with each other is to raise the temperature. However, because of the different strengths of different chemical bonds, the temperature required to break these chemical bonds is different. Sometimes the required temperature is very high, which has certain difficulties in actual operation.

In addition to the method of heating, the other is to use electric and magnetic fields. In Chapter 4, we discussed electric and magnetic fields. They are both vector fields. The electric field is represented by a capital black $\boldsymbol{E}$ and the magnetic field is represented by a capital black $\boldsymbol{B}$. When a charge Q enters the electromagnetic field at a speed $\boldsymbol{V}$, the force $\boldsymbol{F}$ that it receives is called Lorentz force and satisfies Lorentz equation. Note that both speed and force are vectors, so the letters in bold are used:

$$\boldsymbol{F} = Q\,(\boldsymbol{E} + \boldsymbol{V} \times \boldsymbol{B}) \tag{11.6}$$

The above equation is a vector operation. The cross symbol $\times$ in the equation indicates that the direction of the force on the charge is 90° with the direction of the magnetic field. There is no $\times$ in front of the electric field, which means that the direction of the force on the charge is consistent with the direction of the electric field. This equation was first proposed by the Dutch physicist Hendrik Lorentz (July 18, 1853–February 4, 1928). Lorentz equation shows that in an electromagnetic field, a charged particle is accelerated and its direction is changed by a force. This particle can collide with other particles around it, producing the same effect as raising the temperature. Normally, molecules at room temperature are electrically neutral, and a neutral particle is not subjected to the force of an electromagnetic field. Due to thermal motion, gas and liquid molecules still move randomly in all directions. But in a solid, because of the limitation of the crystal lattice, the atoms can only vibrate thermally near the crystal lattice. The semiconductor manufacturing processes, such as RIE and PECVD, primarily use gaseous chemicals to complete etching and deposition. So, in this section we will use gas as an example to illustrate how plasma is generated under the action of Lorentz force.

As mentioned above, gaseous molecules move in all directions at room temperature. Table 11.1 lists the mean (average) thermal velocity of some gases at 20 °C. Mach number in the table is the ratio of the velocity of an object to the velocity of sound, named after Austrian physicist and philosopher Ernst Mach (February 18, 1838–February 19, 1916).

$$M = \frac{u}{v} \tag{11.7}$$

where M is the Mach number, u is the object velocity, and v is the sound velocity.

Table 11.1 Mean thermal velocity of some gases at 20 °C [6].

Gas	Chemical symbol	Mean velocity (m/s)	Mach number
Hydrogen	H_2	1754	5.3
Helium	He	1245	3.7
Water vapor	H_2O	585	1.8
Nitrogen	N_2	470	1.4
Air		464	1.4
Argon	Ar	394	1.2
Carbon dioxide	CO_2	375	1.1

As can be seen from this table, the mean thermal velocity of gases exceeds the speed of sound. Collisions occur between them. Some ions and electrons are generated. In addition, cosmic rays can also generate some ions and electrons. But the concentration of charged ions and electrons produced in these ways is very low, which is why the chemical reaction rate is very slow at room temperature. However, if an electromagnetic field is added and the energy of the electromagnetic field exceeds a certain limit, these low concentration ions, especially electrons, hit other molecules at a faster speed under the acceleration of the electromagnetic field, break more molecular bonds, and excite more electrons. When the concentration of ions and excited electrons exceeds a certain number, the plasma discharge is initiated. This process is called ionization. An example of ionization is lightning in rainy weather, where the high-strength electric field between clouds triggers the plasma by air molecules. The generation of plasma enables chemical reactions to proceed at room temperature or even lower temperature.

In the actual process, we pass the selected gases into the process chamber, apply an electric or magnetic field to the process chamber, and generate plasma in the chamber. According to the discussion in Chapter 4, we revealed that the electric field is related to the capacitor, and the magnetic field is related to the spiral coil inductor. Therefore, when designing a process chamber capable of generating plasma, only two structures can be adopted: a capacitive structure and an inductive structure. These two structures can be used alone or mixed.

References

1 Shimpi, A.L. (2012). Apple A5X die size measured: 162.94 mm^2, Samsung 45 nm LP confirmed. AnandTech, March 16.

2 Wolf, S. (2004). *Microchip Manufacturing*, 67–69. Lattice Press.

3 Hruska, J. (2014). Intel's 14 nm Broadwell chip reverse engineered, reveals impressive FinFETs, 13-layer design. ExtremeTech.
4 Ghoushchi, V.P. (2015). Opto-mechanical design and development of an optodigital confocal microscope. University of Murcia, M.Sc. Thesis. p. 9.
5 Numerical Aperture. Nikon, Microscopy, The source for microscopy education.
6 Pfeiffer Vacuum Technology, 1.2.4 Thermal velocity. pfeiffer-vacuum.com.

12

Photolithography (Lithography)

Photolithography is to make device patterns on the surface of semiconductor substrate. After the patterns are completed, other processes will follow-up, such as etching, deposition, and so on. Figure 12.1 is a simple schematic diagram of the flow of lithography process and etching. The lithography process shown in the figure is like the film photography which we used before, but there are three main differences: (i) the photoresist-coated wafer replacing photographic film, (ii) the photomask replacing the scenery, and (iii) the ultraviolet light replacing the sunlight. During film photography, the film feels different light intensities in the scene, which means that the light intensity received on the film is uneven. During photolithography, ultraviolet light passes through the pattern windows of the mask, and the intensity that the photoresist receives is uniform. Below we will discuss the lithography process based on this schematic diagram.

12.1 The Steps of Lithography Process

When we get a semiconductor wafer or piece and need to do a photolithography process, the steps shown in Figure 12.2 are usually followed:

12.1.1 Cleaning

When we have a wafer or a piece, to avoid contamination of the process, the wafer or piece must be cleaned first. In Chapter 10, we discussed RCA cleaning (Figure 10.15). This is the standard cleaning method for silicon wafer. This method can be used to clean Si wafers, Si covered with SiO_2, and Si covered with Si_3N_4. But for III–V semiconductor materials (refer to Chapter 5) and silicon wafers that have completed metal processing, RCA cleaning cannot be used, because this cleaning solution can corrode compound semiconductor materials and metals. At this moment, the common cleaning method is to use organic

Semiconductor Microchips and Fabrication: A Practical Guide to Theory and Manufacturing,
First Edition. Yaguang Lian.

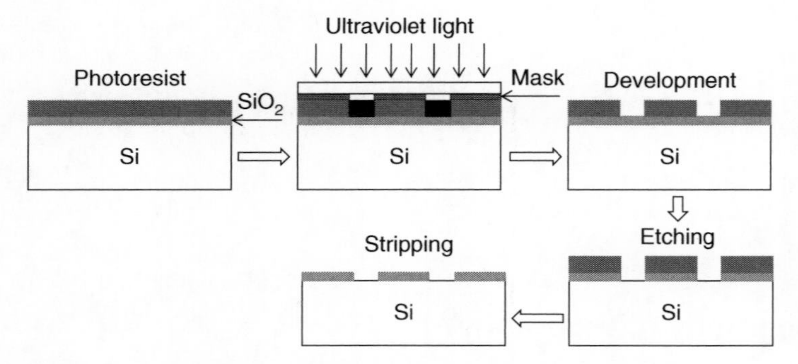

Figure 12.1 Schematic diagram of photolithography process.

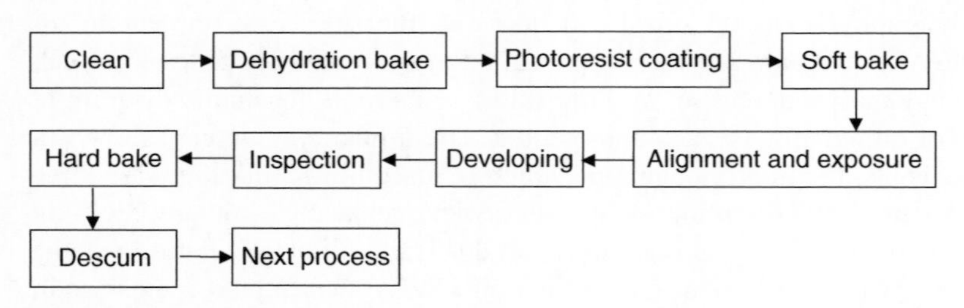

Figure 12.2 Lithography process step diagram.

solvents soaking and washing wafer or piece. We first use acetone, then use alcohol or isopropyl alcohol (IPA), and finally rinse with DI water. In addition, because RCA cleaning is complicated and dangerous to use, organic solvents can also be used to clean silicon wafers, SiO_2/Si and Si_3N_4/Si when the requirements are not strict. No matter what method is used, the final step is to rinse with DI water. After rinsing, use a nitrogen gun to blow dry, or use a spin dryer. Please see Figure 12.3.

12.1.2 Dehydration Bake

The last step of cleaning is to rinse the wafer or piece with DI water. Although it is dried with a nitrogen gun or a spin dryer after rinsing, there are still water traces on the surface of the wafer or piece. We need to dehydrate the surface. There are two main methods of the dehydration. Put the wafer or piece into a vacuum chamber or onto a hotplate to heat up for the dehydration. The hot plate is simple and cheap to use. So, in most cases, the hot plate is used for dehydration bake. The photo of the hot plate is shown in Figure 12.4.

To further understand the necessity of dehydration, we need to briefly introduce the hydrophilicity and hydrophobicity of the material surface. When a solid

Figure 12.3 N_2 gun (a) and spin dryer (b).

Figure 12.4 The picture of hotplate.

surface is in contact with liquid, there are two situations of the surface. One is hydrophilic surface, and the other is hydrophobic surface, see Figure 12.5. For liquid water, SiO_2 is a hydrophilic surface and Si is a hydrophobic surface. If Si stays in the air for a period of time, there will be a natural oxide layer on the surface. So, in general, it is also a hydrophilic surface. Therefore, a thin layer of water molecules is adsorbed on the surface. The photoresist used in photolithography is hydrophobic. When a hydrophilic surface with water molecules meets a hydrophobic material, the adhesion is poor. Due to the hydrophilicity of SiO_2, the water adsorbed makes the surface to have poor adhesion with the photoresist, and the normal photolithography process cannot be performed. So, the dehydration bake is required to drive the water out of the surface. The temperature

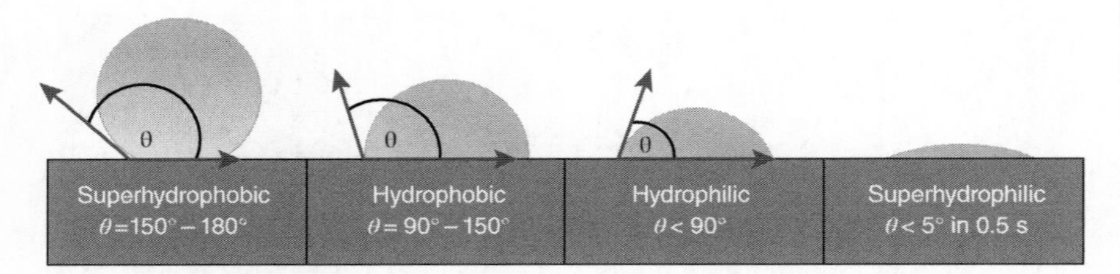

Figure 12.5 Schematic diagram of the hydrophilicity and hydrophobicity of the material surface. Source: [1] Ahmad et al. / Taylor & Francis.

for desorption of surface water molecules is 150–200 °C [2]. A recommended dehydration bake condition is that the hot plate temperature is 150–200 °C and the time is 15–10 minutes.

12.1.3 Photoresist Coating

Photoresist (PR or resist) is a photosensitive organic material used in the photolithography process. There are two commonly used resists-positive photoresist and negative photoresist. If the exposed areas can be removed by the developer, the unexposed areas cannot be removed by the developer, it is positive photoresist. The PR in Figure 12.1 is positive resist. If the exposed areas cannot be removed by the developer, the unexposed areas can be removed by the developer, which is negative photoresist. Sometimes we do not distinguish between positive PR and negative PR, referred to as PR, please see Figure 12.6. In order to further improve the adhesion between the resist and the wafer or piece surface, a layer of primer needs to be coated on the surface before coating the PR on it. The commonly used primer is HMDS (hexamethyldisilazane). The left one of Figure 12.7 is HMDS. The right one is polymethyl methacrylate (PMMA) for e-beam lithography.

Dehydration bake was discussed in the previous paragraph. However, after baking, a small amount of water or hydroxide (OH^-) may remain on the surface of the wafer or piece, which will adversely affect the adhesion of the PR. HMDS can remove residual water and hydroxide radicals on the surface and combine with the surface. In this way, when the photoresist is coated on the wafer or piece surface containing HMDS, its adhesion will be improved.

When the wafer or piece has just completed the growth or deposition of SiO_2 or Si_3N_4, and if the next step is photolithography process, do not leave it in the room for too long, but coat the resist immediately. No need to clean, bake, or coat HMDS, just coat the PR directly. Our commonly used SiO_2 or Si_3N_4 equipment are thermal oxidation, low-pressure chemical vapor deposition (LPCVD), low-temperature oxidation (LTO), and PECVD. The temperature of all these processes is above 250 °C. Just after completion, there will be no traces of water

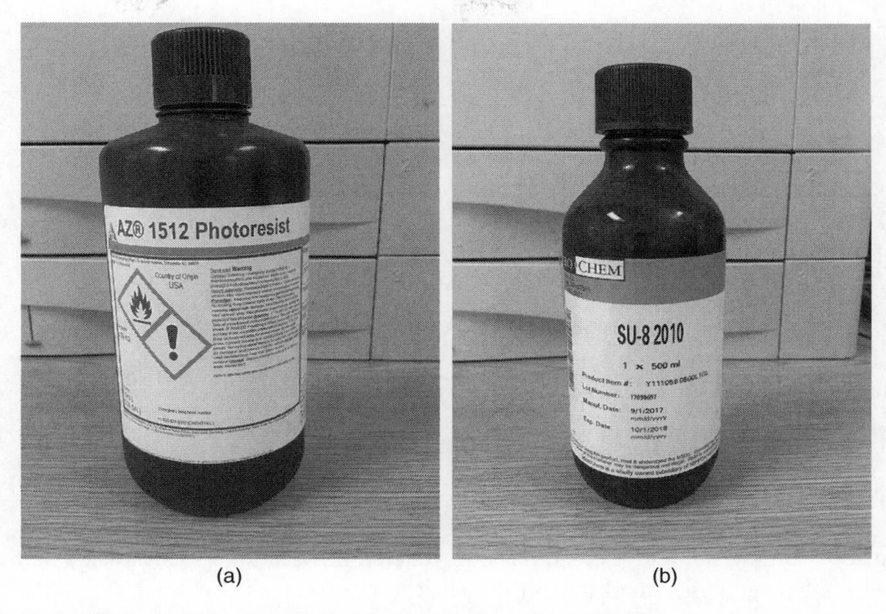

Figure 12.6 Positive photoresist (a) and negative photoresist (b).

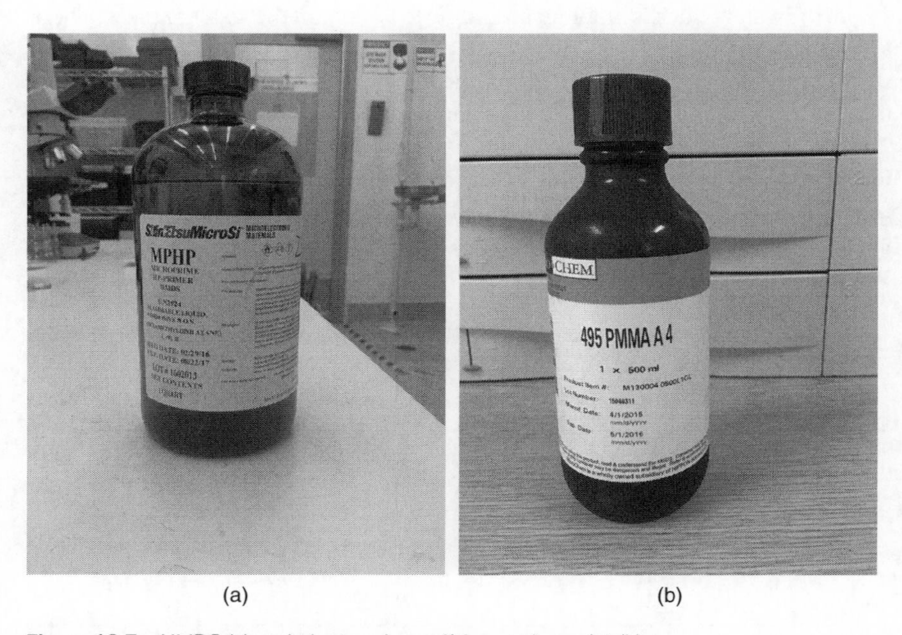

Figure 12.7 HMDS (a) and electron beam lithography resist (b).

Figure 12.8 Spinner.

or OH^- on the wafer or piece surface. In general, the cleanliness level inside the equipment is higher than that in the clean room (refer to Section 7.4), so the pre-PR coating steps can be omitted.

The equipment used for PR coating is a spinner, as shown in Figure 12.8. The size of the wafer chuck is different according to the size of the wafer or piece. There are small holes on the chuck, which are connected to a vacuum pump. Through vacuum, the wafer or piece is held on the chuck. The PR is dropped on the surface of the wafer or piece. A motor accelerates the chuck to the spinning speed of 2000–6000 rpm (revolutions per minute) within a few seconds and keep this speed for 30–60 seconds, so that the resist is evenly spread on the wafer or piece surface. The unit of thickness of PR is usually μm. Table 12.1 shows the relationship between the thickness of AZ 1500 series positive PR and the spinning speed of the spinner. Figure 12.9 shows the silicon wafers before and after photoresist coating.

Table 12.1 The thickness (μm) of the AZ 1500 series positive PR vs. the spin speed (rpm).

Spin speed	2000	3000	4000	5000	6000
AZ 1505	0.71	0.58	0.50	0.45	0.41
AZ 1512HS	1.70	1.39	1.20	1.07	0.98
AZ 1514H	1.98	1.62	1.40	1.25	1.14
AZ 1518	2.55	2.08	1.80	1.61	1.47
AZ 1529	4.10	3.35	2.90	2.59	2.37

Source: Ref. [3] / Clariant.

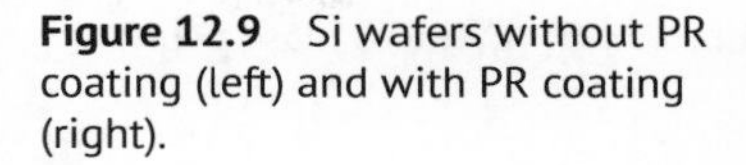

Figure 12.9 Si wafers without PR coating (left) and with PR coating (right).

12.1.4 Soft Bake

After the coating of photoresist, the wafer or piece needs to do soft bake, also called pre-bake. The purpose of this step is to remove most of the solvent in the PR and make it solid for the next step of alignment and exposure. Usually, the photoresist in the bottle contains 65–85% solvent, which makes the PR thinner and can successfully complete the PR coating step. After the coating is completed, the solvent content will drop to 10–30%, but the surface is still very sticky at this time and cannot do alignment and exposure. After the soft bake is completed, the solvent in the PR is reduced to 5% [2], and the liquid-cast film is converted into solid form. So that the alignment and exposure can proceed smoothly. In addition, the soft bake can improve the adhesion between the PR and the surface, which is beneficial to the development after exposure.

In most cases, the soft bake temperature is 90–110 °C. According to the thickness of the photoresist and the temperature of the hot plate, the time is 1–5 minutes.

12.1.5 Alignment and Exposure

After the soft bake is completed, alignment and exposure are performed. Alignment is to align the finished patterns on the wafer or piece with a photomask by using a mask aligner (aligner). In the early days, the aligners were manual alignment. Now most of them use automatic alignment. But in university laboratories and clean rooms, manual alignment is mainly used. Figure 12.10 is a photo of an aligner. Figure 12.11 is a photo of a photomask. The brown area on the mask is opaque, and the ultraviolet light emitted by the aligner cannot pass through this area. The white part is transparent. The ultraviolet light passes through this area to expose the resist on the surface under the mask. The patterns on the mask are transferred to the PR. For the convenience of observation, a mask with large patterns is specially found here as an example to illustrate. Please also refer to Figure 12.1.

Figure 12.10 EVG 620 mask aligner.

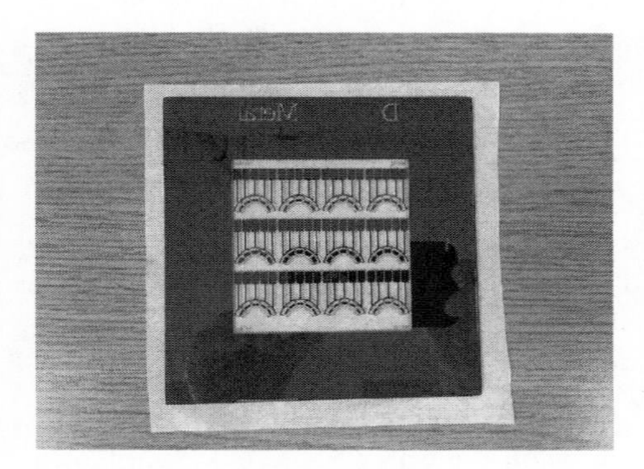

Figure 12.11 A photomask.

Figure 12.10 is a contact aligner, that means the mask and the surface of wafer or piece contact each other. This type of aligner can also perform non-contact proximity exposure. There is also a long-distance non-contact machine. The projection exposure system widely used in production is the long-distance non-contact lithography machine. Figure 12.12 is a schematic diagram of the three exposure modes. Projection photomask is also called reticle. Typical projection systems use reduction optics (4X–5X).There are two kinds of projection systems-stepper and scanner. During exposure, a stepper moves only the wafer-step and repeat, and a scanner moves the reticle and wafer simultaneously step and scan.

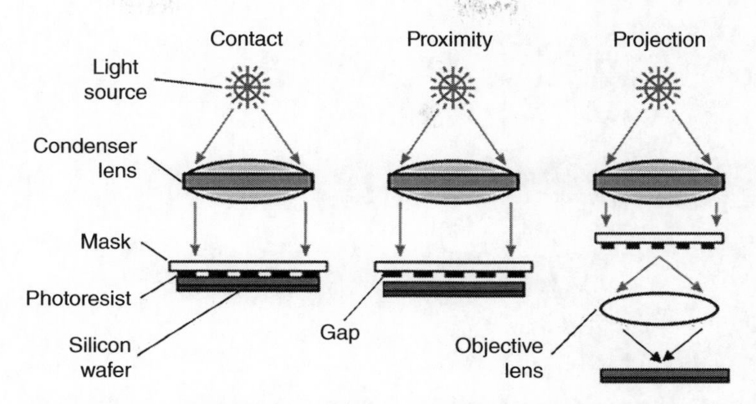

Figure 12.12 Basic operation modes of optical lithography: In contact lithography the mask is in direct contact with the resist, in proximity lithography there is a small gap in between. Projection lithography uses a lens system between the mask and the sample. Source: [4] Haberfehlner.

The contact aligner can get high resolution, but the defects on the surface of photoresist are big. This is because if there are small dirty spots on the surface of PR, once the mask and the wafer or piece contact each other, the damaged area of the small particles will become much bigger. In addition, the contact one is cheap, and the diffraction effect of light is small.

The smallest pattern size that can be obtained by the proximity mode is larger than that of the contact mode. This is because diffraction reduces the resolution of this mode, but the PR surface has fewer defects.

The projection type has higher resolution and lower defects on the resist surface. Take the stepper as an example. When it is working, the tool shrinks the size of patterns on the reticle several times and projects them onto a certain area of the wafer. Then the wafer moves its position, and tool projects the patterns to a new position. With this way, repeat step by step, and finally make patterns all over the surface of the wafer. We can imagine that with a certain area of circuit patterns, the larger the wafer size is, the more circuits are finished with same step–step process. The production cost will be lower, and the price will be cheaper. Therefore, the size of Si wafer continues to become larger, see Figure 7.10.

There are different types of light sources used to generate ultraviolet (UV) light. High-pressure mercury-vapor lamp (HPMVL) was once widely used in semiconductor photolithography processes. Because the light source is cheap and easy to operate, it is still a workhorse in university laboratories and clean rooms. Figure 12.13 is a photo of a HPMVL. When the mercury lamp is lit, the pressure of the ionized mercury gas in the lamp can reach 40 atm [2], which is why we call it a high-pressure lamp. Figure 12.14 is the spectrum of the HPMVL. In the

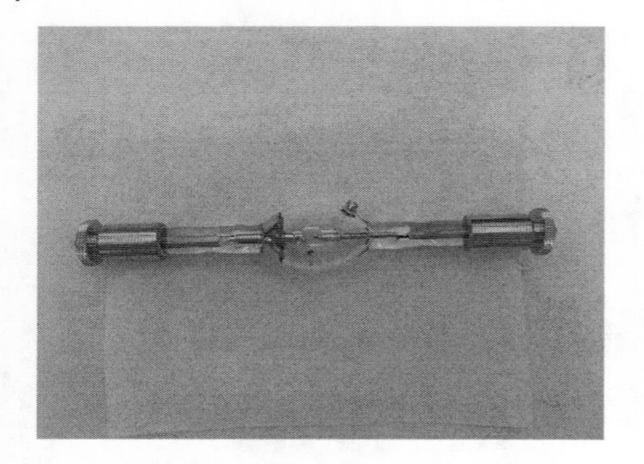

Figure 12.13 High-pressure mercury UV lamp.

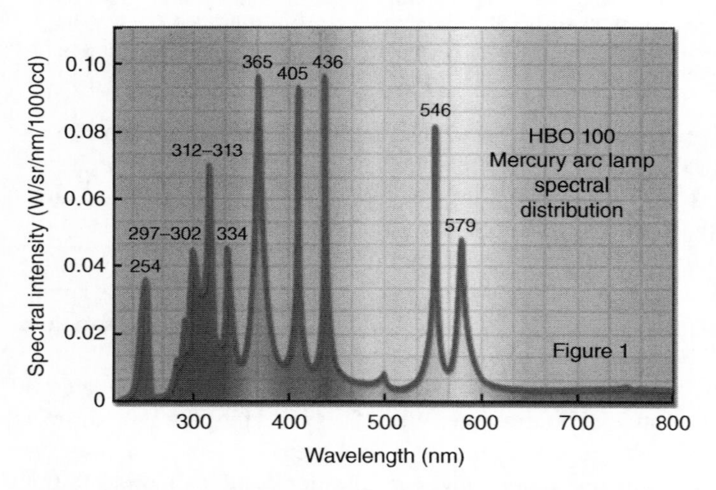

Figure 12.14 The spectrum of the high-pressure mercury lamp. Source: Reprinted with permission of ZEISS.

spectrum, there are three peaks of luminescence. The wavelengths are 436, 405, and 365 nm, respectively. These three wavelengths are named G-line, H-line, and I-line. Among the three spectral lines, the I-line has the shortest wavelength. According to Rayleigh criterion, it has the highest resolution. Therefore, in the current lithography systems with HPMVLs, most of them use I-line. In the mercury lamp spectrum, 546 and 312–313 nm are also two spectral lines with higher peaks. However, their peaks are smaller than G-line, H-line, and I-line. The output power is smaller, and the exposure time is longer. In addition, the long wavelength of 546 nm is not conducive to the production of devices with small feature sizes Therefore, the two spectral lines of 546 and 312–313 nm are not used much in the process. Quartz glass is mostly used to make photomask and reticle.

12.1.6 Developing

After exposure, the wafer or piece is placed in the developer to display the latent patterns and become the device patterns on the photoresist. The following processes are performed on these patterns. Figure 12.15 is negative PR and positive PR developers. The developers for both resists are usually aqueous-alkaline. They are based either on a diluted sodium hydroxide or potassium hydroxide solution. These two types of developers are metal ion containing (MIC) or metal ion bearing (MIB). Another type of developer is metal ion free (MIF), mostly based on organic TMAH (TetraMethylAmmoniumHydroxide). For AZ series photoresists, the positive one can use either AZ MIC developer or MIF developer, the negative one only uses MIF developer. Another kind of developer for negative resist is a special organic solvent solution, for example SU-8 developer is shown in the figure.

When developing, agitate wafer or place in a container for desired time. After development, the positive PR pattern has the developing opening with wide top and narrow bottom (close to the substrate), and the negative PR pattern has the developing opening with narrow top and wide bottom. Refer to Figure 12.16.

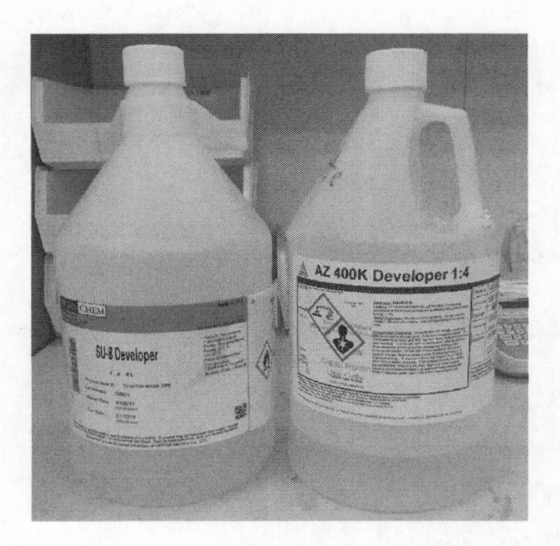

Figure 12.15 Negative PR developer (left) and positive PR developer (right).

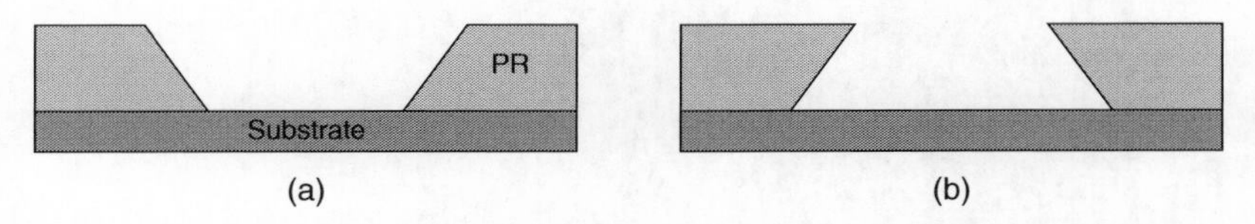

Figure 12.16 The profiles of PRs after development, the (a) is positive and the (b) is negative. Source: [5] Lueke et al. / MDPI / CC BY 4.0.

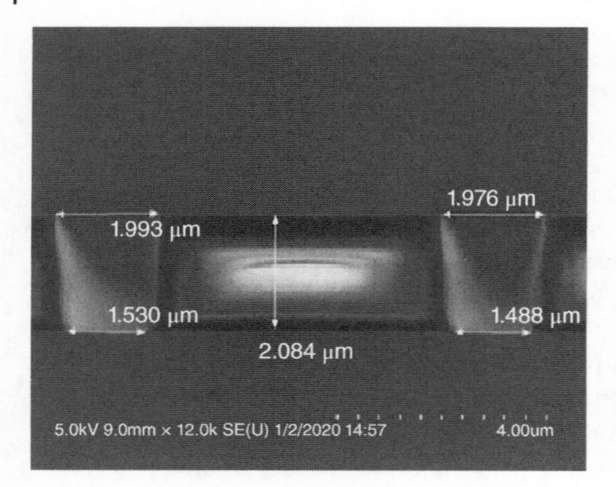

Figure 12.17 The SEM picture of positive PR after development.

Figure 12.17 is a SEM (scanning electron microscope) picture of the openings' side view of the positive PR patterns. If we want to have a steeper profile, a simple way is to appropriately increase the exposure time and shorten the development time.

12.1.7 Inspection

After the development, put the wafer or piece under a microscope for observation to check whether the pattern openings are clean and whether the alignment between different layers meets the accuracy specifications. Figure 12.18 is a photo

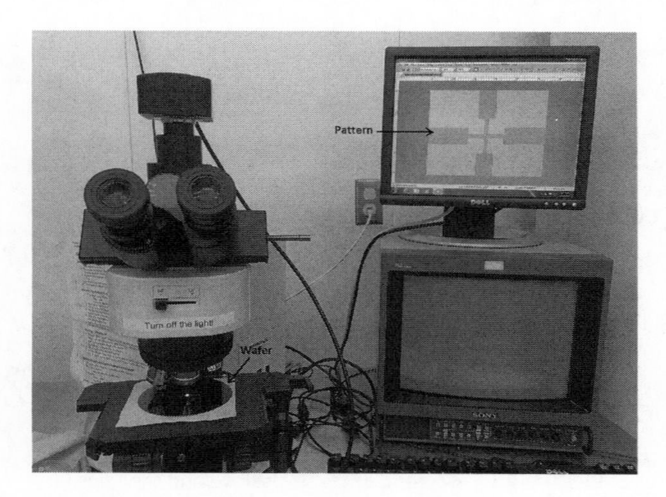

Figure 12.18 Check the patterns of the sample under a microscope.

of a microscope and the sample being observed. If the result is not satisfactory, remove the PR and redo the process. For the most PRs, a simple way to redo the process is to put the sample on a spinner. When the spinner starts to spin, spray the surface with acetone, followed by isopropanol, never use DI water. Wait for the spinner to stop and observe the surface of the sample. If there is no PR residue, we can drop and spin to coat a new resist layer.

12.1.8 Hard Bake

This step is also called post-bake. For most developers, the last step of development is to rinse with DI water. For an organic solvent developer, the last step is generally to rinse with isopropanol. The purpose of the hard bake is to remove the water residue or residual organic solvent, and to make the photoresist to harden continually, laying a good foundation for the next process. But, some photoresists (for example, SU-8 negative PR) do not need to do hard bake.

For positive resists, a common problem in hard bake is reflow. When the temperature of the hot plate is high, the edge of the pattern opening appears to reflow, making the edge round and affecting the next process. Figure 12.19 is a reflow phenomenon of SPR PR. As can be seen from the figure, when the temperature of the hot plate is 95 °C, the edge of the PR opening begins to reflow. Therefore,

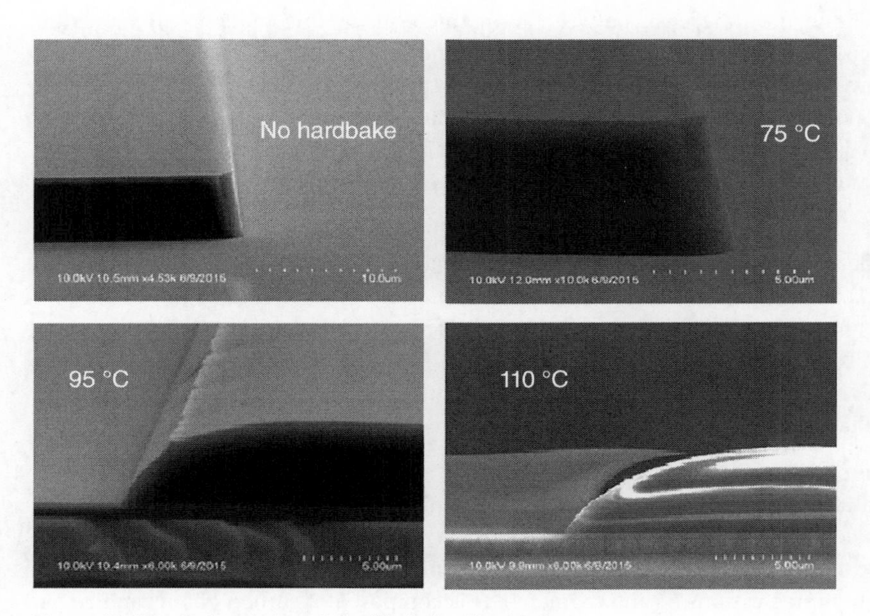

Figure 12.19 Reflow issue of the PR during hard bake.

under normal circumstances, the temperature of hard bake should be lower than the temperature of soft bake, and in the meanwhile, the time should be extended. Negative resists crosslink after exposure. The crosslinking makes them thermally stable, so even elevated temperature will not deteriorate the resist profile [6].

12.1.9 Descum

After the development is completed, observe the openings under the microscope. Although they look clean, there may be a thin layer of PR residues left on the surface of the openings. The residues may be introduced from insufficient exposure or developing time. These residues do not have much impact on dry etching but will have a huge impact on metal processes such as ohmic contact. So, we need to further clean the opening-remove the residues, generally by using oxygen plasma method. This step is called descum.

To explore this issue further, we need to say a few more words about organic matters. Photoresist is an organic matter (organic molecule or compound). Organic matters are compounds containing carbon, but some exceptions, such as CO, CO_2, SiC, etc. Most inorganic matters are molecules not containing carbon. The smallest organic molecule is methane (CH_4). Compared with inorganic molecules, the biggest feature of organic compounds is that all large and macro molecules are composed of organic compounds. If we use daily life example to illustrate the difference between inorganic and organic matters, an inorganic molecule is like a grass, and an organic molecule is like a tree. The difference between a tree and grass is that a tree has a trunk. Because of the trunk, a tree can support its huge body. Through this example, we can think of an organic matter. If a large molecular structure of an organic matter is supported, it must have a structure like a tree trunk, and carbon plays this role. What is so great about carbon? [7]

- Carbon atoms can be linked by strong, stable covalent bonds.
- Carbon atoms can form stable bonds to many other elements (H, F, N, S, P, etc.). Most organic compounds contain a few hydrogens, and sometimes oxygen, nitrogen, sulfur, phosphorus, etc.
- Carbon atoms can form complex structures, such as long chains, branched chains, rings, chiral compounds, complex 3D shapes, etc.

The chiral compound mentioned above is a compound that has an asymmetric center and cannot be superposed on its mirror image. Figure 12.20 shows the molecular structure of Novolak, which is widely used as a photoresist material. The molecule is represented by skeletal formula and carbon atoms are represented by the intersections of line segments. It is composed of carbon chains and rings. The carbon ring exists in the form of benzene ring (right one in the figure).

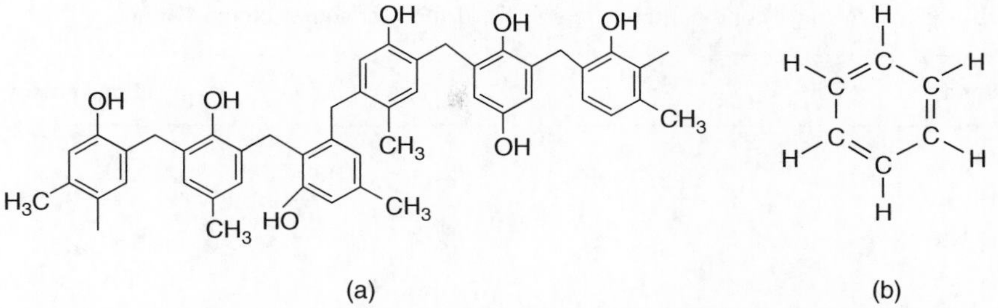

(a) (b)

Figure 12.20 (a) Molecular structure of Novolak, which is a phenol-formaldehyde condensation polymer with a branched structure (Source: [8] Reiser et al. / with permission of Elsevier) and (b) benzene ring.

An important reason why carbon has this characteristic is that it has the second strongest bond in nature. Please see Table 12.2. For the same-element structure, nitrogen has the strongest bond (N≡N), followed by carbon (C≡C). The mole (symbol: mol) is a unit that represents the quantity of a substance. The number of microscopic substances (atoms, molecules, ions, etc.) in a mole is about 6.02×10^{23}. This number is called Avogadro number, in memory of the Italian scientist Amedeo Avogadro (1776–1856). Although nitrogen has the largest bonding energy and the strongest bond, it is unfortunate that nitrogen exists in the form of a gas in a large temperature range. Its boiling point is −195.795 °C. Carbon exists as a solid in a large temperature range. Its sublimation temperature is 3642 °C. Sublimation means that the substance changes directly from a solid to a gas without passing through a liquid state. This allows carbon to be a "trunk" within a large temperature range. Carbon provides a solid foundation for the formation of organic matter. Life on the earth consists of the organic molecules, so life on the earth is carbon-based life.

Now let us return to the descum. As mentioned above, descum uses oxygen plasma to clean PR residues in the PR openings. Why does plasma be used? We have already discussed before, so I do not repeat it here. Why does oxygen be used? This is because of carbon dioxide or carbon monoxide produced by the reaction between oxygen and carbon in the PR. CO_2 or CO is a gas at room temperature. In addition, other materials that make up photoresist can also form volatiles after reacting with oxygen, or after being separated from the molecular structure, they are themselves gaseous. So, the residues can be removed with oxygen. For the same reason, fluorine and hydrogen also can be used for PR descum, because their reactants with carbon, carbon tetrafluoride (CF_4) and methane (CH_4) are gases at room temperature. Compared with oxygen, fluorine and hydrogen are more dangerous, especially hydrogen, which is even more combustible and dangerous gas. Their price is more expensive than oxygen, so we usually use oxygen

Table 12.2 Average bond energies (strengths, kJ/mol) for some common bonds.

Bond	Bond energy
H—H	436
H—C	415
H—N	390
H—O	464
H—F	569
H—Si	395
H—P	320
H—S	340
H—Cl	432
H—Br	370
H—I	295
C—C	345
C=C	611
C≡C	837
C—N	290
C=K	615
C≡N	891
C—O	350
C=O	741
C≡O	1080
C—F	439
C—Si	360
C—P	265
C—S	260
C—Cl	330
C—Br	275
C—I	240
N—N	160
N=N	418
N≡N	946
N—O	200
N—F	270
N—P	210
N—Cl	200

(Continued)

Table 12.2 (Continued)

Bond	Bond energy
N—Br	245
O—O	140
O=O	498
O—F	160
O—Si	370
O—P	350
O—Cl	205
O—I	200
F—F	160
F—Si	540
F—P	489
F—S	285
F—Cl	255
F—Br	235
Si—Si	230
Si—P	215
Si—S	225
Si—Cl	359
Si—Br	290
Si—I	215
P—P	215
P—S	230
P—Cl	330
P—Br	270
P—I	215
S—S	215
S—Cl	250
S—Br	215
Cl—Cl	243
Cl—Br	220
Cl—I	210
Br—Br	190
Br—I	180
I—I	150

Source: Ref. [9] / Rice University / CC BY 4.0.

for the step of descum. From the above discussion, we can see that the descum is actually dry etching, but the etched layer is very thin. We can also conclude that all organic matter will be etched by oxygen. However, if an organic substance contains a certain component, after reacting with oxygen, it produces non-volatile product, the equipment will be contaminated. So, we should pay attention to this aspect.

In the dry etching process, we usually use fluorine-containing gases to etch silicon, silicon dioxide and silicon nitride (refer to the reaction Eq. (11.4)). Fluorine can also etch PR, so a small amount of PR residues has little effect on dry etching. But, if these residues are left between the metal and the semiconductor surface, it will have a significant impact on the semiconductor-metal contact. I measured the impact before. The ohmic contact resistance without the descum can sometimes reach 5–10 times higher than that after the descum.

So far, we have finished the basic process flow of photolithography. Of course, there are some special circumstances, and this basic flow is not immutable. Some processes and photoresists need post exposure bake (PEB). AZ 5214E image reversal process requires PEB, usually 110–130 °C for a few minutes. Another case is that many AZ negative resists need PEB. PEB can carry out the crosslinking which is initiated during exposure rendering the exposed structures insoluble in the developer, typically 110–120 °C for a few minutes [6]. After finishing photolithography, we can do other processes.

12.2 Alignment Mark (Mark) Design on the Photomask

In the lithography process, the design of photomask is also an important topic. In this section, we do not discuss the entire design, because the requirements of different devices and circuits design are different, and the design rules of different companies' production lines are also different. We only focus on the alignment mark on the mask used for manual alignment. Several points need to be considered in the design of the alignment mark, which are often encountered in university laboratories and clean rooms.

The fabrication of a microchip cannot be completed with one step of lithography. It takes multiple lithography steps to complete. Between layers made by different lithography step, their pattern structures need to be aligned with each other, as shown in Figure 12.21. The brown and blue patterns in the figure need to be aligned. The cross patterns on both sides are commonly used alignment marks. The alignment marks are used to align the patterns of different layers of the microchip. The alignment marks can be designed on both sides of the wafer or in the middle of the patterns, depending on the size of the sample (whole wafer or piece) during the lithography process.

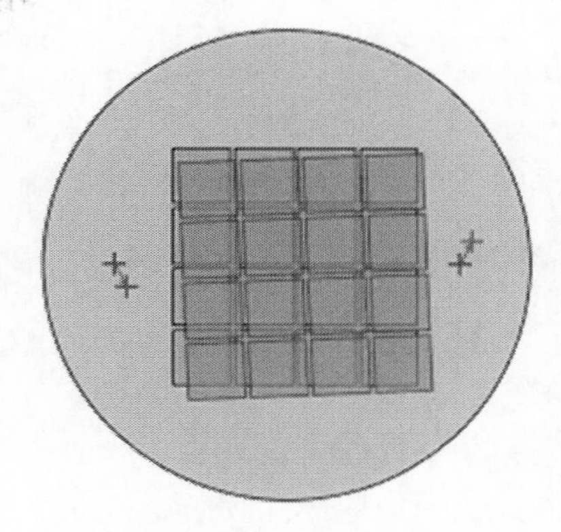

Figure 12.21 Alignment in the photolithography (between brown patterns and blue patterns). Source: Reprinted the image of Rick J. Bojko and used with the permission of N. Shane Patrick, Washington Nanofabrication Facility.

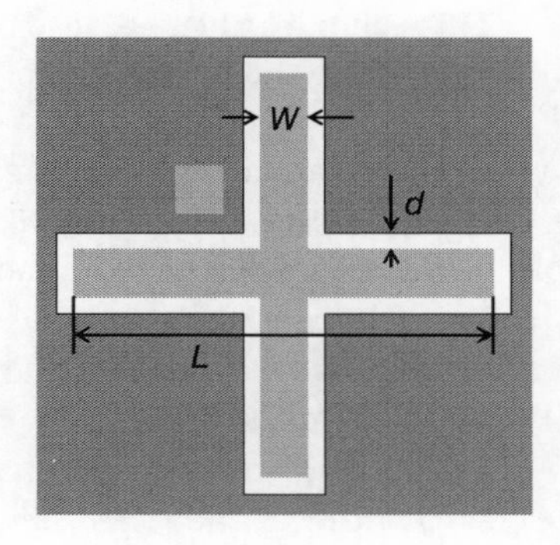

Figure 12.22 Cross alignment mark.

Usually, in the first step of lithography, the first layer of cross alignment mark is made on the wafer or piece. For the convenience of the next alignment step, the cross patterns are preferably made of metal (gold is the best), so that it is convenient to observe later. The substrate can also be etched to make the cross patterns, but the contrast is not as good as that of metal. In the subsequent photolithography of other layers, the alignment can be performed based on the first layer. Figure 12.22 is a set of typical cross alignment mark. The yellow one in the mark is the cross pattern of the first layer, and the green one is the pattern of other layers. The white area is transparent, and the other parts are opaque. L is the length of the mark, W is the width, and d is the alignment overlay space between two layers of the marks. The question discussed in this section is how to determine the values of L, W, and d, and the direction of the alignment mark. We need to consider these issues in the following three cases:

Case 1. The design of L and W. The alignment mark cannot be designed to be too small, which is inconvenient to watch. According to the size of the device area, the length is generally 100–200 μm, that is, $L = 100–200\ \mu m$. The width is generally 20–40 μm, that is, $W = 20–40\ \mu m$.

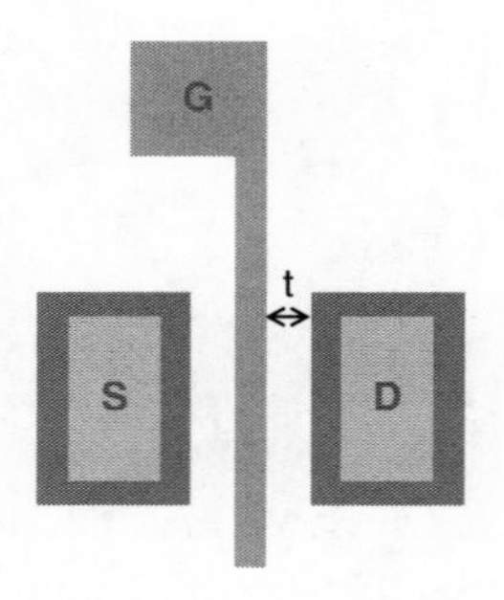

Figure 12.23 The typical structure of MESFET.

Case 2. Design of d. Please note, d cannot be designed to be zero, that is, $d = 0$. This is not feasible, because once d is zero, the alignment situation cannot be seen under the microscope of the aligner. The design of d should consider the following two points:

1. The accuracy of the device: We take a MOSFET as an example to illustrate, please see Figure 12.23. In the MOSFET process, the drain region D and the source region S are fabricated at the same time, and the gate region G is fabricated separately. D and S are ohmic contacts, and these regions need to be heavily doped. The yellows in the figure are the metal in the D and S electrodes, and the blues are the heavily doped areas. The area of the metal is slightly smaller than the doping area. Ideally, the distances between D and G, S and G are same, denoted by t in the figure. One of the design rules of d is less than t, that is, $d < t$. The purpose of this is that in the step of alignment, once the alignment accuracy is not good, even if the two layers of alignment marks touch together ($d = 0$ in a certain direction), t will not be zero, the device can still work, but the characteristics deteriorate.
2. The magnification of the aligner's microscope: Karlsuss aligner in Figure 12.24 is commonly used to contact aligner at universities. Due to the requirements of depth of field (DOF) and field of view (FOV), the magnification of microscope used for the aligner cannot be too large. The maximum magnification of this microscope is 300. Now the devices are getting smaller and smaller, so we hope that d is as small as possible. However, because of the limitation of microscope magnification, if d is too small, it is difficult to see clearly under the microscope, which is not conducive to alignment. For example, if the design is $d = 1\ \mu m$, it is 0.3 mm under the microscope with a magnification of 300 times, which is difficult for the eyes to see clearly. If the design is $d = 2\ \mu m$, it is 0.6 mm under the microscope of 300 times, this scale is not difficult to see clearly with the eyes. So, the second design rule of d is to refer to the magnification of the aligner's microscope.

Figure 12.24 Karlsuss MJB3 contact aligner.

In short, the design of the alignment accuracy *d* should be considered according to the requirements of device design and the situation of contact aligner, rather than designing blindly.

To avoid misalignment problem, modern CMOS technology uses self-alignment technique. The CMOS structures in Figures 11.2–11.4 are finished by the technique of self-alignment. An important reason for using polycide (Section 6.4) instead of metal as electrodes in CMOS devices is to use self-alignment process. This topic will not be discussed in this book. For more details, please read reference [10].

Case 3. Design of direction. If it is processing of whole wafer, we can determine the direction through the flats of the wafer (Figure 10.9). A mask usually has some words on it that can be used to determine the direction (Figure 12.11). But when processing a small piece, as shown in Figure 12.25, this is a common situation in the university clean rooms. Every time these small pieces are placed in an aligner, the direction cannot be determined. In addition, when aligning under the microscope, we only use the alignment marks. If there is no direction indication on the mark, the difference between the direction of a small piece and a mask by 90°, 180°, and 270° cannot be found. So, it is necessary to design a pattern indicating the direction on the mask. We can use different patterns. A simple method is to add a small square to a corner of the cross pattern in the mark, as shown in Figure 12.22. This small square pattern does not need to consider the alignment overlay space, and the size of each layer can be the same. It just gives a sign to avoid the deviation in the direction when a small piece is aligned. From the picture, we can clearly see the edge bead issue in the corner of right one. The left one is cut from a whole wafer after PR coating, so no edge bead on it.

Figure 12.25 Small pieces of semiconductor.

12.3 Contemporary Photolithography Equipment Technologies

So far, the lithography machine we have discussed is only used to make micron-level microchips. With I-line, the pattern with a feature size of 0.5 μm can be finished. However, it is difficult to make smaller size than 0.5 μm by using I-line. According to Rayleigh criterion, to distinguish smaller sizes, we need to make the following improvements in the optical system: large optical lens, high refractive index optical medium and short wavelength of exposure light source.

We can make large optical lens, but due to the manufacturing and other reasons, we cannot make the lens too large. Using a high-refractive index optical medium, the space between final lens and wafer is filled with water which is one of the most used optical medium. This technology used in a contemporary advanced lithography machine is called immersion lithography, please see Figure 12.26. Regarding short wavelength, 193 nm ArF excimer laser is used as exposure light source to replace I-line HPMVL.

Immersion lithography was made by Advanced Semiconductor Material Lithography (ASML) in 2003. When using ArF excimer laser, the NA can reach 1.35. 193 nm is deep ultraviolet (DUV). The refractive index of water is around 1.33 (Section 8.3.2) in the visible light range (400–700 nm) but is 1.44 in DUV of 193 nm. Such a high NA optical systems are over 1.2 m high and weigh more than a metric ton [11]. Figure 12.27 is a photo of ASML scanner with ArF excimer laser light and water immersion stage. The NA of this machine is 1.35. By using this

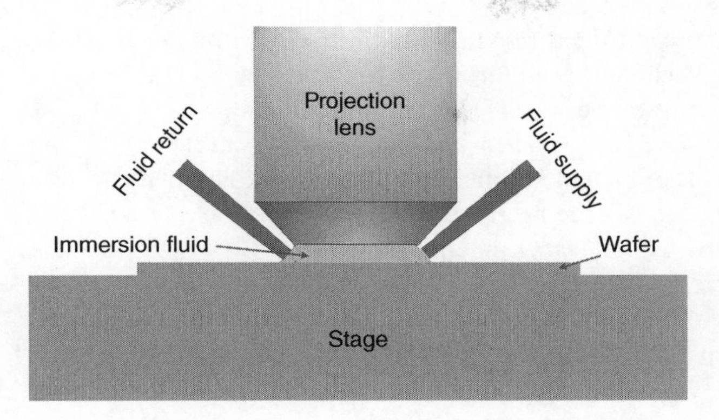

Figure 12.26 Schematic diagram of immersion lithography technology, including the last lens element contacting the immersion fluid (water).

Figure 12.27 ASML TwinScan XT:1950i scanner. Source: [12] Roger et al., 2009 / with permission of Annual Reviews.

system, single exposure can extend resolution down to 3× nm hp (Section 7.3), and double patterning techniques can further extend to 2× nm hp [13]. Here 3× and 2× mean thirties and twenties.

As the feature size becomes smaller and smaller, the wavelength of the light source becomes shorter and shorter. The light source used in the process of 7 nm and smaller is extreme ultraviolet (EUV) with a wavelength of 13.5 nm. Let us return to the Planck energy Eq. (2.1) ($E = h\nu$) to calculate the EUV requirements for optical lenses. Planck constant $h = 6.626 \times 10^{-34}$ J·s in the equation. The relationship between the energy unit electron volt (eV) used in microscopic particles and the commonly used energy unit Joule (J) is 1 eV = 1.602×10^{-19} J. The ν in the equation is the frequency of light, the speed of light $C = 3 \times 10^{8}$ m/s = 3×10^{17} nm/s. The relationship between the speed of light (C), frequency (ν), and wavelength (λ) is:

$$C = \lambda \nu \tag{12.1}$$

Using formulas (2.1) and (12.1), and substituting the above values, the following conclusions can be obtained:

1. The material of the quartz photomask we mentioned above is SiO_2. Its band gap is $E_{SiO_2} = 9$ eV. Using the above formulas and values, we can get the shortest wavelength that can pass through the quartz mask which is $\lambda = 137.87$ nm. From this number, we know that quartz can be used to make optical lenses and photomasks for DUV system.
2. For the light source of $\lambda_{EUV} = 13.5$ nm, quartz glass cannot be used. It is necessary to use a material with a band gap of at least $E_{EUV} = 90$ eV for optical lenses and photomasks. Unfortunately, such material is nonexistent.

Judging from the electromagnetic wave spectrum in Figure 7.20, the EUV has entered the X band and can be absorbed by almost all materials, including air. To avoid absorption, the light system of a EUV lithography machine adopts a reflective mirror structure, and the photomask also adopts this structure. The optical system is placed in a vacuum chamber, and the wafer and the mask are also placed in a vacuum chamber. In addition, it is also very demanding for the machining accuracy of the lithography equipment. Processing a 7 nm chip with a 13.5 nm light source requires carefully designed software for correction. In short, such a lithography system is very difficult to make. Currently, only Dutch ASML company (ASML Holding N.V) can manufacture it worldwide. The price of an ASML EUV projection lithography system is more than US$100 million [14], while the price of an etching system for production is between US$4 million and US$7 million [15]. Figure 12.28 is an ASML EUV scanner lithography system.

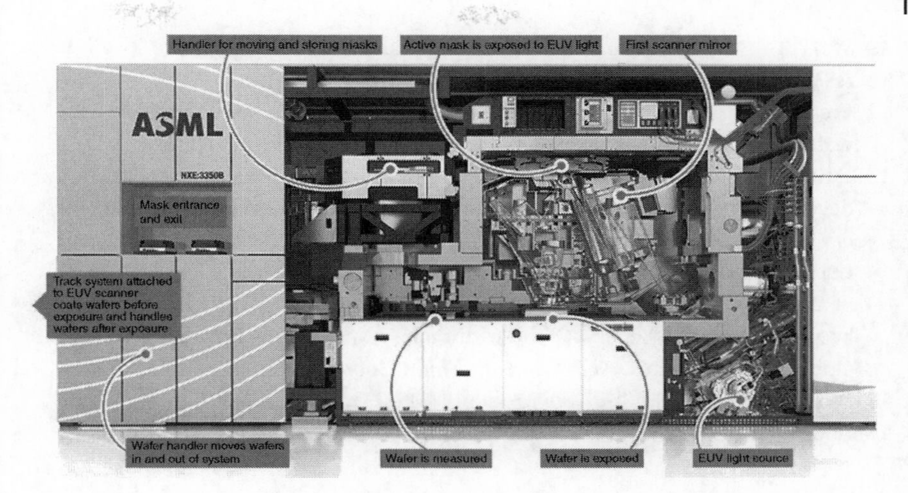

Figure 12.28 ASML EUV scanner lithography system. Source: [16] Courtland et al., 2016 / with permission of IEEE.

References

1 Ahmad, D., van den Boogaert, I., Miller, J. et al. (2018). Hydrophilic and hydrophobic materials and their applications. *Energy Sources Part A: Recovery, Utilization, and Environmental Effects* 40 (22): 2687.

2 Wolf, S. and Tauber, R.N. (2000). Chapter 12: Lithography I: optical photoresist materials and process technology; Chapter 13: Lithography III: optical aligners and photomasks. In: *Silicon Processing for the VLSI Era*, Process Technology, 2e, vol. 1. Lattice Press, p. 510, p. 515, p. 589.

3 Product data sheet, AZ 1500 series, Standard photoresists, Clariant.

4 Haberfehlner, G. 3D nanoimaging of semiconductor devices and materials by electron tomography, Graz University of Technology, ResearchGate, p. 12.

5 Lueke, J., Badr, A., Lou, E., and Moussa, W.A. (2015). Microfabrication and integration of a sol-gel PZT folded spring energy harvester. *Sensors* 15: 12218–12241.

6 MicroChemicals. Resists, Developers and Removers. Revised: 2013-11-07.

7 Boudreaux, K.A. Organic compounds: alkanes, Chapter 1. In: *CHEM 2353 Fundamentals of Organic Chemistry Organic and Biochemistry for Today*. Angelo State University.

8 Reiser, A., Huang, J.P., He, X. et al. (2002). The molecular mechanism of novolak-diazonaphthoquinone resists. *European Polymer Journal* 38: 619–629.

9 Rice University Chemical bonding and molecular geometry. In: *Chemistry*, Chapter 7. 7.5 Strengths of ionic and covalent bonds.

10 Wolf, S. (2004). *Microchip Manufacturing*. Lattice Press.

11 ASML, Lithography principles, Lenses & Mirrors. (2019).

12 French, R.H. and Tran, H.V. (2009). Immersion lithography: photomask and wafer-level materials. *Annual Review of Materials Research* 39: 93–126.

13 Furukawa, T., Terayama, K., Shioya, T., and Shima, M. (2013). Material development for Arf immersion extension towards sub-20 nm node. *Journal of Photopolymer Science and Technology* 26 (2): 225–230.

14 Clark, D. (2021). The Tech Cold War's 'Most Complicated Machine' That's Out of China's Reach. A $150 million chip-making tool from a Dutch company has become a lever in the U.S.-Chinese struggle. It also shows how entrenched the global supply chain is. *The New York Times* (July 4).

15 王聪/张天闻. (2018). "国内刻蚀机供应商崛起在望", 摩尔芯闻. .

16 Courtland, R. (2016). The molten tin solution. *IEEE Spectrum* 31.

13

Dielectric Films Growth

As insulating layers and optical films, dielectrics are widely used in the manufacturing of semiconductor microchips. Among these dielectrics, SiO_2 and Si_3N_4 are the two most used films. The electrical and mechanical stabilities of the Si–SiO_2 interface make this structure the cornerstone of MOSFETs. Si_3N_4 has five main applications:

1. The last protective film (layer) of the integrated circuit, which is called passivation film (layer).
2. A barrier layer for partial isolation of oxidation process.
3. The optical refractive index of SiO_2 is 1.46, and that of Si_3N_4 is 2.05. Due to the difference in refractive index, they and their combination are important components of optoelectronic devices.
4. The stress of Si_3N_4 film can be easily adjusted. So it has found a place in the manufacturing of many types of devices, such as cantilever, micro, and nano tube (Figure 7.19), and other structures.
5. SiO_2 and Si_3N_4 can be used as mask layers in III–V dry etching.

The growth methods of the two films are carried out by chemical reactions. Different machines use different temperatures, which we will discuss below. At these temperatures, the resulting dielectric film has an amorphous structure. This structure is composed of many tiny crystal grains. One of their main defects is pinholes. Pinholes are small holes in the film. When metal is grown, metal particles can pass through these pinholes, which weakens the insulation of the film. When the temperature is increased, the tiny crystal grains that make up the film can diffuse in the film at a faster speed, filling pinholes, reducing the pinhole density, and improving the insulation of the film. Conversely, when the temperature is decreased, the pinhole density increases, reducing the insulation of the film. So, for electrical insulation, high-temperature process is a good choice. From this point of view, the higher temperature, the better quality of the film.

Semiconductor Microchips and Fabrication: A Practical Guide to Theory and Manufacturing, First Edition. Yaguang Lian.

However, due to the following two reasons, not all materials and processes can be finished with high temperature methods for these two dielectric films:

1. Material restrictions. For silicon, the film growth temperature can be as high as 1200 °C (Chapter 10). However, III–V materials cannot withstand such high temperature.
2. The limitation of the process. After the metal is deposited on the semiconductor surface, in order to achieve a good contact, the metal needs to be heated up for annealing (sintering). In most cases, the annealing temperature is 400–500 °C.

From the above discussion, we have a conclusion: In order to meet the process requirements, we should have film growth technology which temperature is less than 400 °C. The different equipment developed based on this point uses different temperatures for the growth of SiO_2 and Si_3N_4. In this chapter, we mainly introduce the growth processes of these two films.

13.1 The Growth of Silicon Dioxide Film

The Si–SiO_2 system is a milestone invention. Based on this invention, the silicon planar technology was created (Figure 7.15). For SiO_2 growth, there are three main methods: thermal oxidation, LTO-low temperature oxide, and PECVD. The temperature of thermal oxidation is 900–1200 °C, the temperature of LTO is 400–750 °C, and the temperature of PECVD is 250–350 °C. It can be seen from this that the SiO_2 growth process uses three different temperature ranges. Thermal oxidation gives the best quality film. PECVD can complete film growth below 400 °C. LTO gives a range between these two. In these three processes, thermal oxidation is the direct reaction of oxygen and silicon in the surface of the substrate to produce SiO_2, LTO generates SiO_2 by chemical reaction through thermal decomposition, and PECVD uses two different gases to generate SiO_2 through chemical reaction with the aid of plasma. For special requirements, such as growing SiO_2 on photoresist, we can continue to lower PECVD temperature, but the quality of the film is decreased. In addition, this section also introduces another deposition technique with even lower temperature.

13.1.1 Thermal Oxidation Process of SiO_2

Thermal oxidation is to put Si wafers directly into an environment of oxygen (dry oxidation) or water vapor (wet oxidation). The temperature is increased to

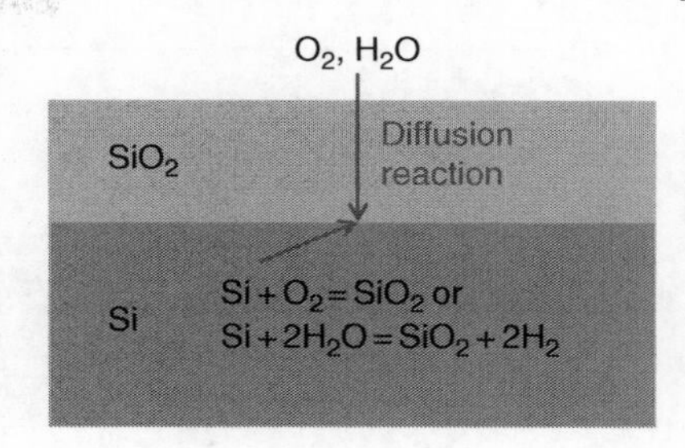

Figure 13.1 Schematic diagram of thermal oxidation process [1]. Source: Chinese Technical Books.

900–1200 °C, so that oxygen and silicon react directly to form silicon dioxide. The chemical reaction formulas in this temperature range are

$$\text{Dry : Si(solid)} + \text{O}_2\text{(gas)} = \text{SiO}_2\text{(solid).}$$
$$\text{Wet : Si(solid)} + 2\text{H}_2\text{O(gas)} = \text{SiO}_2\text{(solid)} + 2\text{H}_2\text{(gas)} \tag{13.1}$$

SiO_2 is a solid, and it stays on the surface of the Si substrate to form SiO_2 film. The above chemical reactions occur at the interface of SiO_2–Si. After the film is formed, oxygen and water will pass through the SiO_2 film with thermal diffusion and reach the SiO_2–Si interface. After reacting with silicon, a new SiO_2 layer is formed, as shown in Figure 13.1. Thermal oxidation is a unique process of silicon, and it has the following characteristics:

1. It can reach the highest temperature in the processes.
2. The quality of the grown SiO_2 film is the best.
3. The growth process is a direct chemical reaction at the interface of SiO_2–Si to form a SiO_2 film, which is unique among all film growth processes. During oxidation, oxygen first reacts with the contaminations and defects on the surface of silicon, so that these contaminations and defects are removed from the silicon surface. This feature allows the thermal oxidation process to obtain a fresh and clean silicon surface, which is not achieved by other film growth processes.
4. The growth rate of the wet method is higher than that of the dry method, and the density of the film grown by the dry method is better than that of the wet method, which means that the quality of dry oxidation is better than that of wet oxidation.
5. Since oxygen and water vapor must diffuse through the grown film to reach the SiO_2–Si interface, as the thickness of the film becomes larger and larger, the time required for diffusion becomes longer and longer. Therefore, the growth rate decreases with time until it becomes saturated. Please see Figure 13.2. Saturation means that when the thickness exceeds a certain value, SiO_2 will not grow again. This is not the case with other methods. Once the conditions

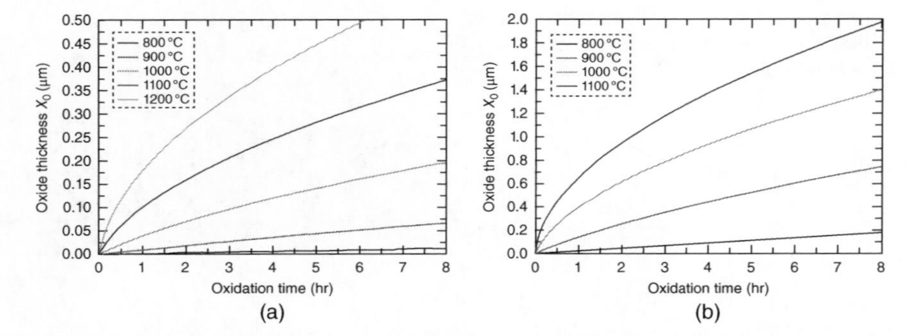

Figure 13.2 The relationship between the thickness of the thermal oxidation SiO_2 film and time. The lattice plane is (100) (refer to Figure 10.10). (a) dry oxidation and (b) wet oxidation. Source: Reprinted with permission of IuE, TU Wien.

are determined, the growth rate is fixed and will not change with time and thickness.

6. The proportions of silicon consumed are 44% by the dry and 41% by the wet [1]. Assuming to grow 1000 Å of SiO_2, 440 Å of Si is used in the dry, and 410 Å of Si is used in the wet. Other methods do not consume silicon.

Thermal oxidation is done in an oxidation furnace. The main structure of the furnace is a quartz tube with heating resistance coils around it. Put the silicon wafers or pieces in the quartz boat and push it into the quartz tube. For dry oxidation, UHP (Section 7.4) oxygen is input the furnace and passed over the wafers or pieces. For wet oxidation, oxygen is input a bubbler containing DI water at 95 °C. The oxygen carries water vapor and goes to the furnace over the wafers or pieces inside. The heating resistance coils are powered to heat the tube. SiO_2 film is generated in the surface of silicon substrate by chemical reaction of silicon and oxygen or silicon and water. Figure 13.3 is a schematic diagram and photos of the oxidation furnace and its accessories. The quartz is mainly composed of silicon dioxide, which has a melting point of over 1600 °C and is one of the commonly used materials in high-temperature processes. As many kinds of metals can react with chlorine to produce volatile chlorides (discuss later), similarly sometimes, HCl or Cl_2 can be mixed in the gas flow to remove metallic impurities and improve the quality of SiO_2 film.

13.1.2 LTO Process

The second method of growing SiO_2 is LTO. The growth temperature of LTO is lower than thermal oxidation temperature and is chemical vapor deposition (CVD). This growth method does not consume silicon substrate, nor does it

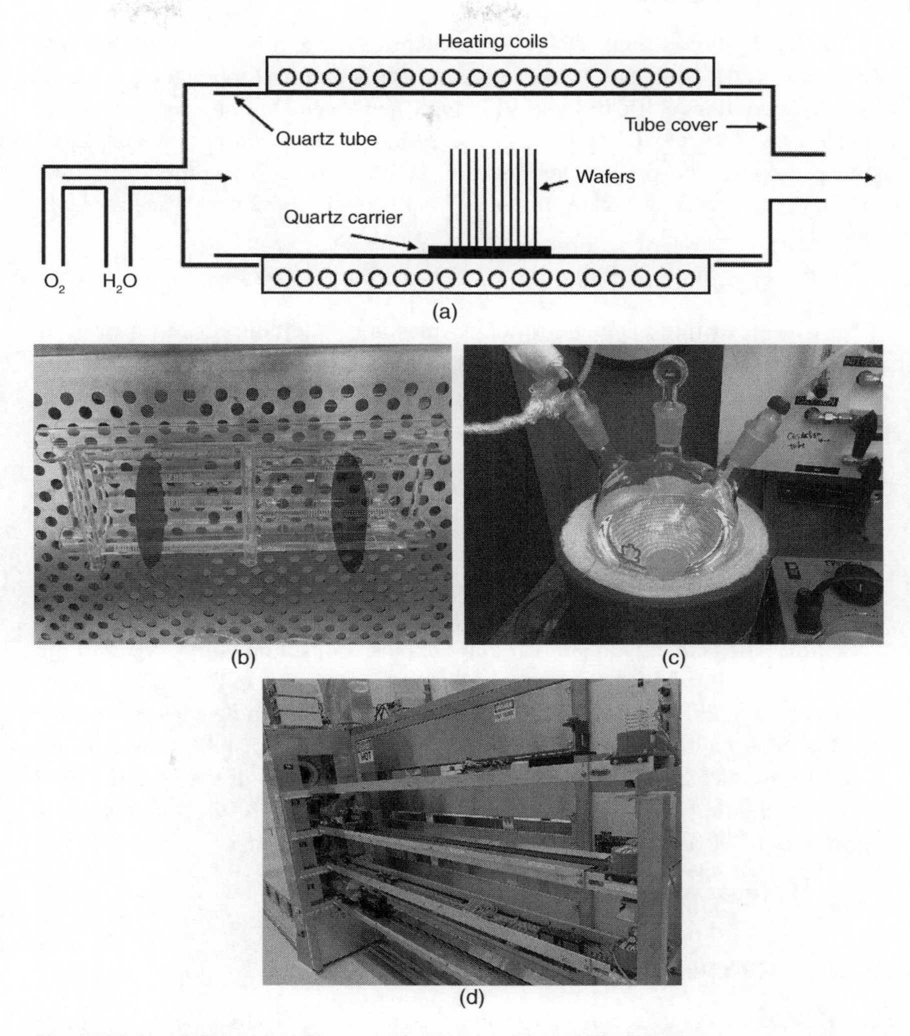

Figure 13.3 (a) Schematic diagram of oxidation furnace. (b) Si wafer and quartz boat. (c) Water bubbler. O_2 flows into the bubbler and takes away water vapor. (d) A furnace. Source: (a) [2] / Bhargav Veepuria (b–d) Lindberg / Tempress 8500.

consume other types of substrates. So the substrate can be silicon, or other semiconductors and nonsemiconductor materials. The equipment used is like the thermal oxidation furnace. It also uses a quartz tube and has a heating resistance wire, but the chemical gases and the temperature are different. There are two main types for LTO oxide layer depositions:

Type 1. TEOS deposition. TEOS is tetraethoxysilane, the chemical formula is $Si(OC_2H_5)_4$, and the other name is Tetraethyl OrthoSilicate, TEOS is the abbreviation of this chemical. TEOS generates thermal decomposition reaction at 650–800 °C, and the product SiO_2 is deposited on the surface of the substrate. The process temperature is 720–750 °C. The pressure is over 100 mTorr. For comparison, 1 atm = 760 Torr. The reaction process is as follows [2]:

$$\text{TEOS} \rightarrow \text{SiO}_2 + \text{Gaseous organic molecules} + (\text{SiO} + \text{C}) \qquad (13.2)$$

The oxygen in the reaction formula comes from TEOS itself and cannot be introduced from the outside. The oxygen introduced from the outside causes the quality of film to deteriorate. The content of SiO and C in the reaction formula should be minimized. This can be achieved by adjusting the process parameters.

Type 2. Silane deposition. When silane (SiH_4) and oxygen meet, they will have chemical reaction to generate SiO_2 that deposits on the substrate surface. SiH_4 is a gas at room temperature. Its structure is very unstable, because silicon and hydrogen are both metal-like elements, the chemical bond between them is very weak. It can decompose and release active hydrogen and silicon at room temperature. The active hydrogen can have explosive reaction with oxygen in the air to form water, and the silicon also reacts with oxygen to form SiO_2. Active hydrogen and silicon mean that they exist as atoms not molecules. For safety, SiH_4 should be diluted with high-purity nitrogen to 5% or less, or with other inert gases, such as helium. In order to ensure the quality of the film, the temperature should be heated above 400 °C, generally 400–450 ° C. The pressure is from 150 to 300 mTorr [3]. The chemical reaction equation is

$$\text{SiH}_4(\text{gas}) + 2\text{O}_2(\text{gas}) \rightarrow \text{SiO}_2(\text{solid}) + 2\text{H}_2\text{O}(\text{gas}) \qquad (13.3)$$

13.1.3 PECVD Process of Silicon Dioxide

The PECVD process uses plasma technique to deposit SiO_2. In order to complete the deposition, a PECVD system must have the following main parts:

1. Process chamber. This is the core part of the whole system because the samples are deposited in the chamber.
2. In order to ensure the quality of the film, the samples must be heated. As mentioned above, the temperature should be lower than 400 °C, generally set to 250–350 °C. The temperature is usually tested and controlled by thermocouple.

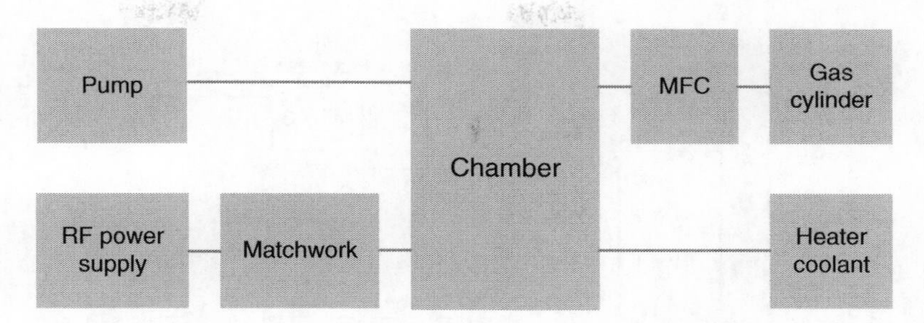

Figure 13.4 Schematic diagram of PECVD system. MFC, mass flow controller.

3. Maintaining a certain process pressure. If the pressure is too low, the growing film is not dense enough. If the pressure is too high, large particles are likely to be produced, and the quality of the film will deteriorate. The pressure is generally set up between 500 mTorr and 5 Torr. Therefore, a vacuum pump system is needed to control the pressure in the process chamber.
4. It is necessary to input gas into the process chamber, so a gas supply system is required.
5. A power supply (source) is required to generate an electric or a magnetic field. To make it easier for gas molecules to collide with each other to achieve the purpose of activating chemical reactions, the power supply generally uses a high-frequency power source. The most used frequency is 13.56 MHz. This frequency is in the RF range, please see Figure 7.20. So this power supply is also called RF power supply.

In fact, the abovementioned main parts are standard configurations of many kinds of semiconductor process equipment. The schematic diagram of the structure is shown in Figure 13.4. We will explain each part of the system in the dry etching in Chapter 14.

In PECVD SiO_2 system, for further safety, it is not only SiH_4 that is diluted with nitrogen or other inert gases but also reacts with N_2O instead of O_2. N_2O is laughing gas.

13.1.4 TEOS + O_3 Deposition Using APCVD System [4]

An Atmospheric Pressure Chemical Vapor Deposition (APCVD) system is implemented for SiO_2 film deposition. TEOS and ozone (O_3) are mixed into the APCVD system. The deposition temperature range is 125 to 400 °C. Figure 13.5 is the schematic diagram of the APCVD system.

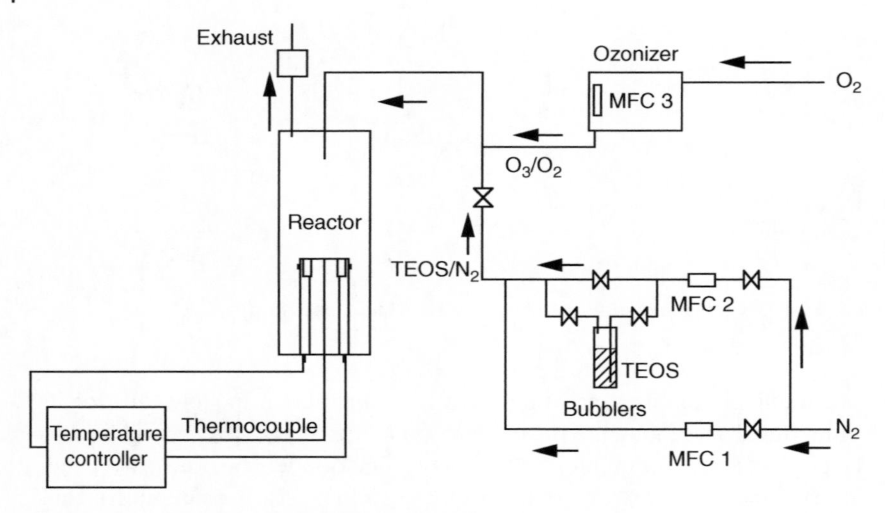

Figure 13.5 Schematic diagram of the APCVD system.

13.2 The Growth of Silicon Nitride Film

There are two main growth methods for silicon nitride, Low Pressure Chemical Vapor Deposition (LPCVD) and PECVD. The pressure of LPCVD is generally set to 0.15–2 Torr. Compared with this process, the thermal oxidation process is processed at 1 atm. There are also high-pressure processes, such as the High-Pressure Oxidation – HIPOX process, which has a pressure of 10–25 atm. It is characterized by increasing the growth rate and is widely used in LOCOS and STI techniques [5], see Section 11.1.

13.2.1 LPCVD

The basic structure of LPCVD equipment is like that of thermal oxidation furnace. Its main components are quartz tube and heating resistance wire. Thermal oxidation furnace does not have vacuum pumps. LPCVD is connected to vacuum pumps. Figure 13.6 is a schematic diagram and photos of an LPCVD system. The diagram shows three heating zones. The wafers are put into the center zone to do the process. Thermal oxidation furnace has the same heating zones. The boats in the picture are placed in a cart to protect them. The material of the boat can be quartz or graphite. The equipment used for the LTO processes (TEOS and silane depositions) is the same as the LPCVD equipment. In fact, one LPCVD machine can have several different quartz tubes for the growth and deposition of different films. The polysilicon for gate we mentioned earlier is completed with LPCVD equipment. In the process of depositing silicon dioxide by the silane method,

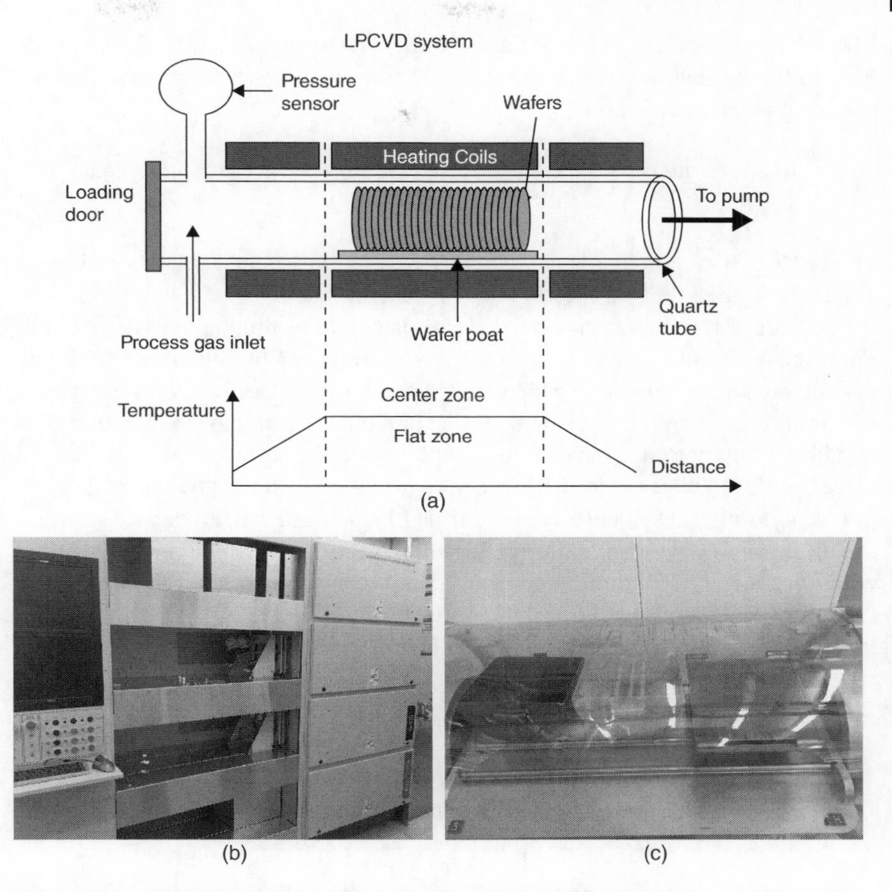

Figure 13.6 LPCVD system. (a) schematic diagram Hong Xiao [6], (b) is a machine (firstnano) and (c) are wafer boats.

without oxygen, a polysilicon film will be deposited on the surface of the device in the temperature range of 600–700 °C. In PECVD, there is no oxygen source, and silicon film can also be deposited. Due to the lower temperature, the deposited film by PECVD is amorphous silicon. The difference between polycrystalline and amorphous is that the grain size of polycrystalline is much larger than that of amorphous.

For the deposition of silicon nitride film, the commonly used gas is Dichlorosilane – DCS. Its molecular formula is H_2SiCl_2, which provides a silicon atom. Another gas is ammonia, the molecular formula is NH_3. The reason why pure nitrogen is not used is that nitrogen has the largest bonding energy (see Table 12.2) and is not easy to break, so ammonia is used instead of nitrogen. In the

PECVD system, ammonia is also used instead of nitrogen to deposit silicon nitride film. In the temperature range of 750–850 °C, the two gases produce the following chemical reactions:

$$3H_2SiCl_2\,(gas) + 4NH_3\,(gas) \rightarrow Si_3N_4\,(solid) + 6HCl\,(gas) + 6H_2\,(gas) \tag{13.4}$$

The LPCVD Si_3N_4 film has a very dense structure. One of the applications is used as a mask in the LOCOS process.

LOCOS is a traditional integrated circuit isolation technique, and is shown in Figure 13.7. The first step in the process is to grow thermally a thin layer of silicon dioxide on silicon substrate, which is called pad-oxide. Use LPCVD to deposit a silicon nitride film on this oxide pad. Use photolithography to make patterns on this film. Use dry etching to etch the opening until the surface of silicon. Finally, place the silicon wafer in the high-pressure oxidation furnace, and thermal oxidation is performed. The dense silicon nitride film acts as a barrier mask to prevent the diffusion of oxygen or water vapor; so oxidation only occurs in the opening without silicon nitride film. "Bird's beak" in the figure is an inherent problem in LOCOS technique. Its occurrence is caused by lateral oxidation in the process, so it can only be used for the devices with a feature size greater than 0.25 μm. Please see Section 11.1.

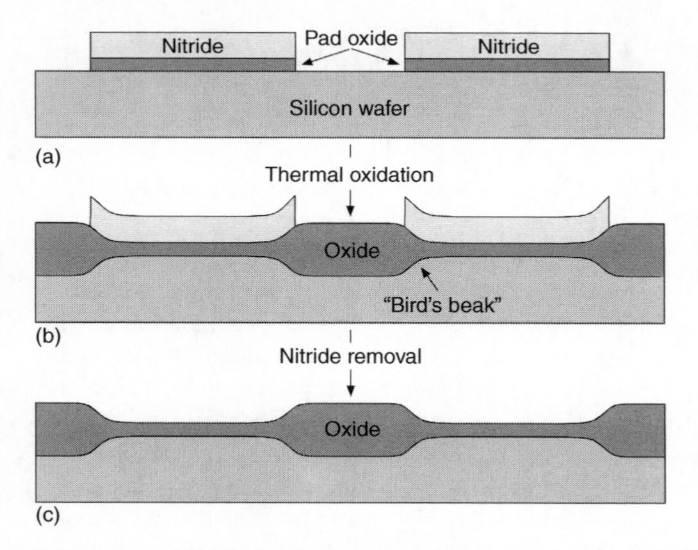

Figure 13.7 Schematic diagram of LOCOS process flow. Source: Reprinted with permission of IuE, TU Wien. (a) Opening nitride and oxide until Si surface. (b) Thermal oxidation through the opening. (c) Removing nitride film.

The growth of Si_3N_4 film by using thermal nitridation is extremely slow and self-limiting. This is because silicon nitride is so dense that when nitrogen and silicon react to produce a Si_3N_4 film, this film will prevent the diffusion of nitrogen and stop the growth of Si_3N_4. Through the chemical reaction with ammonia, thickness of the film grown by thermal nitridation is only 3–4 nm [5]. Due to this limitation, this process is only used in the manufacture of some special devices. The reaction of Si and NH_3 shows as following:

$$3Si\,(solid) + 4NH_3 \rightarrow Si_3N_4\,(solid) + 6H_2\,(gas) \quad (13.5)$$

13.2.2 PECVD Process of Silicon Nitride

PECVD Si_3N_4 deposition equipment and PECVD SiO_2 equipment are the same. But look closely, they are still somewhat different:

The first difference is that different gases are passed into the process chamber to deposit different films. SiO_2 is dilute silane and laughing gas, while Si_3N_4 is dilute silane and ammonia.

The second difference is that SiO_2 equipment usually needs one power source, namely 13.56 MHz, which is called a high-frequency (HF) power supply. Si_3N_4 equipment has two sets of power sources, one is a power source with a frequency of 13.56 MHz, and the other is a power source with a frequency of 300–400 kHz [7], which is called a low-frequency (LF) power supply. On a silicon substrate, when the deposited Si_3N_4 film is completed by a high-frequency power supply, the film shows tensile stress, causing the silicon wafer to be convex. When the deposited Si_3N_4 film is completed by a low-frequency power supply, the film exhibits compressive stress, which causes the silicon wafer to be concave, see Figure 13.8. Figure 13.9 is a photo of a PECVD machine used to deposit Si_3N_4. It can be used to deposit Si_3N_4 or SiO_2 film, depending on the gases input.

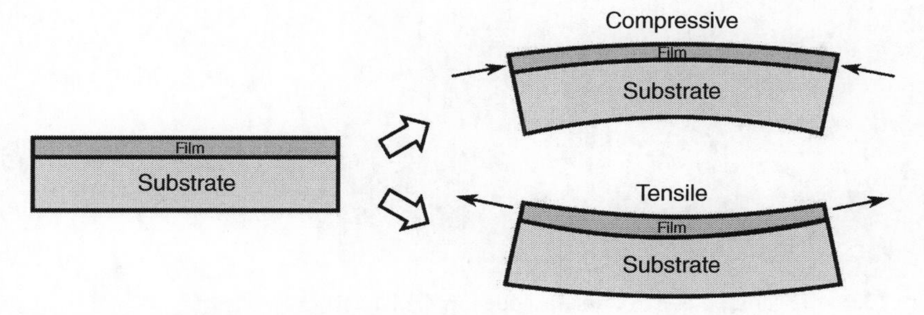

Figure 13.8 Changes of compressive stress and tensile stress substrate. Source: Reprinted with permission of IuE, TU Wien.

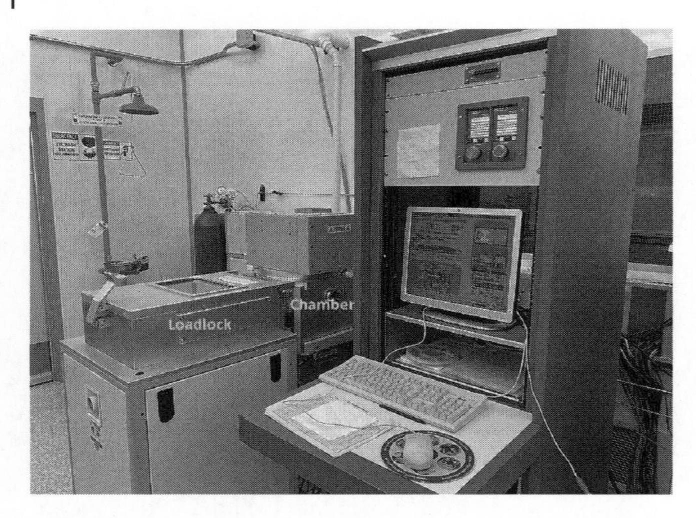

Figure 13.9 The photo of a PECVD machine for depositing Si_3N_4 film.

As mentioned above, with HF power, we can make Si_3N_4 film with tensile stress; with LF power, we can make Si_3N_4 film with compressive stress. If the film is done by using HF and LF powers at the same time, we can make Si_3N_4 film with low stress. The stress of Si_3N_4 has been found in many applications in the devices' manufacturing, especially in micro-electromechanical systems (MEMS). Figure 13.10 is a cantilever made by Si_3N_4 film with tensile stress. The device was made by Professor Chang Liu's research group at the

Figure 13.10 Scanning electron microscope (SEM) picture of cantilevers.

University of Illinois. The μm-sized transformer shown in Figure 7.19 is also made using the stress technique of by Si_3N_4 film.

As can be seen from Figure 13.8, if a compressive stress film is deposited on Si surface, the substrate and the film tend to stick together. If a tensile stress film is deposited on Si surface, the substrate and the film tend to separate. Therefore, if a Si_3N_4 film is used as the final passivation film, a compressive stress film should be used, and the reliability of the device is higher than that of a tensile stress film used.

In the above processes, we use combustible gases. For example, DCS and silane are both combustible gases (they easily separate out hydrogen). DCS is also a toxic gas (chlorine is separated out). We classify combustible and toxic gases as hazardous gases. When using hazardous gases, the exhaust gases removed by pump from the equipment must first be processed by CDO and exhaust gas treatment equipment before they can be discharged into the atmosphere. CDO is the abbreviation of "Controlled Decomposition Oxidation," which means controlled oxidative decomposition. The word for exhaust gas treatment equipment is "scrubber," which is used to remove harmful materials from the exhaust gases. Fluorine-containing gases CF_4, CHF_3, and SF_6 are not combustible and toxic. So they may go to scrubber directly without passing CDO. Figure 13.11 are photos of CDO and scrubber.

(a) (b)

Figure 13.11 CDO (a) and scrubber (b).

13.3 Atomic Layer Deposition Technique [8]

The film growth techniques we discussed above are widely used in the manufacturing of microchips. As the feature size of the devices gets smaller and smaller, these techniques have two common issues that are difficult to figure out: thin films deposition and achieving a conformal layer on a high aspect ratio structure. In semiconductor manufacture, an aspect ratio is primarily the ratio of the height and width of an etching structure, please see Figure 13.12. Atomic Layer Deposition (ALD) is a promising system to solve two issues. Let us use CVD system as an example to describe ALD system in this section.

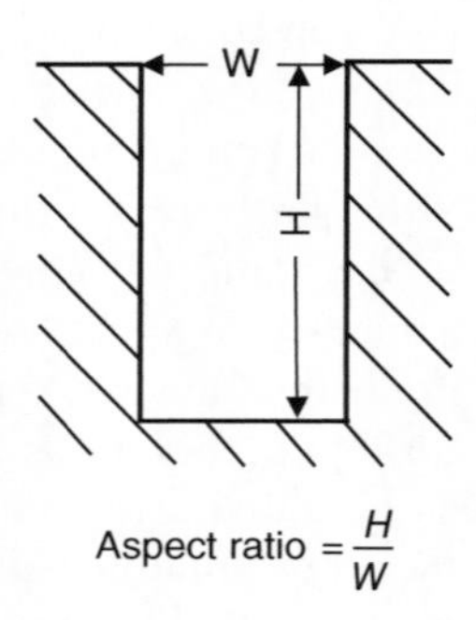

Figure 13.12 Schematic diagram of aspect ratio.

Typically, CVD processes are used in flow-type reactors with simultaneous and continuous reactants injection and continuous removal of reaction by-products. These processes are mainly characterized by the following parameters: deposition temperature (T, °C), deposition pressure (P, mTorr-Torr), deposition time (t, min, or s), and gas flow

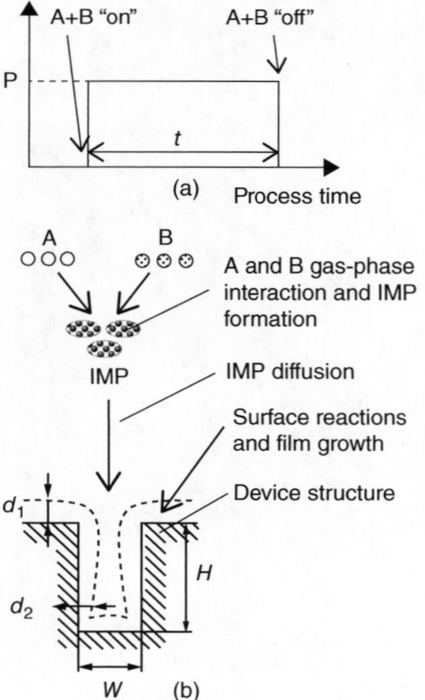

Figure 13.13 A simple scheme of the CVD process with the formation of intermediate products (IMP) from the reaction of A and B in the gas phase, diffusion and forming the film on the structure surface.

rate (sccm, talk later). However, ALD technique shows completely different process implementation and features. Unlike CVD processes, ALD processes are organized cyclically with sequential injection of the precursors and other reactants separated by an inert gas (mostly N_2). There are two types of ALD system: thermal activated ALD (TA-ALD) and plasma-enhanced ALD (PE-ALD). Figure 13.13 is the diagram of CVD processes with reactants A and B. Figure 13.14 is the diagram of a cyclic process with precursor A and reactant B and represents TA-ALD process. In the case of PE-ALD, the plasma discharge is usually switch-on some short time after the start of the reactant pulse and switch-off at the end of the pulse.

The step coverage (film conformality) can be expressed as the ratio of the film thickness at the bottom of the step (or gap) to the film thickness at the top of this step (or gap), d_2/d_1 (%). Depending on CVD process and its conditions, the ratio from a few percent to almost 100%. As the conformality is not good, and as the film thickness increases, the top opening of the structure will close-off, leaving a void inside the film. Unlike CVD, ALD completely excludes reactant interaction in the gas phase, and the formation and diffusion of IMPs (intermediate products). As compared to CVD, all processes of ALD are localized strongly on the surface of objects. A key advantage is layer-by-layer deposition (therefore, it is called ALD). Due to the self-limited chemical reaction, which is beneficial to the thin films' deposition. In addition, ALD is a strongly surface-limited process and has

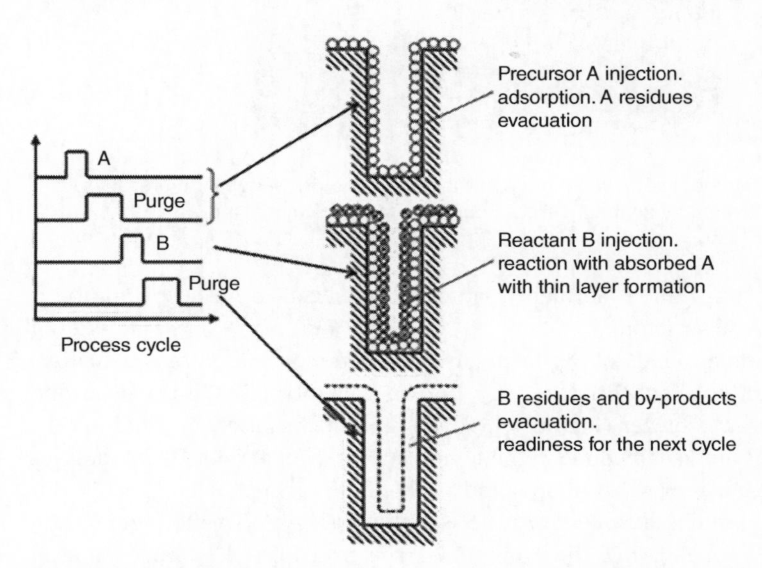

Figure 13.14 Schematic diagram of a cyclic process with precursor A and reactant B of ALD forming the film on the device structure surface.

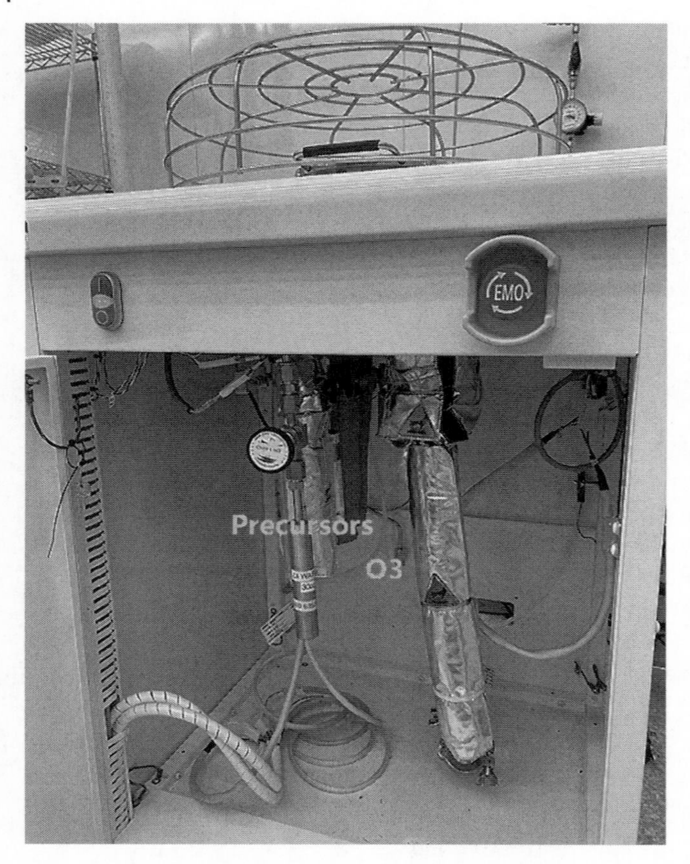

Figure 13.15 A photo of TA-ALD. The process chamber in on the top. The precursor cylinders are covered by heating jackets. Source: Veeco Instruments Inc.

an excellent thin film conformality, which is beneficial to do deposition on a high-aspect ratio structure.

There are many kinds of Si precursors. One commonly used one is BDEAS {bis(diethylamino)silane $SiH_2[N(C_2H_5)_2]_2$}. The reactant for SiO_2 is usually ozone produced by an ozonizer (Figure 13.5). The reactant for Si_3N_4 is usually NH_3. The processes are usually done in the temperature range 200–400 °C. Monolayer films and high conformality of SiO_2 and Si_3N_4 can be created by using ALD. But the deposition rate is slow. Figure 13.15 is a photo of TA-ALD, and Figure 13.16 is a schematic of PE-ALD. In Figure 13.16, spectroscopic ellipsometer is used for in situ measurement of the film thickness. VAT valves are widely used in the manufacturing equipment of semiconductor.

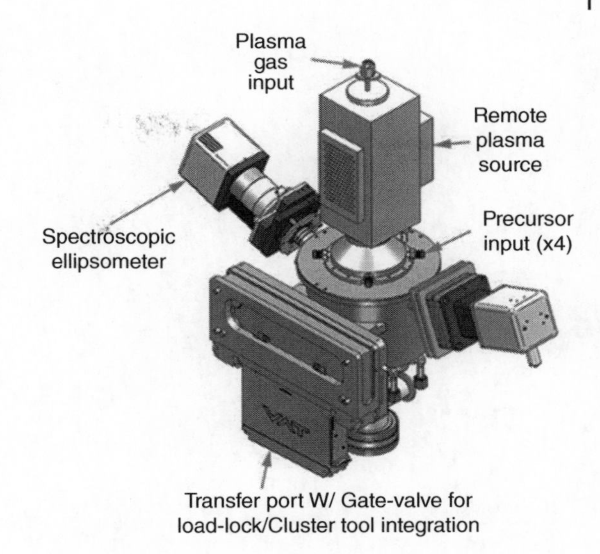

Figure 13.16 Schematic diagram of ALD reactor with plasma enhancement. Source: [9] Gaines.

References

1 电子工业生产技术手册 7, 半导体与集成电路卷，硅器件与集成电路, 184, 国防工业出版社，1991年页。

2 高等学校教学参考书, 半导体器件工艺原理, 厦门大学物理系半导体物理教研室编, 人民教育出版社, 44–45 页, 1977年。

3 Alamariu, B. tube6c-LTO standard operation procedure, MIT.

4 Juárez, H., Pacio, M., Díaz, T. et al. (2009). Low temperature deposition: properties of SiO_2 films from TEOS and ozone by APCVD system. *Journal of Physics: Conference Series* 167: 012020.

5 Wolf, S. and Tauber, R.N. (2000). *Silicon Processing for the VLSI Era: Process Technology*, 2e, vol. 1, p. 283, p. 299. Lattice Press.

6 Xiao, H. *CVD and Dielectric Thin Film*, Chapter 10, 22.

7 Van de Ven, E.P., Connick, I.W., and Harrus, A.S. (1990). Advantages of dual frequency PECVD for deposition of ILD and passivation films. *VMIC Conference*, (12–13 June 1990). IEEE. pp. 194–201.

8 Vasilyev, V.Y. (2021). Atomic layer deposition of silicon dioxide thin films. *ECS Journal of Solid State Science and Technology* 10: 053004.

9 Gaines, J.R. (2017). KJLC® awarded a patent for its atomic layer deposition system and process.Kurt J. Lesker, November 28. US patent 9,695,510.

14

Introduction of Etching and RIE System

Etching is an important process in the manufacture of semiconductor microchips. After photolithography is completed, the device patterns produced by the lithography can be permanently transferred and prepared on the substrate under the PR by etching. In the transferring and preparation, most of the microchip's manufacturing processes require accurate transfer with constant pattern size. The process can be divided into two types: wet etching and dry etching. Wet etching uses liquid to etch the substrate, and dry etching uses gas to etch the substrate.

14.1 Wet Etching

Wet etching involves many things. For example, the etching rate of the etchant used for silicon is different on different lattice planes, but this issue is not involved here. This section discusses the wet etching of SiO_2, also including Si_3N_4. They are the most used dielectrics in semiconductor processing.

The commonly used silicon dioxide (oxide) etchant is hydrofluoric acid (HF) and buffered oxide etch (BOE) in which water and ammonium fluoride (NH_4F) are added to hydrofluoric acid. This kind of etchant is known as HF-based etchant. The concentration of commonly used HF liquid is 49%, and its etching rate is too fast. So it is necessary to add DI water to dilute HF during the process, or use BOE. BOE is also called buffered HF or BHF. Figure 14.1 is the photos of HF and BOE. HF etches SiO_2, the principle is that fluorine and silicon react to produce a volatile, see Eq. (11.4). Nevertheless, the etching process of wet and dry techniques is different. As mentioned in Section 11.3, in order to complete dry etching at room temperature, plasma technique is needed in most cases. Wet etching does not require plasma, but the key is that the etchant must be adsorbed on the surface being etched. Since the HF-based etchant is an aqueous solution, the surface to be etched must be hydrophilic (see Figure 12.5). The surface of SiO_2 is hydrophilic, so this etchant can etch SiO_2. The surface of Si is hydrophobic, so this etchant

Semiconductor Microchips and Fabrication: A Practical Guide to Theory and Manufacturing, First Edition. Yaguang Lian.

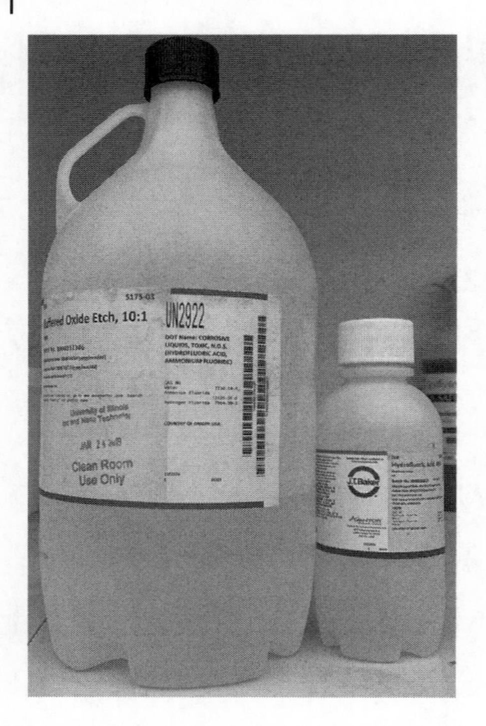

Figure 14.1 HF (small bottle on the right) and BOE (big bottle on the left).

cannot etch Si. This is very interesting because according to the chemical reaction formula, SiF_4, the reactant of F and Si is a volatile. In theory, HF can etch Si, but in fact, it does not work because the surface of Si is hydrophobic.

Wet etching of silicon nitride is divided into two situations: (i) Si_3N_4 deposited by PECVD can be etched by HF-based etchant, but the etching rate is lower than that of SiO_2; (ii) Si_3N_4 deposited by LPCVD can be etched by HF-based etchant with slow etching rate. The commonly used method is in a phosphoric acid (H_3PO_4) solution at 150–170 °C.

There are some differences between dry etching and wet etching. Dry etching is implemented through chemical reaction, physical sputtering, or the combination of chemical and physical actions. Wet etching is primarily achieved by chemical reaction. Dry etching of silicon and silicon dioxide is divided into three situations: (i) It etches Si and SiO_2; (ii) It only etches Si but not SiO_2; (iii) It only etches SiO_2 but not Si. Dry etching of silicon and silicon nitride is divided into two situations: (i) It etches Si and Si_3N_4; (ii) It only etches Si but not Si_3N_4. These situations will be discussed in detail in Chapter 15.

Compared with dry etching, wet etching has two main advantages: (i) low cost. The main equipment for wet etching is a fume hood (see Figure 14.2). The price is lower than the equipment for dry etching. (ii) High-production efficiency. Up to 25 wafers can be placed in a wafer cassette and placed in the bath for etching at the same time (see Figure 14.3). The bath showed in the figure can be heated. HF

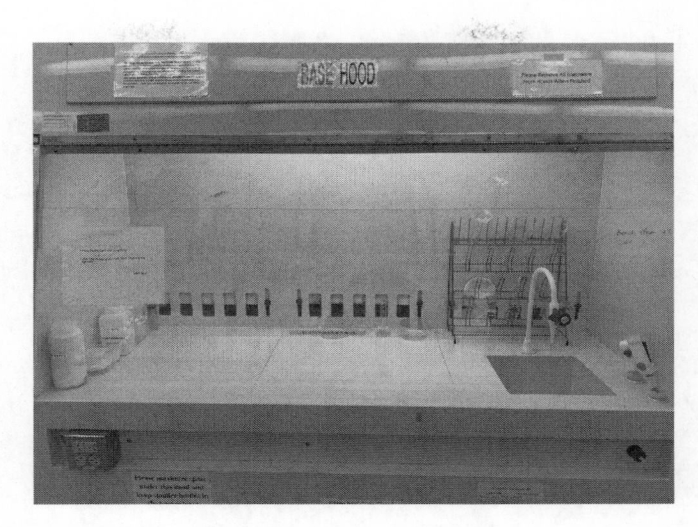

Figure 14.2 A fume hood with simple structure for wafer cleaning, wet etching, and developing.

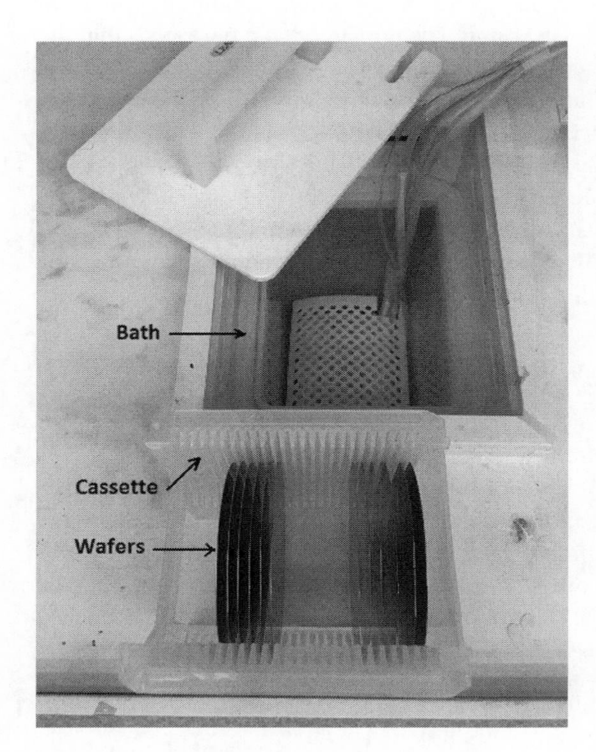

Figure 14.3 Etch bath, cassette, and wafers.

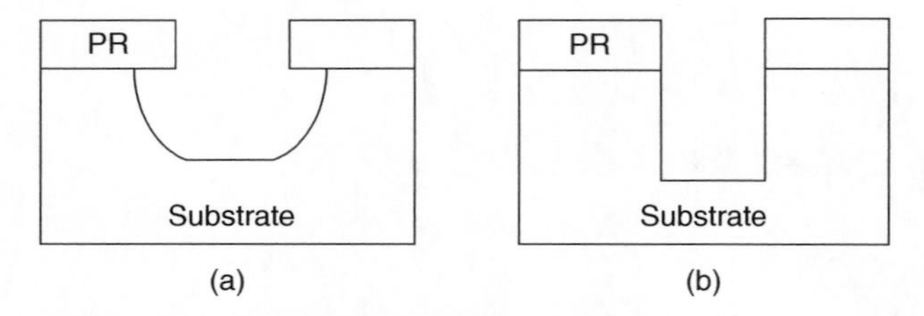

Figure 14.4 The profiles of isotropic etch (a) and anisotropic etch (b).

bath has a similar structure but no heating wires inside. However, wet etching has two main disadvantages: (i) It is difficult to perform small size etch. (ii) It can only perform isotropic etch, that is, the etching rate is the same in all directions. In the manufacture of microchips, anisotropic etch is main target. Figure 14.4 is a schematic diagram of isotropic and anisotropic etching profiles.

14.2 RIE System for Dry Etching

As we said before, from a chemical point of view, dry etching and chemical vapor deposition are essentially the same. The fundamental difference is that the product is volatile after chemical reaction, which is the etching; the product is nonvolatile after chemical reaction, which is the deposition. RIE is the most used dry etching equipment. An RIE machine and a PECVD machine are very similar in appearance. The basic structure of the two machines is shown in Figure 13.4. In a process chamber, either deposition or etching can be done, depending on the gases input. We will discuss RIE equipment below related to the Figure 13.4. Due to the similarities between RIE and PECVD, PECVD will often be mentioned for comparison in the following discussions.

14.2.1 RIE Process Flow and Equipment Structure

Let us use the dry etching of SiO_2 to illustrate the RIE process flow. The diagram of the flow is shown in Figure 14.5. The first step is to grow or deposit a SiO_2 dielectric film on the wafer surface. The second step is to complete the photoresist patterning on the film, where the photoresist is used as a masking film to protect the unexposed areas from etching. The third step is to put it into the machine to complete the etch of the openings without the PR protection. In the third step, "che + phy" refers to chemical reaction and physical sputtering (bombardment); "radicals" refers to the atoms, or the collection of atoms that exist in the plasma. They are electrically neutral, but in the states of incomplete chemical bonding, making them very chemically reactive.

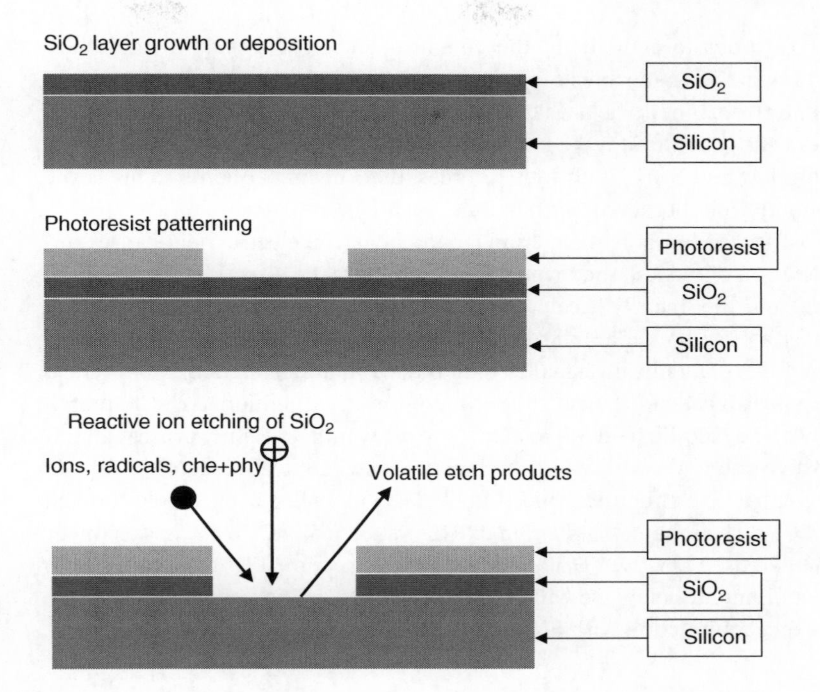

Figure 14.5 Schematic diagram of RIE dry etching process.

Let us use CF_4 as an example. The most abundant ionic specie found in CF_4 plasma is CF_3^+. In addition to CF_4 molecules, ionic species, and electrons, there are a large number of radicals formed. In CF_4 plasma, the most abundant radicals are CF_3 and F. In general, radicals are thought to exist in plasma in much higher concentration than ions. For more details, please refer the reference [1] on page 670. These particles are also subject to the collision of electrons and ions in the plasma. Due to the similarity, we use the basic structure of PECVD shown in Figure 13.4 to discuss RIE system. In RIE, the core part is the process chamber, because the samples are processed here.

As discussed in Section 11.3, a plasma process chamber can only adopt two structures: capacitance or inductance. What structure should RIE adopt? To determine this issue, we need to discuss dry etching. In the process, the two most important materials to be etched by dry etching are SiO_2 and Si_3N_4, which are chemically the same as silicon etching. They are carried out by the reaction of fluorine and silicon to form a volatile. When the fluorine-containing gas is split and activated by plasma, the active radical F can directly react with Si to produce volatile SiF_4. However, F cannot directly react with Si inside SiO_2 and Si_3N_4 because silicon atoms in these two materials are bound by oxygen and nitrogen. In order to etch them, the chemical bonds of these two molecules must be broken. Gas molecules can collide with each other and split through the action of electromagnetic field. SiO_2 and Si_3N_4 are solids, and collision between molecules is impossible. How to break the bonds? This requires bombarding the surface of SiO_2 and Si_3N_4 with ions to break their chemical bonds. This is the physical sputtering (bombardment) of ions, as shown in Figure 14.5.

When a charged ion enters electromagnetic field, the electric field causes linear acceleration of the ion, the magnetic field changes the direction of movement of the ion, and the linearly accelerated ion will physically bombard the surface of the substrate. Although a radical can also randomly bombard the surface, the sputtering force of a radical is smaller than that of an accelerated ion, so physical sputtering is also called physical sputtering of ions. As mentioned in Chapter 4, capacitance and electric field are related. So if we want to design a process chamber, in which there are both chemical reaction and physical sputtering, we can only use a capacitive structure. This kind of plasma is called capacitively coupled plasma (CCP). The schematic diagram of RIE system using CCP process chamber is shown in Figure 14.6. PECVD also uses a similar structure. Figure 14.7 is a photo of a RIE equipment. Below we will describe each part of the system according to Figure 14.6 in conjunction with Figure 13.4.

14.2.2 Process Chamber

As mentioned above, the process chamber (chamber) in an RIE machine uses a capacitive structure. When the chamber is opened, the structure is shown in Figure 14.8. The top (upper) electrode in the figure is also called showerhead because this electrode has many small holes (like showerhead used in

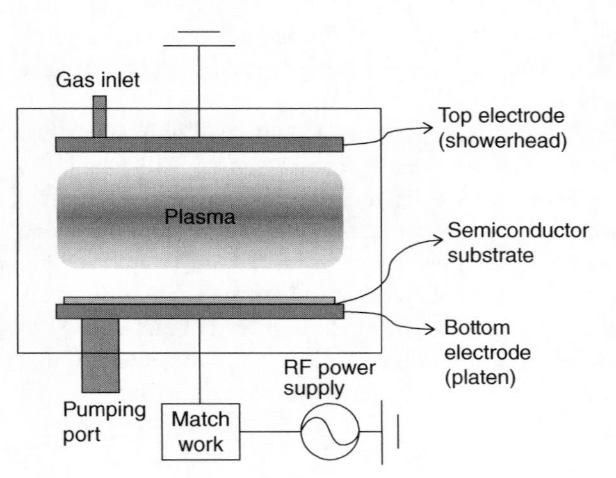

Figure 14.6 Schematic diagram of the overall structure of the RIE system.

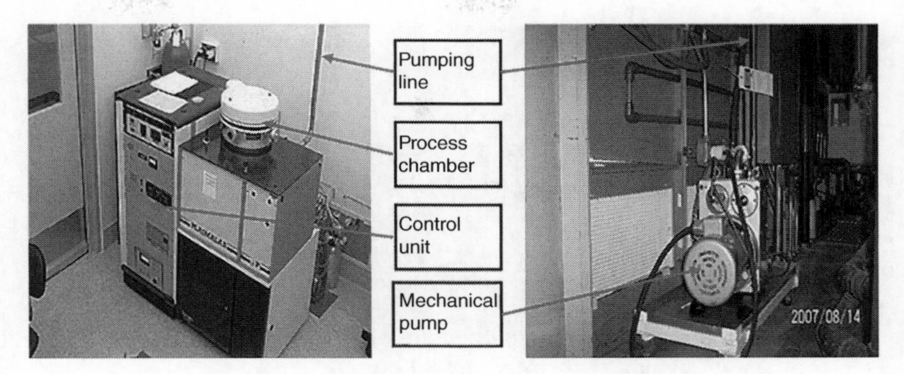

Figure 14.7 Photos of an RIE machine. Source: PlasmaLab RIE.

Figure 14.8 The structure and names of a process chamber. Source: PlasmaLab RIE.

the bathroom) through which etching gases enter the chamber. The bottom (lower) electrode has another name platen. An O-ring is inlaid on the outer edge of the showerhead. The O-ring ensures that the chamber can maintain a certain vacuum. The wafer or piece is placed on the platen. The etching gases are introduced into the chamber. Under the action of the electric field, plasma

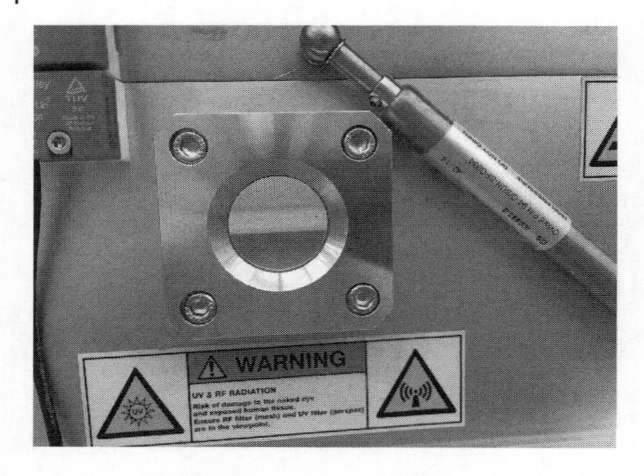

Figure 14.9 The color of plasma glow.

discharge is generated, and photons are emitted. The light emitted by different gas has different wavelength and different color. There is a small window on the sidewall of the chamber to observe the luminous color, as shown in Figure 14.9. A PECVD machine has a process chamber with the same structure and has similar light emission.

14.2.3 Vacuum Pumps

The vacuum pumps (pumps) used in different systems are different. Figure 14.10 shows commonly used pumps. The left is a turbomolecular pump (turbo pump), and the right is a Roots blower and mechanical (backing) pump. If the process

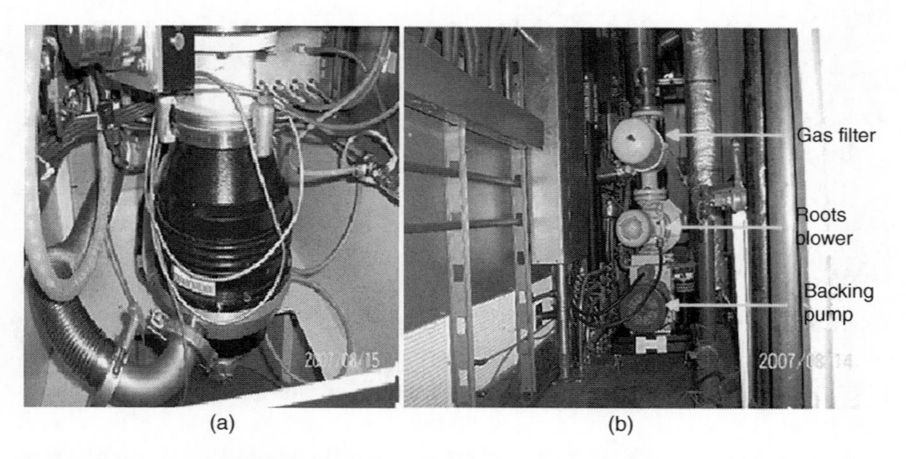

Figure 14.10 A turbomolecular pump (a), Roots blower and mechanical pump (b).

pressure is 100 mTorr or higher, and the gas flow is not large, a mechanical pump can meet the requirement. However, if high vacuum is required, or the gas flow rate is large, it is necessary to assist other types of pumps. RIE requires a high vacuum, the turbo pump can meet this requirement. The structure of the pump uses a combination of turbo and mechanical pumps. PECVD hopes to have a low vacuum, but the gas flow is large, mainly because the diluent gas flow is large. The pump structure used in PECVD is a combination of a Roots blower and a mechanical pump. The "Gas filter" showed in the figure is frequently used in PECVD system. This is because PECVD is a process that produces tiny particles, and the filter can block the particles that may enter the pump, extending the service life of the pump. The pumping lines are used to connect pumps to the process chamber through pumping port, as shown in Figure 14.6.

14.2.4 RF Power Supply (Source) and Matching Network (Matchwork)

As mentioned in Section 13.1.3, the frequency of most power supplies used for the process is 13.56 MHz. Other frequencies are sometimes used in the power supplies. 300–400 kHz supplies are used for PECVD Si_3N_4 deposition and 2 MHz supply is used for inductively coupled plasma (ICP) dry etching (discuss in Chapter 15). Theses frequencies are in the radio frequency (RF) range, so they are called RF power supplies. In addition to RF supplies, we also use DC and microwave (2.45 GHz) power supplies in the process. Figure 14.11 is the photo of a 13.56 MHz power supply. To understand the power supply, impedance matching needs to be discussed. For RF power supply, it satisfies impedance Eq. (4.9), which contains the components of capacitance and inductance. Different size of RIE chamber, the capacitance is different. So with different sizes and frequencies, the

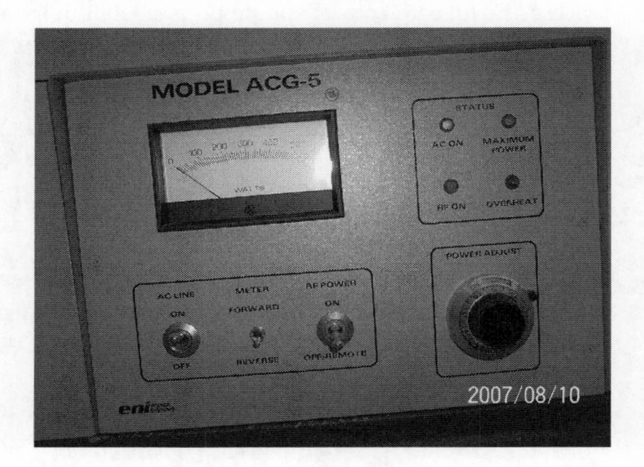

Figure 14.11 RF power supply.

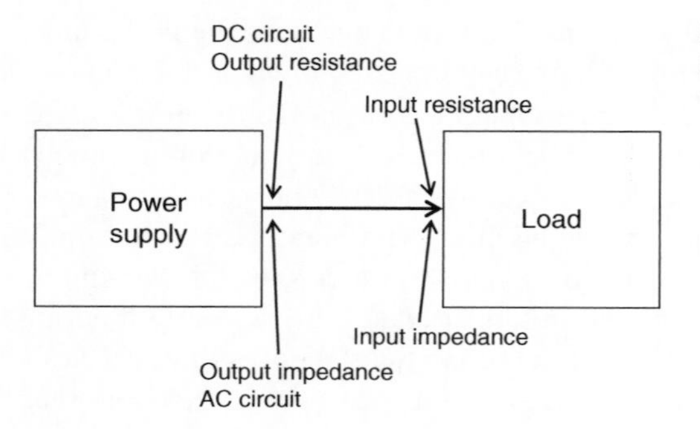

Figure 14.12 The difference between DC circuit and AC circuit.

impedance of the system is also different. Because the working frequency of the supply is very high, when a supply is connected to a process chamber, it is easy to cause a problem called impedance mismatch. Figure 14.12 is a schematic diagram of the connection between a supply and a load. For a DC circuit, output resistance of the supply and input resistance of the load are the same. For an AC circuit, output impedance of the supply and input impedance of the load are different in many times. This is called impedance mismatch. As a product, the output impedance of the power supply is generally designed to be 50 Ω. This number has a special name, characteristic impedance. In our actual system, the load is the process chamber. It is difficult for the input impedance of the chamber to achieve a characteristic impedance of 50 Ω, which causes impedance mismatch problem. Therefore, on a RF power supply, there are two power readings (Figure 14.11): FORWARD is the incident power, which refers to the power from the supply to the chamber; REVERSE is the reflected power, which refers to the power from the chamber to the supply. In the system, FORWARD is sometimes written as INCIDENT, and REVERSE is sometimes written as REFLECTED. During the process, the reflected power should be as small as possible, preferably zero. If the reflected power is large, the following three problems will occur:

1. The power entering the process chamber is smaller than the power designed in the recipe. The etching result deviates from the expected value, which is lower than the designed value.
2. It is difficult in ignition of the plasma discharge. 3. It may cause damage to the power supply.

If the matching between the supply and the chamber is not good, and there is a large reverse power, we should use a matching network (matchwork) to eliminate the mismatch problem, please see Figures 14.13 and 14.14. Matching is a very important issue in the design of RF and microwave circuits. Different structures are used for different systems and frequencies. For RF power supply used in semiconductor technology, a matching network is mainly composed of

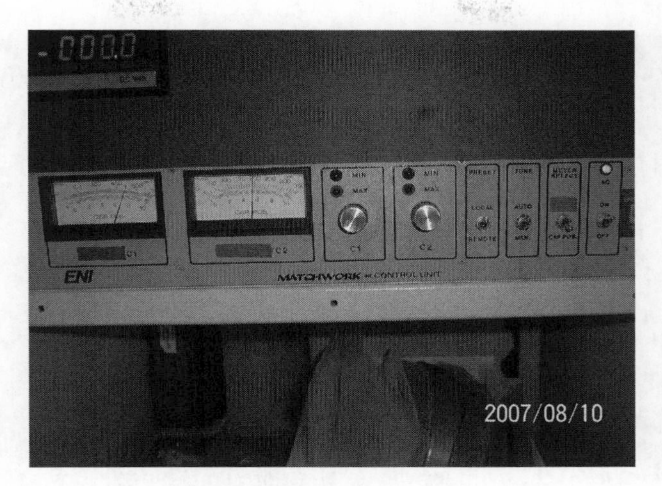

Figure 14.13 Matchwork controller.

(a) (b)

Figure 14.14 The variable capacitors used in matching networks. (a) one uses two vacuum capacitors and (b) one uses two rotary capacitors.

two variable capacitors and an inductor. The capacitors are called capacitor 1 (C_1) and capacitor 2 (C_2), or load C and tune C, or magnitude C and phase C, please see Figure 14.14. The matchwork is placed between the supply and the chamber, see Figures 13.4 and 14.6.

14.2.5 Gas Cylinder and Mass Flow Controller (MFC)

The chemical gases used in the process need containers for storage. Cylinders are the containers to store the gases (see Figure 14.15a). Since most gas cylinders store high-pressure gases and cannot be directly connected to the equipment, we need to use a regulator to reduce the pressure from a high-pressure cylinder

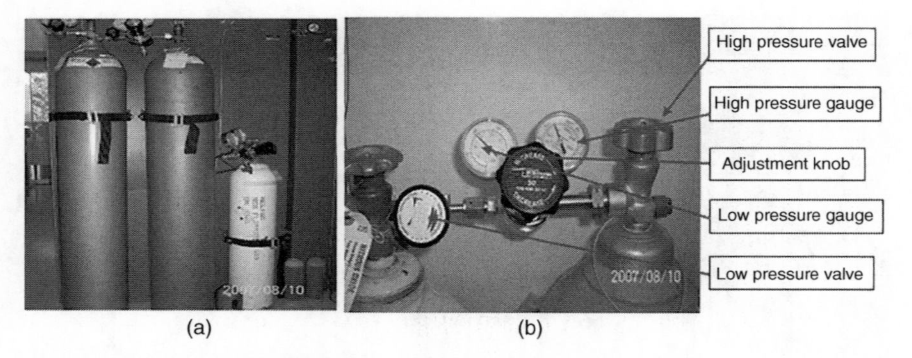

Figure 14.15 Gas cylinders (a) and a regulator (b).

to a safe working pressure for the equipment use (see Figure 14.15b). In the figure, "High-pressure valve" is the cylinder valve, "High-pressure gauge" shows the pressure inside the cylinder, "Adjustment knob" is a pressure-reducing valve, "Low-pressure gauge" reads the pressure after the adjustment knob, and "Low-pressure valve" is the valve after the adjustment knob. In the following chapters, the high-pressure and low-pressure gauges will be referred to as pressure gauge for short.

The boiling point or sublimation temperature of each chemical gas is different, and the pressure of the cylinder that stores it is also different. The lower the temperature, the higher the cylinder pressure. The higher the temperature, the lower the cylinder pressure. Some etching gases are liquid or close to liquid at room temperature. At this time, the pressure of the cylinder is very low, and it can be directly connected to the system without a regulator. Figure 14.16 shows the values displayed by several high-pressure gauges of different gases. The unit of the pressure gauge in the figure is pounds per square inch (psi), 1 atm = 14.7 psi. The pressure of CF_4 in the picture exceeds 1500 psi, CHF_3 is close to 600 psi, and SF_6 is only 300 psi. These cylinders have been used for some time, and the pressure has decreased. The pressure of the new cylinder is 2000 psi for CF_4, 635 psi for CHF_3, and 320 psi for SF_6 [2]. Their corresponding boiling points are shown in the figure. From the pressure of the cylinder, we can understand which gas has a low boiling point and which gas has a high boiling point. BCl_3, one of the most used etching gases for III–V semiconductors, has a boiling point of 12.6 °C and a cylinder pressure of 4.4 psi [2]. In this case, the cylinder can be directly connected to the equipment and no regulator is required.

The three gases mentioned above, CF_4, CHF_3, and SF_6, are the three most used gases for dry etching of Si, SiO_2, and Si_3N_4. The first two gases contain carbon, so sometimes we call CF_4 as Freon 14 and CHF_3 as Freon 23. The three gases are very stable. Plasma method is required to ionize them for the dry etching. There are many differences between the three gases, but in the process, the main difference

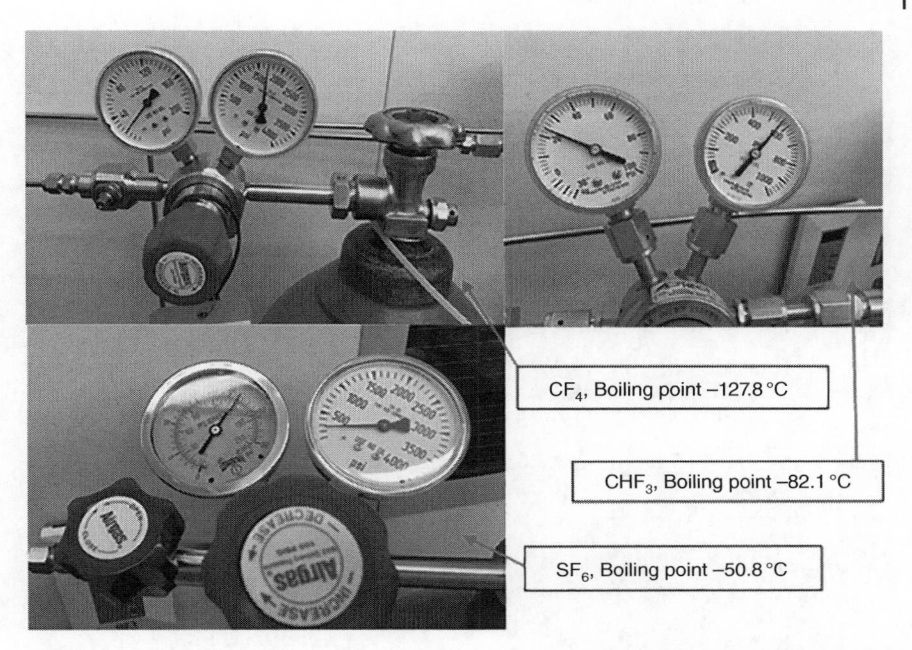

Figure 14.16 Different boiling points correspond to different pressures of the cylinders with unit psi.

between them is the content of fluorine. Sulfur hexafluoride (SF_6) contains 6 F, Freon 14 has 4 F, and Freon 23 has only 3 F. Therefore, under a same condition, SF_6 etching rate is the highest, CF_4 is in the middle, and CHF_3 is the lowest. Here are three results of PECVD Si_3N_4 etching rate: SF_6 is around 1564 Å/min, CF_4 is around 835 Å/min, and CHF_3 is around 284 Å/min.

When a cylinder is opened, the high-pressure gas passes through the regulator to reduce to low pressure, generally about 20 psi. The low-pressure gas then enters the gas flow control device along the gas line. This device is mass flow controller (MFC), as shown in Figure 14.17. The most used gas flow rate unit in the process is Standard Cubic Centimeters per Minute (SCCM). One cubic centimeter (cm^3) is 1 milliliter (ml). About learning, Confucius said 2500 years ago: "learn one point, understand three points." We just learned about SCCM. If we encounter other flow rate units, we should be able to read them out according to this method. Figure 14.18 is the other two flow rate units, SCFH is Standard Cubic Feet per Hour and SLPM is Standard Liter per Minute. The "standard" refers to the standard conditions of temperature and pressure–temperature is 0 °C and pressure is 1 atm.

At this point, a complete gas supply system has been established, see Figure 13.4. One gas connects with one MFC. If more than one gas used in the etching, they are merged to one gas line after passing through the MFCs and enter the process

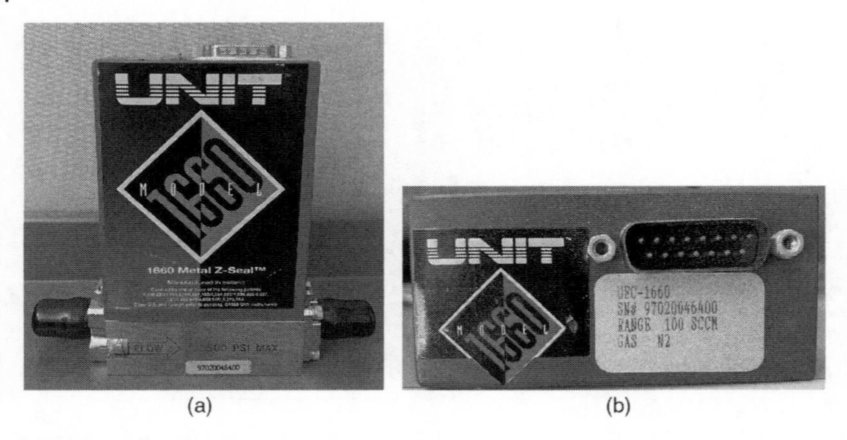

(a) (b)

Figure 14.17 MFC (a) and the unit of gas flow (b).

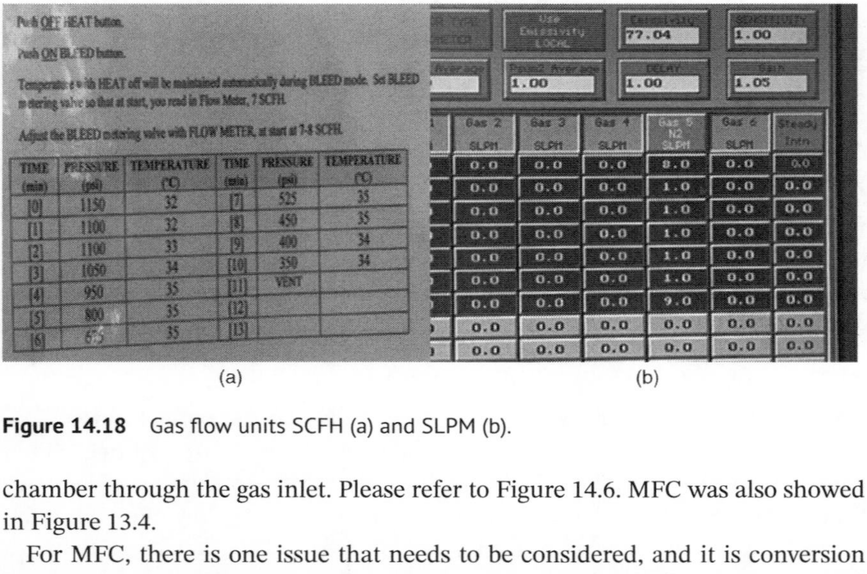

(a) (b)

Figure 14.18 Gas flow units SCFH (a) and SLPM (b).

chamber through the gas inlet. Please refer to Figure 14.6. MFC was also showed in Figure 13.4.

For MFC, there is one issue that needs to be considered, and it is conversion factors (correction factors) between different gases. The gas labeled on MFC in Figure 14.17 is N_2. It can be considered that this MFC is calibrated with nitrogen. Its range is 0–100 sccm. This means that when using this MFC to control N_2, the minimum flow rate is 0 sccm, and the maximum flow rate is 100 sccm. Corresponding to different gases and needs, the calibration gas of MFC is different, and the range is also different (see Figure 14.19). When the upper MFC is used for chlorine, the maximum flow is 50 sccm; when the lower one is used for silane, the maximum flow is 150 sccm.

Figure 14.19 Cl_2, 50 sccm MFC and SiH_4, 150 sccm MFC.

Table 14.1 Gas conversion factors with respect to N_2.

Gas name	Conversion factor
N_2, Nitrogen	1.00
Air	1.00
Ar, Argon	1.39
H_2, Hydrogen	1.01
He, Helium	1.45
O_2, Oxygen	0.99
Cl_2, Chlorine	0.86
N_2O, Nitrous oxide	0.71
CH_4, Methane	0.72
NH_3, Ammonia	0.73
SiH_4, Silane	0.60
CHF_3, Trifluoro methane (Freon23)	0.50
CF_4, Carbon tetrafluoride (Freon14)	0.42
BCl_3, Boron trichloride	0.41
$SiCl_4$, Silicon tetrachloride	0.28
SF_6, Sulfur hexafluoride	0.26
C_4F_8, Octafluorocyclobutane	0.16

Source: Adapted from Ref. [3].

The question now is, if an MFC is broken, for example a 100 sccm CF_4 MFC is broken, there is no same MFC at this time, but there is a 100 sccm N_2 MFC. We can install this N_2 MFC on the gas line of CF_4 to control CF_4. 100 sccm N_2 MFC is used for CF_4. Is the maximum flow rate also 100 sccm? The answer is no. We must multiply by a number called conversion factor or correction factor to convert N_2 to CF_4 (see Table 14.1). The gases listed on the table are commonly used gases in the process. In this table, all gases are converted based on N_2. The factor of CF_4 is 0.42, so if 100 sccm N_2 MFC is used to control CF_4, the flow rate of CF_4 at full scale is $100 \times 0.42 = 42$ sccm. The conversion of other gases can be deduced by analogy.

14.2.6 Heater and Coolant

In PECVD, the platen needs to be heated to about 300 °C, so there must be a set of heating and control systems in the PECVD equipment. RIE usually does not require heating, and photoresist can be used as a mask for etching in the process.

Figure 14.20 Water chillers, the left is a small one and the right is a large one.

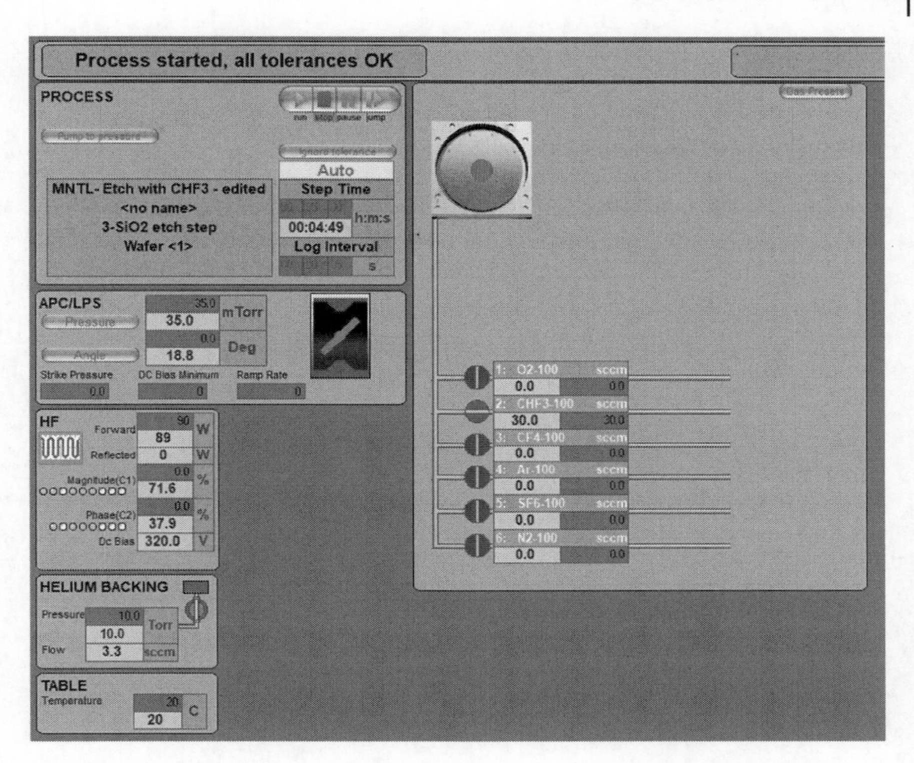

Figure 14.21 The computer screen of operating software. Source: Oxford Instruments, PlasmaPro 80 RIE.

The plasma in the process generates a lot of heat, which requires cooling. In addition, a large power supply also needs cooling. So in RIE and PECVD systems, cooling is an essential part. In cooling technology, water is the most used coolant, and other types of coolants can also be used. The system can be directly connected to the chilly water system in the clean room, or water chillers (chillers) can be used, see Figure 14.20.

So far, a whole RIE system is introduced. Most of the pictures I took from old model machines. New model machines have become compact. It is not easy to find these parts from the outside. But we can find such information on the computer screen of the operating software (see Figure 14.21). From this figure, we can find MFC, flow rate, pressure, forward power, reflected power, matchwork, and DC bias (talk in Chapter 15).

References

1 Wolf, S. and Tauber, R.N. (2000). Chapter 14, Dry etching for ULSI fabrication. In: *Silicon Processing for the VLSI Era*, Process Technology, 2e, vol. 1, 670. Lattice Press.
2 Matheson Tri·Gas, Gases and Equipment, pp. 199, 200, and 225.
3 "Gas Correction Factors for Thermal-based Mass Flow Controllers", MKS Instruments, Inc.

15

Dry Etching

In Chapter 14, we mainly discussed the RIE system. In the dry etching process, there is still much to be understood. In this chapter, we will discuss the etch profile, speed, the selection of etching gases for different materials, and inductively coupled plasma (ICP).

When doing RIE etching process, we need to record the following four parameters:

1. Gas flow rate (flow), the unit is sccm.
2. The process pressure (pressure), the unit is mTorr.
3. The process power (power), the unit is W.
4. DC bias, the unit is V.

For the PECVD deposition process, the first three parameters to be recorded are the same as RIE, except that the fourth parameter is different. This parameter is temperature.

Below we will discuss the abovementioned aspects.

15.1 The Etch Profile of RIE

For dry etching, the profile after etching is one of our main concerns. If the influence of other chemical components in the process is not considered, and fluorine etching of Si, SiO_2, and Si_3N_4 is only examined, the profile of the substrate opening after RIE etching is shown in Figure 15.1. The photoresist cross-section (profile) in the schematic diagram is in an ideal situation. The actual resist cross-section is shown in Figure 12.16. We use the ideal photoresist cross-section to discuss the profile of dry etching. Figure 15.1 uses silicon as an example. SiO_2 and Si_3N_4 have similar conclusions.

For dry etching, the resulting profile has two extreme cases:

Semiconductor Microchips and Fabrication: A Practical Guide to Theory and Manufacturing,
First Edition. Yaguang Lian.

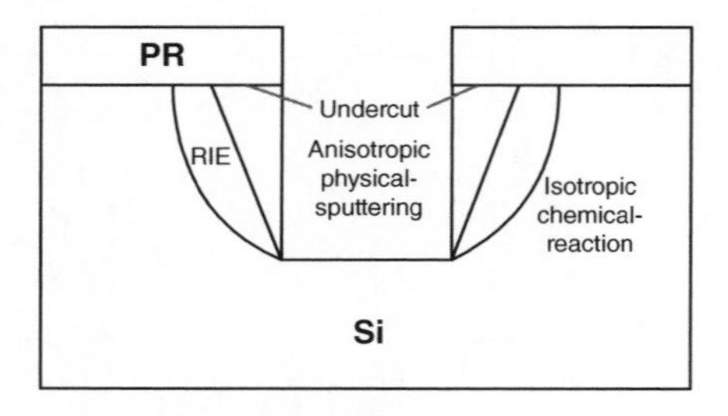

Figure 15.1 Schematic diagram of Si opening profile after RIE dry etching.

15.1.1 Case 1

Completely isotropic etch, which is obtained by a complete chemical reaction. As this reaction is nondirectional, the reaction rate in all directions is the same, and the resulting cross-section is an isotropic profile.

What kind of equipment can be used to obtain such a profile? As mentioned in Chapter 11, the purpose of using plasma technique in the process is to make stable molecular structures unstable through intermolecular collisions; even the chemical bonds are broken so that chemical reactions can proceed smoothly. In the RIE equipment, the process chamber uses a capacitive structure. Capacitor is related to voltage, see formula (4.1). In this way, when RF power enters the chamber, a DC bias voltage (DC bias) is established between the upper and lower electrodes (see Figure 14.21). The RIE system is designed in such a way that the direction of the electric field is from shower head to platen (we will elaborate on it in Section 15.5). The ions are accelerated by the electric field (DC bias) and bombarded the surface of the substrate. This is the origin of physical sputtering of RIE. Therefore, RIE does not produce a complete chemical reaction and obtain a complete isotropic profile. From these discussions, we can conclude that if there is a fluorine-containing gas, it is easy to split and does not need a power supply to generate plasma, and there is no physical bombardment. Such a tool can be used to produce a complete chemical reaction, and the etched profile is isotropic. This dry etching machine is a xenon difluoride (XeF_2) etcher (see Figure 15.2). Figure 15.3 shows the structure of the etcher's process chamber. XeF_2 is a very strange

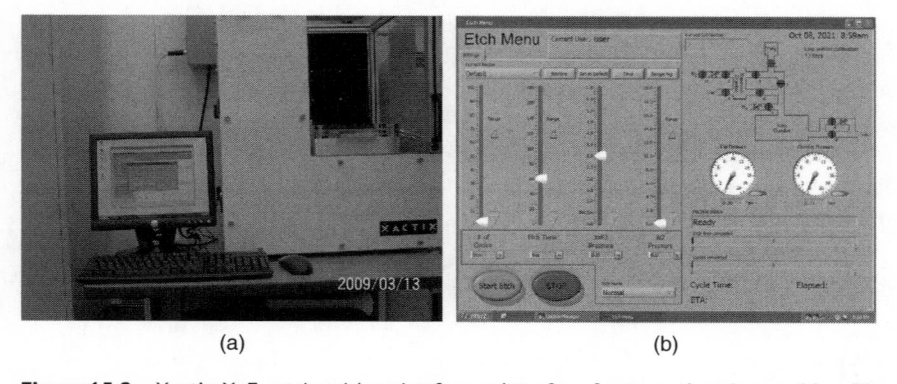

Figure 15.2 Xactix XeF_2 etcher (a) and software interface for operating the machine (b).

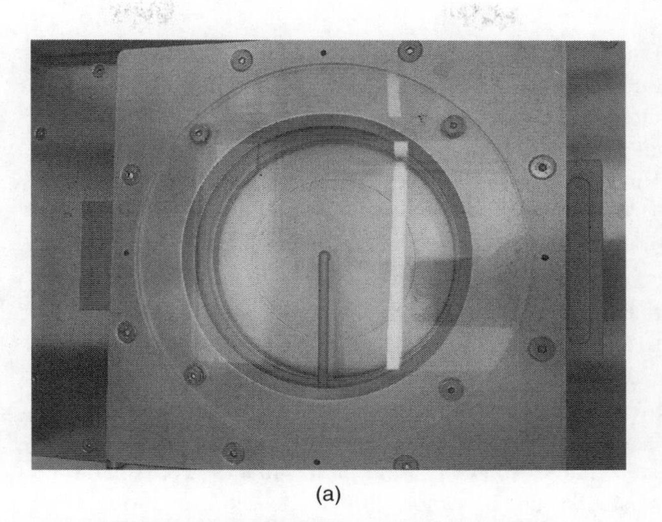

(a)

(b)

Figure 15.3 The process chamber of the etcher (a) is the picture of the chamber, (b) is the structure when the chamber is opened.

substance. Xenon (Xe) is an inert gas, and inert gases should not chemically react with other elements. But XeF_2 is a product of chemical reaction between xenon and fluorine! Nevertheless, it is very unstable and easy to break. This substance has two states, solid and gaseous. It is stored in a gas cylinder. When the pressure exceeds 4 Torr, it maintains a solid state. When the pressure is less than 4 Torr, it sublimates from solid state to gaseous state, separate and free atomic fluorine. Fluorine can react with silicon to form a volatile product, which can be pumped away to achieve the purpose of etching. Since the XeF_2 etcher does not require a power supply, no plasma, and no physical sputtering, it can etch Si, but not SiO_2 and Si_3N_4, or photoresist. For the process chamber of this equipment, we can only call the showerhead and platen. Compared with RIE, XeF_2 etcher is simple, cheap, and does not generate a lot of heat. It is used to etch a single-atom material if this material produces volatile chemical reactions with fluorine, such as silicon, germanium, and tungsten. The existence of XeF_2 etcher makes it possible to do isotropic dry etching of silicon, but it is more difficult to achieve isotropic dry etching of SiO_2 and Si_3N_4. Figure 15.4 is showing two SEM pictures of Si etching profiles. Left one uses SiO_2 as etching mask with a small opening and right one is the etching result with a huge opening. We can see from the pictures that with a small hole, XeF_2 gives almost perfect isotropic etch like a spherical surface. With a big opening, the etching profile is the same as the schematic diagram shown in Figure 14.4. XeF_2 etcher can be used to etch Si only, but not SiO_2 and Si_3N_4. As the etching rate of SiO_2 is extremely slow, when using XeF_2 tool, we should remove the native oxide by using BOE from Si surface to get better etch result. Figure 15.5 shows the etched openings with BOE and without BOE dip. From the photos, we can find that the etched surface with BOE dip is much smoother than the surface without BOE dip. The average etching rate without BOE is around 8% lower than that with BOE. So the PR shadow with BOE is larger than that without BOE.

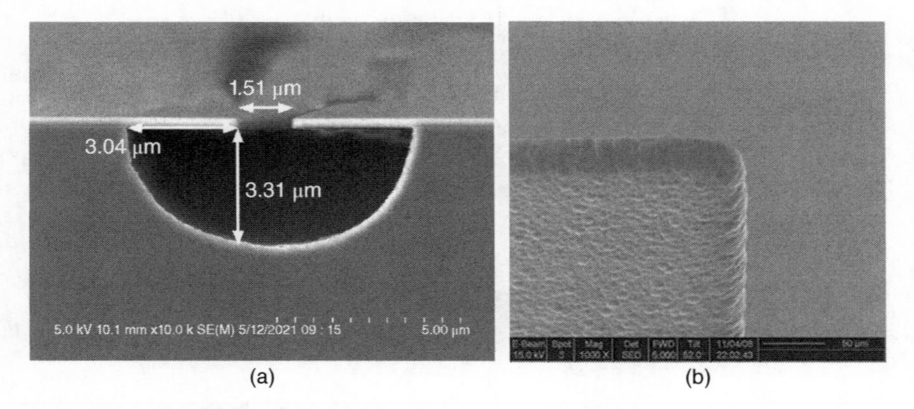

Figure 15.4 SEM pictures of Si etched profiles by using XeF_2 etcher. (a) XeF_2 etches Si through a small opening, (b) XeF_2 etches Si through a large opening.

Figure 15.5 Si etched openings with BOE dip (a) and without BOE dip (b).

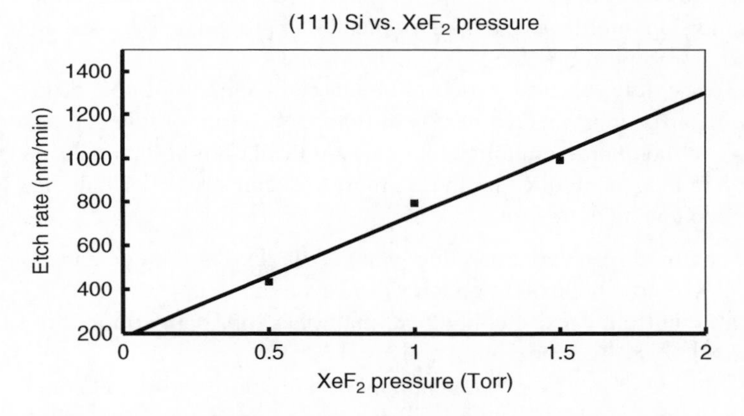

Figure 15.6 The relationship of (111) Si etching rate and XeF_2 pressure.

XeF_2 gives only chemical reactions. The higher the pressure, the faster the etching rate (discussed later in Section 15.5). Figure 15.6 is the curve of Si etching rate vs. XeF_2 pressure. Here I want to talk more about the tool I used for the experiment. Many universities have Xactix XeF_2 etchers. There are two chambers in the tool-expansion chamber and main chamber (picture b in the Figure 15.2). The etching is finished in the main chamber. The pressure we set in the recipe is the pressure in the expansion chamber. When the etch starts, a valve between two chambers opens and XeF_2 source is shared by expansion and main chambers; so the process pressure in main chamber is around half of the pressure in the recipe.

15.1.2 Case 2

Completely anisotropic etch is obtained by complete physical sputtering. The names of physical bombardment and physical sputtering are often used interchangeably. The physical sputtering is produced by the ion bombardment of the sample surface accelerated by the electric field. Its directionality is strong, and it only strikes along the direction of the electric field. Therefore, it is possible to obtain complete anisotropic etch that only etches longitudinally but not laterally. If the gas does not produce a chemical reaction, RIE can achieve this etch. What kind of gas can play this role? The answer is argon (Ar). Argon is an inert gas and does not join any chemical reactions (see Figure 5.1 of the periodic table). Inert gas is the 18th group. In this group, argon is the heaviest mass that can be used easily and on a large scale. The inert gas above it is light in mass, and the natural content below it is low. Ar is easily ionized into Ar^+, so it is widely used in physical bombardment in the semiconductor processes. When only argon is used in RIE, complete physical bombardment and anisotropic etch can be achieved. The RF power entering the process chamber is concentrated into a small area, which increases the power density and the rate of Ar^+ physical sputtering. This technology is called ion milling. When used for metal film growth, it is also called a metal sputtering process, and we will discuss in Chapter 16.

We may ask, since only Ar can be used to obtain complete anisotropic etch, which is also one of the most desired results in the manufacture of microchips, can we only use Ar^+ bombardment for etching? Theoretically, it is yes, and this is why we have ion milling technique. But in fact, there are some problems in doing so, mainly in the following two aspects:

1. This kind of bombardment cannot produce volatile matter. The sputtered matter accumulates around the pattern openings and even the sample surface.
2. Since bombardment can cause etching to all materials, it brings certain difficulties to the selection of etch masking films.

   Therefore, during dry etching, we generally do not add Ar to the etching recipe unless there are special requirements (discuss in Section 15.2). The selected gases produce volatile matter with the etched materials. At the same time, the ions produce ion bombardment on the etched openings, but the bombardment force is lower than Ar^+ bombardment. RIE is a etch process involving chemical reaction and physical sputtering. The profile obtained after etching is neither completely isotropic nor completely anisotropic, but a shape between the two, as shown in Figure 15.1. In Figure 15.1, undercut is an inevitable phenomenon of RIE dry etching. To reduce the undercut, it is necessary to increase the physical bombardment composition and reduce the chemical reaction composition. To increase the undercut, it is necessary to reduce the physical bombardment composition and increase the chemical reaction composition. In the process parameters, to increase the physical composition and reduce the chemical composition, we need to reduce the pressure and increase the power; to increase the chemical composition and reduce the physical composition, we need to increase the pressure and reduce the power. Reducing the pressure means increasing the mean free path of the etching gas (discuss in Section 15.5) and the DC bias voltage. Increasing the power means increasing the DC bias voltage. These are all helpful to improve

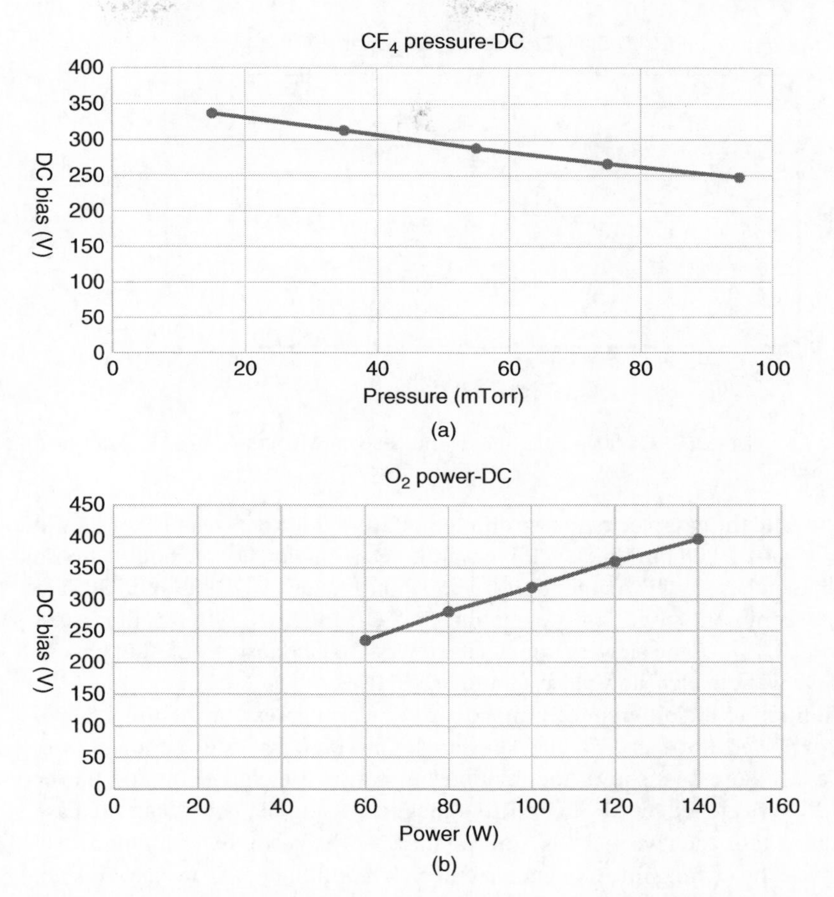

Figure 15.7 The relationships of pressure and DC bias for CF_4 (a), power and DC bias for O_2 (b).

the composition of physical bombardment, thus making the etch profile more anisotropic, and vice versa. Figure 15.7 shows the relationships between pressure and DC bias, power and DC bias. But as the power increases, the ion density increases, and the chemical composition becomes larger. Therefore, be careful when adjusting the power to improve the profile.

15.2 Etching Rate of RIE

Another important consideration in RIE process is etching rate. Only when the rate is known can the time be set to reach the desired depth. The depth of etch is

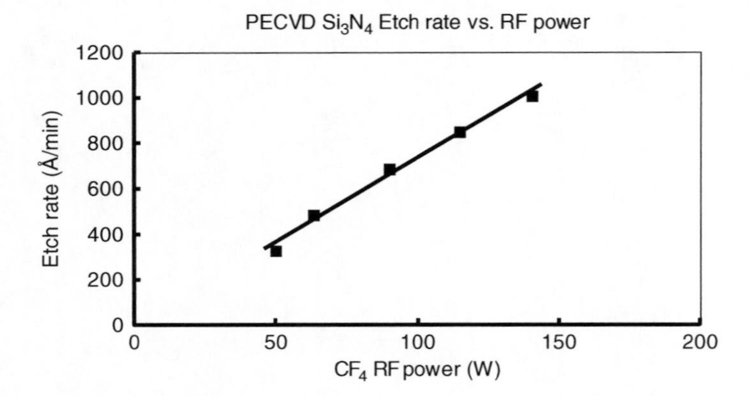

Figure 15.8 Using CF_4 as etching gas, the relationship between the rate and power of PECVD Si_3N_4.

different, and the gas selected is also different. If we want to etch deeply, we should choose a gas that etches quickly. If we want to etch shallowly, we should choose a gas that etches slowly. As mentioned in Section 14.2.5, CF_4, CHF_3, and SF_6 are the three commonly used gases for etching Si, SiO_2, and Si_3N_4. SF_6 has the fastest rate and CHF_3 has the slowest rate. So if we want to etch fast, we should choose SF_6. If we want to etch slowly, we should use CHF_3.

In addition to the different etching rates of different gases, the etching rates of different powers are also different. They are high power, fast rate and low power, slow rate. Figure 15.8 shows the relationship between etching rate and power when Si_3N_4 is etched by CF_4 in an RIE. Since increasing the power can increase the etching rate, then we may ask: Can we increase the power to a very high level to increase the etching rate to meet our fast etch requirements? The answer is no. There is a problem in actual operation if the power is large, the DC bias voltage is high. When the DC bias is higher than a certain value, the process chamber of capacitive structure will face the danger of breakdown. There was a RIE machine I used before, when the DC bias reached 500 V, the system would alarm. Therefore, in the clean room of the university, the RF power supply of RIE tool is generally between 300 and 500 W.

Using the same parameters, the etching rate is different for different substrate materials. Table 15.1 shows the etching rate of CF_4 on some materials. It can be seen from the table that under the same conditions, the rate of Si_3N_4 is faster than that of SiO_2. Due to this reason, Si_3N_4 and SiO_2 are the two commonly used dielectric materials, and discussing this issue will help us understand the influence of chemical bond on etching rate. To illustrate this phenomenon, we need to introduce Arrhenius equation in chemical reactions [1]:

$$\hbar = A \exp\left(-\frac{E}{RT}\right) \tag{15.1}$$

Table 15.1 The etching rate of CF_4 (flow 30 sccm, pressure 35 mTorr, power 93 W).

The materials etched	Etching rate (Å/min)
LPCVD Si_3N_4	376
PECVD Si_3N_4	706
Thermal SiO_2	211
PECVD SiO_2	230
Silicon	421

The k in the equation is called rate constant of the reaction. The larger the k value, the faster the chemical reaction rate. A is pre-exponential factor or frequency factor because it is related to collision frequency of molecules. E is activation energy, T is absolute temperature, and R is ideal gas constant. $R = 8.314\,\text{J/mol/k}$, where mol is the mole (refer to Section 12.1.9); k is absolute temperature Kelvin (refer to Section 5.2). The equation was proposed by Swedish scientist Svante Arrhenius (February 19, 1859–October 2, 1927) in 1889.

Because of the exponential, E and T in the equation have a great influence on the reaction rate. The smaller the E value, the faster the rate. The larger the T value, the faster the rate. For every 10 °C increase in temperature, the reaction rate increases by about two to four times [1], and the etching rate will increase. The greater the bond strength of the molecule, the greater the activation energy required, the slower the chemical reaction rate, and the lower the etching rate. In thermodynamics, we use the standard enthalpy of formation to indicate the energy released or consumed when 1 mol of a substance is created from its constituent elements. It is usually expressed by ΔH_f. ΔH_f of SiO_2 is 911 kJ/mol, and Si_3N_4 is 745 kJ/mol [2]. The ΔH_f of SiO_2 is greater than that of Si_3N_4. The activation energy of Si_3N_4 is smaller, so the etching rate is faster. The activation energy of SiO_2 is larger, so the etching rate is slower.

The enthalpies of different etched materials are important reference values for designing etching recipes. All other conditions are the same, such as the etching gas, the boiling point of the reactant, etc. Using same recipe, in most cases, the etching rate of the material with large value of enthalpy is slow, and the etching rate of the material with small value of enthalpy is fast. If the material to be etched has a larger value of enthalpy, to make the etch going well, a small amount of argon gas (refer to 15.1.2) should be added to the etching recipe. Ar^+ can produce the strongest physical sputtering among dry etching gases. The strong physical bombardment breaks the strong chemical bond and enables the chemical reaction to proceed. For example, the ΔH_f of Al_2O_3 (sapphire) is 1676 kJ/mol [2], and its chemical bond is so strong that when etching this material, we need to add argon in the etching recipe.

15.3 Dry Etching of III–V Semiconductors and Metals

As mentioned in Chapter 11, we use fluorine-containing gases to etch Si, SiO_2, and Si_3N_4. This kind of gases cannot etch III–V semiconductors and most metals. We take GaAs and Al as examples. The boiling point of fluorine and gallium reactant GaF_3 is 1000 °C, and the melting point of aluminum reactant AlF_3 is 1290 °C. Obviously, these reactants are nonvolatile. However, when reacting with chlorine, the boiling point of $GaCl_3$ is 201 °C and the sublimation temperature of $AlCl_3$ is 180 °C. The boiling point and sublimation temperature of these two reactants are much lower. The values given are at 1 atm. At low pressure, these values are even smaller. Therefore, the etching gases for III–V materials, most metals and their oxides are chlorine-containing rather than fluorine-containing. The Al_2O_3 mentioned above must be etched with chlorine-containing gases plus argon.

The boiling point of a substance is the temperature at which the vapor pressure of a liquid equals the pressure surrounding the liquid and the liquid changes into a vapor. The sublimation temperature of a substance is the temperature at which the vapor pressure of a solid equals the pressure surrounding the solid and the solid changes into a vapor. From the principle of physics, the higher the pressure, the higher the boiling point and sublimation temperature; the lower the pressure, the lower the boiling point and sublimation temperature. The relationship between boiling point and pressure follows the Clausius–Clapeyron equation named after Rudolf Clausius (January 2, 1822–August 24, 1888) who was a German scientist and Benoît Clapeyron (January 26, 1799–January 28, 1864) who was a French engineer [3]:

$$\ln \frac{P(T_2)}{P(T_1)} = -\frac{\Delta_{vap}H}{R}\left(\frac{1}{T_2} - \frac{1}{T_1}\right) \quad (15.2)$$

In the equation, ln is natural log, $P(T_1)$ and $P(T_2)$ are vapor pressures at two temperatures, $\Delta_{vap}H = H^V - H^L$ is molar enthalpy change of vaporization, H^V is molar enthalpy of formation for vapor phase, and H^L is molar enthalpy of formation for liquid phase. The relationship between sublimation temperature and pressure also follows this equation.

Molar enthalpies of formation for $GaCl_3$ and $AlCl_3$ are listed in NIST Chemistry WebBook. For example, $\Delta_{vap}H$ of $GaCl_3$ is 72.7 kJ/mol [4]. Let us use $GaCl_3$ to calculate the boiling point under the process pressure. The boiling point is 201 °C = 474.15 K at 1 atm. In Eq. (15.2), $P(T_1) = 1$ atm and $T_1 = 474.15$ K. The most used pressure range for RIE is 5–50 mTorr. This range converted to the standard atmosphere is 6.58×10^{-6}–6.58×10^{-5} atm. The highest process pressure of 6.58×10^{-5} atm is used to do calculation.

Input $P(T_1) = 1$ atm, $P(T_2) = 6.58 \times 10^{-5}$ atm, $T_1 = 474.15$ K, $\Delta_{vap}H$ = 72.7 kJ/mol, and $R = 8.314$ J/mol/k to Eq. (15.2), we can get $T_2 = 311.5$ K = 38.35 °C.

Compared to the boiling point T_1 at 1 atm, the boiling point T_2 is reduced by more than 160 °C at the RIE pressure of 50 mTorr. When the pressure is 5 mTorr, the boiling point will be lower. The situation of $AlCl_3$ is the same. In addition, the plasma can increase chamber temperature during dry etching, which is why the platen of RIE is usually cooled by water. So $GaCl_3$ and $AlCl_3$ are volatile at the processing pressure. In GaAs, arsenic can also react with chlorine to produce arsenic trichloride ($AsCl_3$). The boiling point of $AsCl_3$ is 130.2 °C at 1 atm. From these discussions, we conclude that GaAs and Al (including Al_2O_3) can be etched by chlorine-containing gases, but not fluorine-containing gases. This rule fits most III–V materials and metals.

Indium phosphide (InP) is one of the III–V materials that is widely used to make high-speed transistors. After chemical reaction with chlorine, the boiling point of $InCl_3$ is 800 °C. Even under the process pressure, the reactant with such a high-boiling point cannot be pumped out from the chamber. $InCl_3$ remains in the sample surface as a hard shell that will hinder the further progress of the chemical reaction between In and Cl. To figure out this issue, three methods can be used:

1. Heating up the platen to around 300 °C. As mentioned in Section 14.2.6, RIE is usually kept at room temperature. For InP dry etching, we should buy a RIE that its platen can be heated.
2. Adding Ar in the recipe. During dry etching, Ar^+ provides strong sputtering and removes the surface hard shell from the sample. So that the chemical reaction of In and Cl can keep going.
3. Using CH_4 as etching gas. After chemical reaction of In and CH_4, the reactant is $In(CH_3)_3$. Its boiling point is 134 °C, which is much lower than that of $InCl_3$. $In(CH_3)_3$ can be pumped out from the chamber under the process pressure. But with CH_4 it is easy to create polymer and contaminate the chamber.

If we use Cl-containing gases to etch InP, in addition to adding Ar, we can also add H_2 to the recipe. The chemical reaction of hydrogen and phosphorus creates phosphine (PH_3). The boiling point of PH_3 is −87.7 °C. So phosphorus can be removed from the chamber. CH_4 has hydrogen and does not need to add H_2.

15.4 Etch Profile Control

In the etching process, photoresist is the most-used masking film (mask). In wet etching, the selected etchant has a small etch to the PR. The adhesion between the PR and the substrate is a main factor affecting the quality of etching. In the dry etching, the gases selected should have a small etch of the mask. The ratio of the etching rate of substrate (R_s) to the etching rate of mask (R_m) is called selectivity (S):

$$S = \frac{R_s}{R_m} \tag{15.3}$$

It is better for a larger S value. The ideal situation is infinite. That means that the masking film is not etched at all, which does not exist in the actual process. In the fluorine-containing etching process, the etching rate of the commonly used positive photoresist AZ5214 is about two times faster than that of silicon dioxide. The thickness of AZ5214 is 1.3–2.0 μm. The etching depth of SiO_2 is usually 1000–5000 Å (0.1–0.5 μm). So, after the etch, there is still a certain thickness of PR left on the substrate surface.

In most cases, the thickness of PR has little effect on the dry etching process, but the shape of the opening has a great influence on the etch profile. The shape of the etched profile of the substrate has a great influence on the manufacture of many kinds of microchips. Mostly, it is desirable that the cross-section is a steep 90° angle, which is difficult to achieve. The adjustment of the cross-section angle (control of lateral etching) is related to the settings of process pressure and power as described in Section 15.1. It is also related to the following two factors:

15.4.1 Influence of the PR Opening Shape on the Etch Profile

For positive photoresist, an ideal opening should be a steep 90° angle up and down, as shown on the left in Figure 15.9. In this case, if it is completely anisotropic etch, the etching cross-section is also a steep 90° angle. However, the actual shape of the PR is shown on the right (refer to Figures 12.16 and 12.17). Because the PR has an oblique angle, the thin area is etched during the process, making the upper opening of the profile to be larger than the lower opening, and a complete anisotropic etch profile becomes an incomplete anisotropic etch profile. From this point of view, to achieve the ideal substrate cross-section, the first step is to do a good job in photolithography.

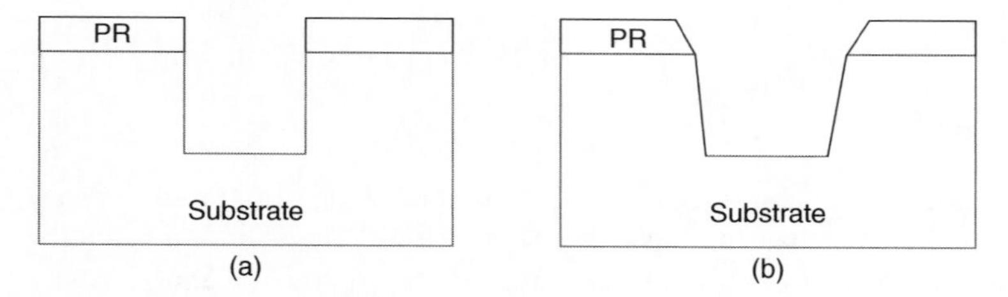

Figure 15.9 The influence of PR profile on the profile of substrate. (a) ideal and (b) actual.

15.4.2 The Effect of Carbon on Etching Rate and Profile

The etching gases Freon 14 (CF_4) and Freon 23 (CHF_3) both contain carbon. CHF_3 also has hydrogen. Another gas used in the process C_4F_8 has a higher carbon content. Carbon plays an important role in RIE. In the etching process, silicon and carbon are competing to produce a chemical reaction with fluorine, and hydrogen also participates in the reaction. The complete reaction of carbon and fluorine forms CF_4, which is a gas, can be drawn out of the process chamber by pump. However, the incomplete reaction of carbon and fluorine forms a Teflon polymer, which stays on the surface of the substrate and the openings to prevent the etching. This process is called passivation. Among the above three gases, C_4F_8

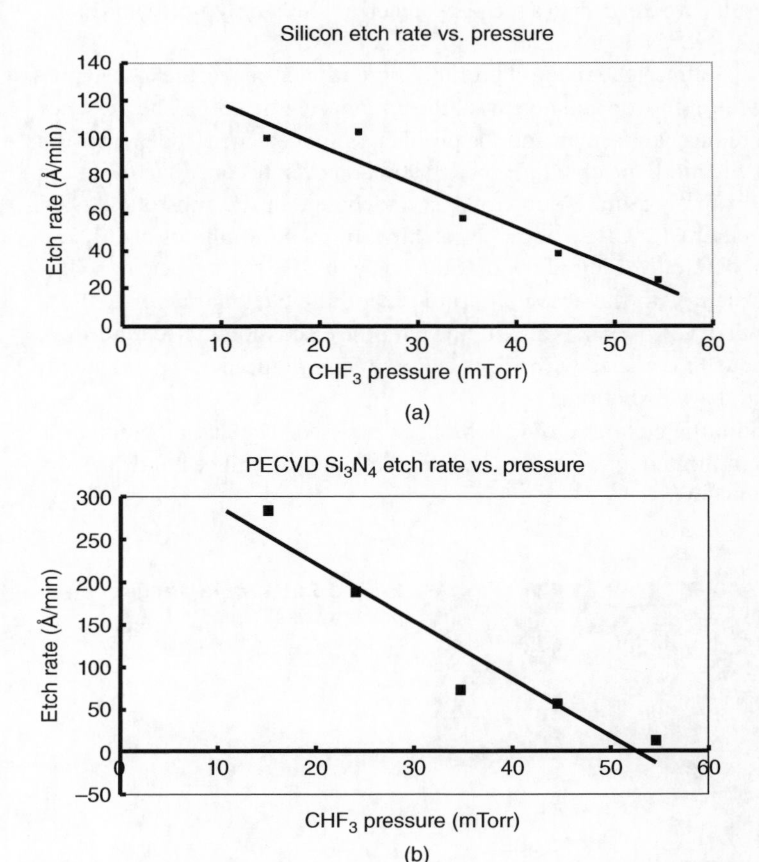

Figure 15.10 The relationship of Si (a) and Si_3N_4 (b) etching rates and pressure using CHF_3.

has the strongest passivation because it has four carbons; CHF_3 is the second because hydrogen can participate in the competition with silicon to react with fluorine, and CF_4 is the weakest. The passivation effect of carbon-free chemicals is less than that of carbon-containing gases, such as SF_6.

Taking CHF_3 as an example, as the pressure increases, the passivation effect increases, and even the etch stops and becomes the deposition of Teflon film. Figure 15.10 shows the relationship curves of Si and Si_3N_4 between etching rate and pressure using CHF_3. The figure shows that as the pressure increases, the etching rates of Si and Si_3N_4 decrease until they reach zero. SiO_2 etching rate has no big change when the pressure is lower than around 55 mTorr because oxygen can react with hydrogen to reduce the competition between hydrogen and silicon. By using these results, we can design a recipe that it etches SiO_2 only, not Si and Si_3N_4 (or etching rate is very slow).

Passivation also affects the shape of profile. As mentioned above, the etched profile of RIE is wide at the top and narrow at the bottom (Figure 15.1). The presence of carbon can reduce lateral etch and the profile becomes steeper. The greater the passivation is, the more the opening tends to 90°, and even become narrow at the top and wide at the bottom. Figure 15.11 is the change of the substrate etched opening angle after CF_4, CHF_3, and C_4F_8 etching. In the SEM photos of CF_4 and CHF_3 etching SiO_2, our results show that the angle of CHF_3 is steeper than CF_4, which is consistent with the above description; and the etching rate of CHF_3 is slower than that of CF_4, which is also in line with the discussion in Section 14.2.5. In another photo, Si is etched with $SF_6 + C_4F_8$, which produced an opening with a narrow top and a wide bottom.

In this section, the etching profile is briefly discussed. The actual situation is much more complicated. If we understand the principle, we will be handy in solving process problems.

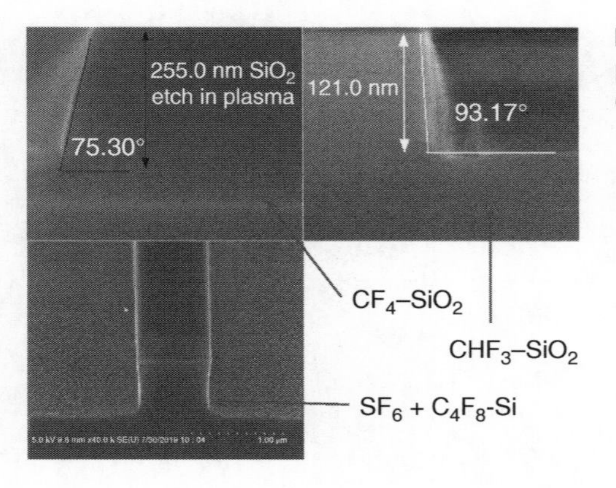

Figure 15.11 Angle change of the profile with different carbon-containing gases.

15.5 Other Issues

We have discussed the etching rate, profile, III–V materials, and metal etch in the previous sections. In RIE, other issues need to be considered. Only by clarifying these can we understand the ICP technique to be discussed in Section 15.6. Although RIE and PECVD equipment are similar in that they both complete the process in a chemical reaction, but they still have some differences. Let us start with these differences to further explore other aspects of RIE.

15.5.1 The Differences Between RIE and PECVD

Difference 1. As mentioned in Section 14.2, from a chemical point of view, there is no essential difference between RIE and PECVD. They are both chemical reactions. However, in RIE, the chemical reaction product is volatile and can be pumped out of the process chamber; in PECVD, the chemical reaction product is nonvolatile and remains on the sample surface. So the difference 1 is that the reactants are volatile and nonvolatile.

Difference 2. In Section 13.1.3, it was mentioned that the temperature of PECVD platen is set to 250–350 °C. The RIE platen is generally kept at room temperature; therefore, photoresist can be used as a mask for etching; so difference 2 is that the platen temperature is different.

Difference 3. In Section 13.1.3, it was mentioned that the process pressure range of PECVD is 500 mTorr–5 Torr. Such a high pressure makes the plasma density high, and the particles produced during the film deposition are dense. In addition, high pressure makes the DC bias low (Figure 15.7) and reduce the bombardment on the surface of deposited film. The pressure of RIE is generally 5–50 mTorr. The reason why the pressure of RIE is low is that during dry etching, the etch gas is expected to have a larger mean free path so that the gas can reach the sample surface with no or less collisions with each other. The collisions can change the direction of linear motion of etch gas, which will have an adverse effect on etching. The pressure is too low, the number of etch gas molecules in the chamber is too small. Too small a number of molecules make the chance of collision with each other small and the plasma difficult to ignite. Now let us talk about the mean free path. The mean free path is the average distance traveled by a moving microscopic particle (such as a molecule, an atom, and a photon). Two particles collide not completely but bounce off when they approach each other at a certain distance. Taking this distance as the boundary, we give the effective diameter d of the particle. When the pressure is p, the expression of the mean free path l of the particle is as follows:

$$l = k_B T / \sqrt{2}\pi d^2 p \tag{15.4}$$

The k_B in the formula is Boltzmann's constant, and T is the absolute temperature. The expression shows that the mean free path is inversely proportional to the pressure. The higher the pressure, the smaller the mean free path; the lower the pressure, the bigger the mean free path.

Figure 15.12 is a schematic diagram of the mean free path of a particle. The distance of the particle travels from the start point to the finish point. When the vacuum is high, the particle can reach the finish point from the start point without collision. When the vacuum is low, the particle has multiple collisions with the other particles from the start point to the finish point. For dry etching, if the sample is placed at the finish point (platen), the etch gases come the chamber from the start point (showerhead), we hope that the gases directly reach the surface of the sample to avoid collisions with each other. So the RIE process pressure should be set low. To understand the issue more, see Figure 15.13 that shows the relationship between the mean free path and pressure of an argon in a sputtering tool (discuss in Chapter 16). For comparison, at room temperature, the mean free path of an air molecule at 1 atm (760 Torr) is 68 nm.

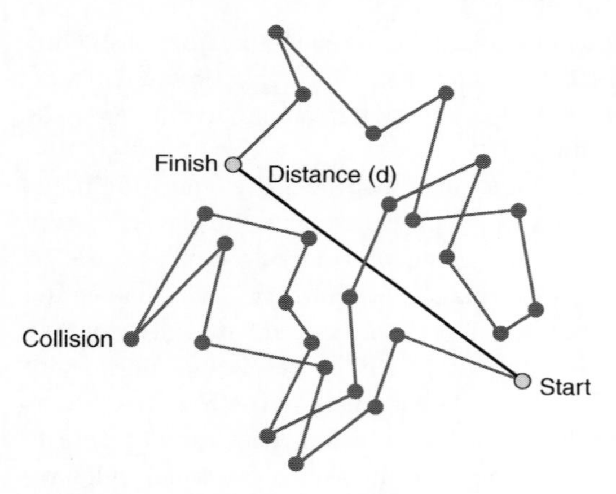

Figure 15.12 Schematic diagram of the mean free path of a particle. Source: Reprinted with permission of school physics.

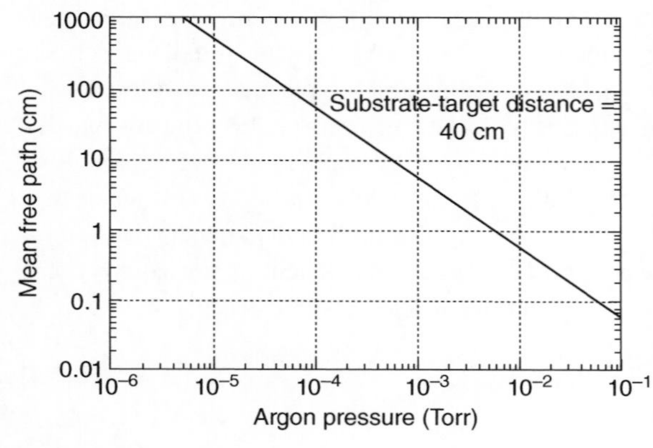

Figure 15.13 The relationship of the mean free path of an Ar in the sputtering tool. Source: [5] Radzimski et al. / AIP Publishing.

Difference 4. The RF power supply of RIE is connected to the platen. The direction of the electric field in the process chamber is from the shower head to the platen, which is the same as moving direction of the etch gas and has a strong physical bombardment to the sample. The RF power supply of PECVD is connected to the shower head. The direction of the electric field in the process chamber is from the platen to the shower head, which is opposite to moving direction of the deposition gas and the physical bombardment of the sample is weak as shown in Figure 15.14. High pressure and connecting power supply to the showerhead make small bombardment on the film during deposition. This is good for the quality of the film.

Difference 5. RIE equipment displays DC bias voltage. PECVD equipment does not display DC bias voltage. As shown in Figure 15.15.

Difference 6. RIE power is generally set to 50–200 W, which has strong physical bombardment. PECVD power is generally set to 20–100 W, which has weak physical bombardment.

Difference 7. For RIE, the chemical reaction happens in the sample surface, and the gas just needs to flow into the chamber; so RIE has less holes in the showerhead (Figure 14.8). For PECVD, the chemical reaction happens in the space between the showerhead and platen. To meet the requirement of creating uniform gas dispersion for deposition, PECVD has more holes in the showerhead (see Figure 15.16).

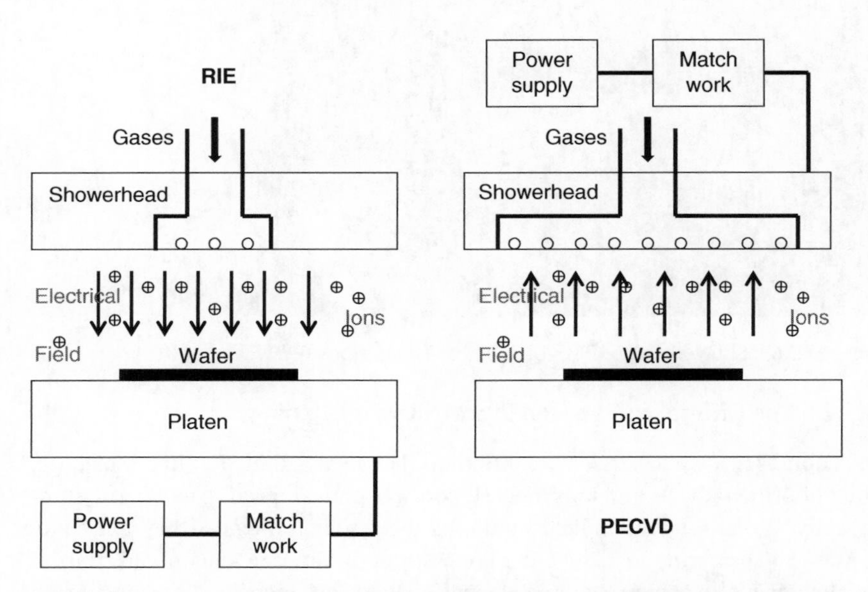

Figure 15.14 The connection position of the RF power supplies in RIE and PECVD, and the direction of the electric field in the process chambers.

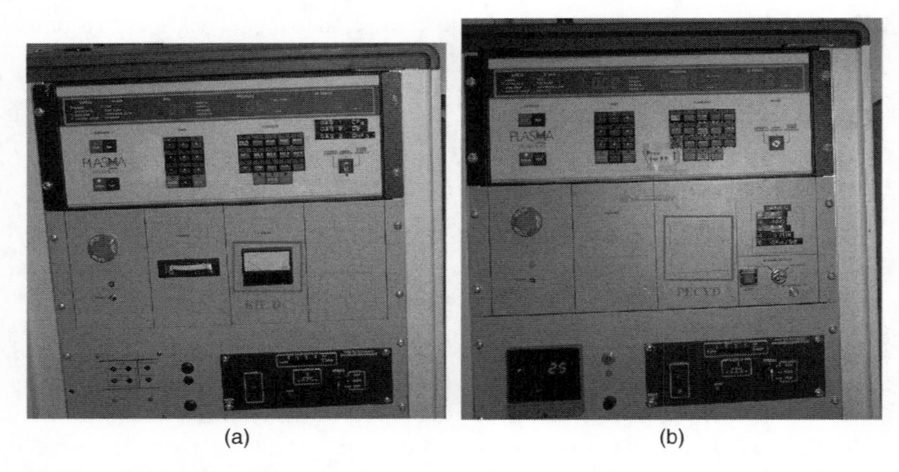

(a) (b)

Figure 15.15 On the left (a) is the RIE console, which displays DC bias. On the right (b) is the PECVD console, which does not display DC bias. Source: PlasmaLab RIE.

Figure 15.16 Showerhead of PECVD. Source: STS Multiplex PECVD.

15.5.2 The Difference Between Si and SiO_2 Dry Etching

In Section 14.2.1, we discussed the difference between Si and SiO_2 in dry etching. When fluorine radicals and ions meet silicon, a chemical reaction can occur automatically. When fluorine radicals and ions meet silicon dioxide, they must rely on physical sputtering to reduce the bond strength between silicon and oxygen, or even break it, to ensure a chemical reaction between silicon and fluorine. From this point of view, for the convenience of considering the process, we can think that the etching rate of silicon depends on the chemical reaction and is linked to the

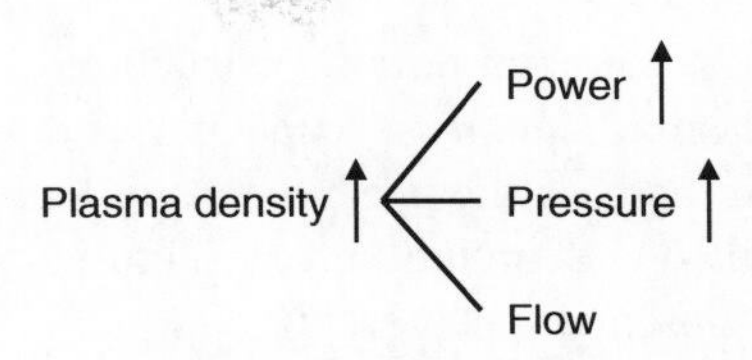

Figure 15.17 The methods to increase the plasma density.

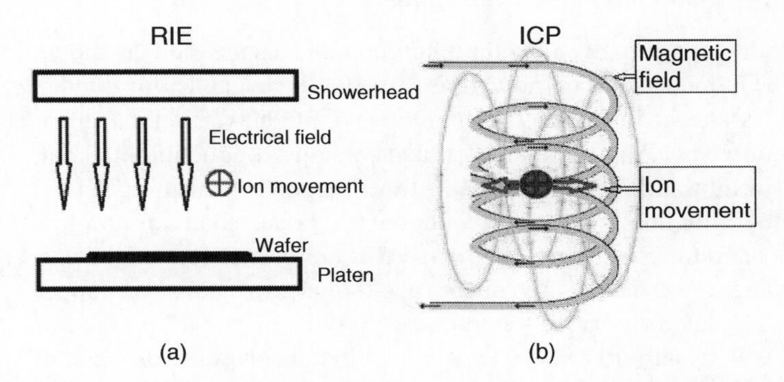

Figure 15.18 Schematic diagram of the force on positive ions in the RIE (a) and ICP (b).

plasma density. The higher the density, the faster the etching rate. This assumption is consistent with Figure 15.6. The etching rate of silicon dioxide depends on the physical sputtering and is related to the power of the plasma. The greater the power, the faster the etching rate. This assumption is consistent with Figure 15.8. When MEMS (see Chapter 13) appeared, the deep etching of silicon became an important issue. From the discussion just now, we can see that in order to increase the etching rate of silicon, it is necessary to increase the plasma density. From the perspective of RIE process, the plasma density can be increased from two aspects: increasing the pressure and increasing the power, as shown in Figure 15.17. We have just discussed that RIE is a low-pressure process, and increasing the pressure is not conducive to dry thing. In addition, because RIE is a capacitive structure, increasing power also increases DC bias voltage, which makes the process chamber to face the danger of breakdown. So the power of RIE cannot be increased much, as described in Section 15.2, generally between 300 and 500 W.

15.6 Inductively Coupled Plasma (ICP) Technique and Bosch Process

As mentioned earlier, an RIE system cannot increase the plasma density by increasing the pressure and power. Since RIE does not make it, a new system must be developed to increase the plasma density and increase the etching rate of silicon. RIE is CCP. As mentioned in Section 11.3, there are only two structures that can be used in a process chamber for plasma generation: capacitor and inductor. The RIE of capacitive structure cannot increase the plasma density, so the question is can the process chamber be designed with an inductive structure? The answer is yes. We are discussing this in the subsequent paragraphs.

15.6.1 Inductively Coupled Plasma Technique

Let us discuss inductor again. An inductor is composed of a spiral coil, as shown in Figure 4.5. The inductance of an inductor is equal to the magnetic flux divided by the current, as shown in Eq. (4.2). There are no DC bias and breakdown concerns. It is conceivable that if this coil is made of superconducting material, then it can flow infinite current and generate infinite magnetic field energy in the coil. But the actual situation is that superconductivity is realized at an ultra-low temperature, its operating cost is very high, and it is not cost-effective for semiconductor equipment. We mostly use copper tube to make this coil. The copper tube is cooled with water. After this treatment, 3000 W or higher RF power can be added to the coil, usually 1000–3000 W; so the inductor solves the problem of high power.

Let us look at the effect of the coil on ions. Figure 15.18a shows the distribution of the RIE electric field. The direction of the electric field is consistent with the moving direction of positive ions such as fluoride ions. This makes it easy to push the ions to hit the sample (physical bombardment) and the platen. Under normal circumstances, the RIE process chamber is made of stainless steel or aluminum. The area of the platen is larger than that of the sample. Therefore, some ions are bombarded onto the platen surface. Once the ions are in contact with sample and metal, they are absorbed. From this point of view, the RIE structure suppresses the increase in plasma density. However, once an ion enters the magnetic field, according to Lorentz Eq. (11.6), the direction of the force acting on the ion by the magnetic field is at an angle of 90° with the direction of the magnetic field. Due to the use of RF power supply, the ion oscillates inside the coil under the action of the magnetic field, as shown in Figure 15.18b. In this case, the plasma density is easily increased.

Based on the abovementioned two reasons: high power and ion oscillation, the process chamber made of coil can increase the ion density to a high level. This process chamber is called inductively coupled plasma (ICP). The actual structure is a vacuum chamber made of ceramics, and a coil made of copper tube around the chamber. Figure 15.18 is a schematic diagram of ion movement and force in RIE and ICP. Replacing showerhead with ICP in an RIE produces an ICP system. Since this system is a combination of ICP and RIE, it is also called ICP RIE. Figure 15.19

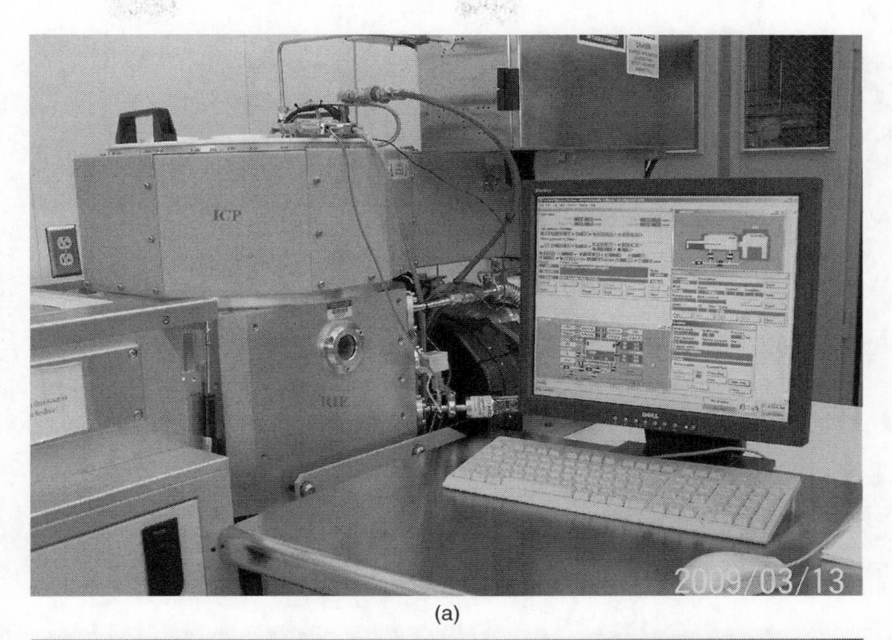

(a)

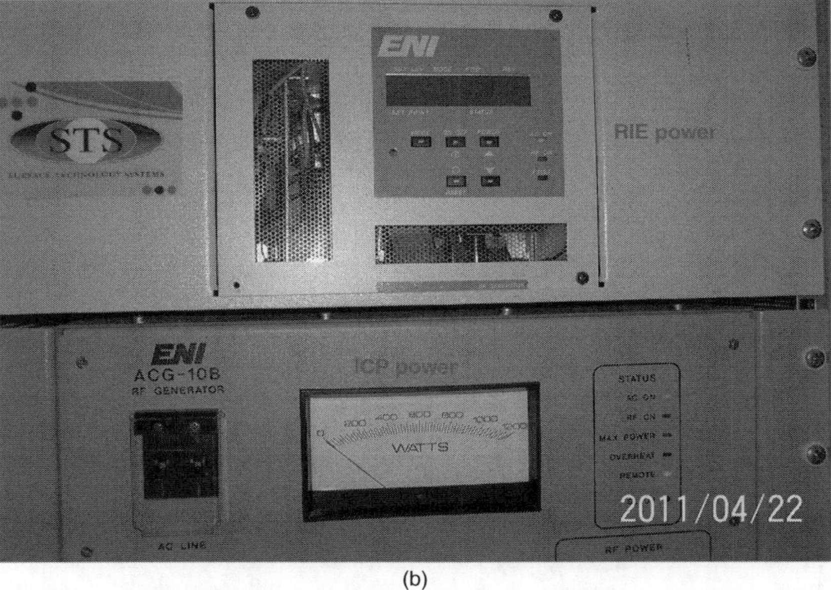

(b)

Figure 15.19 Photos of ICP RIE (STS ASE). The top picture (a) points out the relative position of ICP and RIE. The bottom picture (b) is 300 W small power supply for RIE and 1000 W big power supply for ICP.

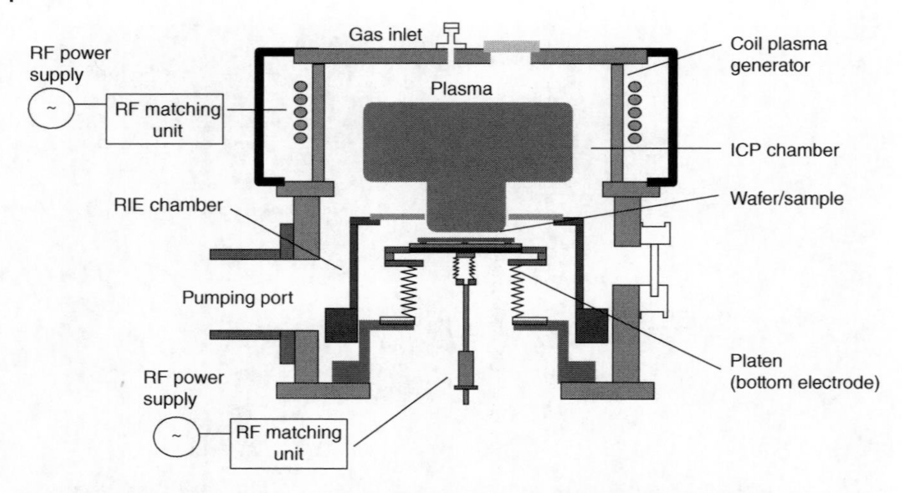

Figure 15.20 Schematic diagram of the internal structure of the ICP RIE process chamber (STS ASE). The frequency for both power supplies is 13.56 MHz. Some ICP machines use 2 MHz supple for ICP chamber.

is a photo of an ICP RIE machine. Figure 15.20 is a schematic diagram of the internal structure of the ICP RIE process chamber. We can compare it with Figure 14.6 to see the difference in structures of RIE and ICP. Compared with RIE, ICP has the following three features:

Feature 1. ICP can provide high-density plasma (HDP) under low pressure. RIE cannot produce HDP. HDP is the fundamental difference between ICP and RIE. It should be emphasized here that in semiconductor manufacturing, there are other types of equipment that can produce HDP. HDP can be used for dry etching or other processes.

Feature 2. For silicon, germanium, and other single element materials that react with fluorine to produce volatile reactants, ICP can significantly increase the etching rate. For compound materials, such as SiO_2, although the etching rate can also be increased, the effect is not as obvious as that of single-element materials such as Si. For example, in an RIE, the etching rate of silicon using SF_6 is usually less than 1 μm/min, but in an ICP, the etching rate can easily exceed 3 μm/min. In general, the etching rate that depends on the chemical reaction increases significantly, and the etching rate that depends on the physical sputtering increases nonsignificantly. So when an ICP RIE is used to etch single-element materials like Si, the system is called Deep RIE (DRIE).

Feature 3. The power can be set separately for ICP and RIE to increase the selectivity S (Eq. (15.3)). Here we take Si and SiO_2 as examples to illustrate this issue.

Figure 15.21 Comb actuator driven microstage.

If the power is only set in the ICP and the RIE power is zero, the DC bias is zero, and there is no physical bombardment. The machine can only etch Si but not SiO_2 and the etching profile of Si is almost isotropic. In real process, due to the thermal movement of particles, the effect of magnetic field on ions, etc., there is a certain amount of physical bombardment. So there is certain etch to SiO_2, but the rate is slow. From this view of point, we can see that increasing the ICP power can increase the etching rate of Si; reducing the RIE power can reduce SiO_2 etching rate. With this adjustment, we can increase the S value of Si and SiO_2 in the fluorine-containing gas. Using SF_6 in an RIE, the S value of Si and SiO_2 is 20–30, but in an ICP, the S value can exceed 100. Due to the high selectivity of ICP, we can use SiO_2 as a masking film to deeply etch Si. Figure 15.21 is a comb actuator driven microstage made by ICP. The device was made by Professor P. Ferreira's research group at the University of Illinois using STS advanced silicon etch (ASE). The device's suspension structure made by using high S value and Bosch process (please see below).

15.6.2 Bosch Process

For the deep etching of silicon, most devices require steep 90° profile, as shown in Figure 15.21. Since RIE contains two parts, chemical reaction and physical sputtering, the etch profile is shown in Figure 15.1, which is wide at the top and narrow at the bottom. Although the etch angle can be adjusted by selecting the carbon content in the etching gases, as shown in Figure 15.11, but due to the limitations

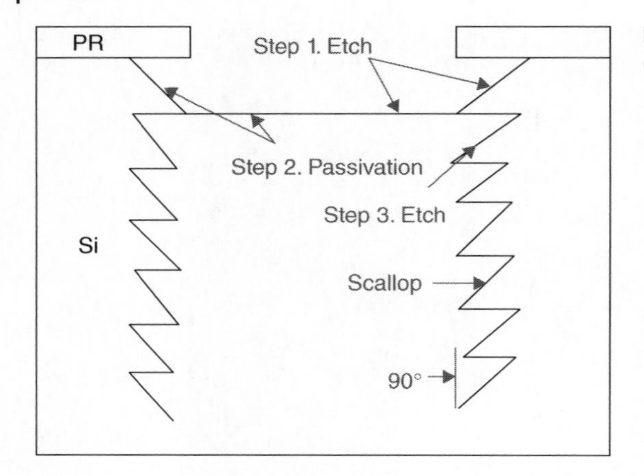

Figure 15.22 Schematic diagram of Bosch process.

of the etching rate and selectivity, this technique has not been widely used in the deep etching of silicon. Bosch process is a process developed in response to this issue. This technique is named after Bosch because it was first developed by a German company, Robert Bosch GmbH. In Bosch process, etching and passivation are performed alternately to achieve a controlled angle (90°) and ensure a certain etching rate.

The following is an example of Bosch process etch recipe (STS ASE):

$$\text{Etching: } SF_6 + O_2, 130\,\text{sccm} + 13\,\text{sccm, ICP } 600\,\text{W, platen } 12\,\text{W}, 12\,\text{s}.$$

$$\text{Passivation: } C_4F_8, 85\,\text{sccm, ICP } 600\,\text{W, platen } 0\,\text{W}, 8\,\text{s}.$$

The entire etch process repeats this cycle – etching and passivation. Figure 15.22 describes the cycle in detail.

Step 1. Etching: $SF_6 + O_2$ is used for etching. SF_6 is the fastest etching gas. Adding O_2 can increase the ability to remove Teflon film. In this step, the power of ICP is 600 W, the power of RIE (Platen) is 12 W, and the time is 12″. RIE has power, DC bias, directionality, physical bombardment, and the etched profile is wide at the top and narrow at the bottom.

Step 2. Passivation: Passivation uses C_4F_8 gas, the freon gas with the most carbon. In this step, the power of ICP is 600 W, the power of RIE (Platen) is 0 W, and the time is 8″. RIE has no power, no DC bias, no directionality, and no physical bombardment. The purpose of this step is to complete the total chemical reaction to produce Teflon film, and evenly cover the etched opening.

After that, the first step of etching is repeated. Due to the low-pressure process and the electric field corresponding to the DC bias, the ions basically reach the bottom of the etched opening in a straight line, as shown in Figure 14.5. The bottom part bears a large momentum, while the sidewall bears a small momentum (see Figure 8.10). The Teflon film at the bottom can be removed by physical bombardment, but the sidewall cannot be removed. So repeat the first step of etching, and the same etch profile can be obtained. Then passivation and etching… then stop until reaching the depth of etch. The obtained profile angle can reach 90°, but the sidewalls are not smooth, resulting in a jagged surface. When the size becomes smaller, those sharp angles will become rounded, making the sidewalls look like scallops, so the word "scallop" is used to describe this kind of sidewall. Figure 15.23 shows some structures made by ICP RIE (STS ASE) using Bosch process in the clean room at the University of Illinois. The photo on the top shows that using the same recipe, the smaller the structure size, the slower the etching rate. This is not difficult to understand. Because of small size, it is more difficult for etching gas to enter the structure, and it is also more difficult for the reaction product to come out from the structure. The wall-like structure in the middle clearly shows the scallop-like sidewalls. The pillar structure on the bottom, through the adjustment of the recipe, makes the scallops on the sidewall less undulating.

In Bosch process, the entire etching process is performed by the competition between silicon and carbon. These two materials compete to react with fluorine. If the reaction of silicon is dominant, the etching profile tends to be RIE structure and becomes wide on the top and narrow on the bottom. If the carbon reaction is dominant, more Teflon film is produced, causing a problem called micro-grass, which is shown in Figure 15.24a. When the problem of micrograss occurs, it is necessary to reduce the passivation step and increase the etching step.

The method is in the etching step; increase the RIE power and the etching time, which can be increased together or individually. In the passivation step, decrease the ICP power and passivation time, which can be decreased together or individually. The adjustment of these two steps can be carried out together, or one step can be selected separately. Using same etch recipe, the small opening is etched normally, the opening surface is shiny. But the large opening is etched abnormally, and the opening surface is black. This is a situation that often encounters the micrograss problem in Bosch etch process. If we have this problem, we should solve it according to the method just mentioned. The adjustment should be made according to the specific situation in the process. Figure 15.24b is the adjusted etched profile. The photo shows that the angle of the profile is basically 90°.

In most cases, the devices prefer smooth etched sidewall rather than undulating. The sidewalls showed in Figure 15.23 cannot meet the needs of most device manufacturing. To solve the issue of scallops, we can do thermal oxidation after

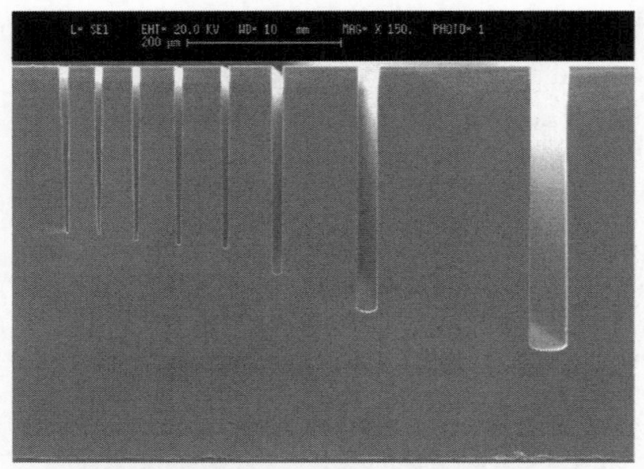

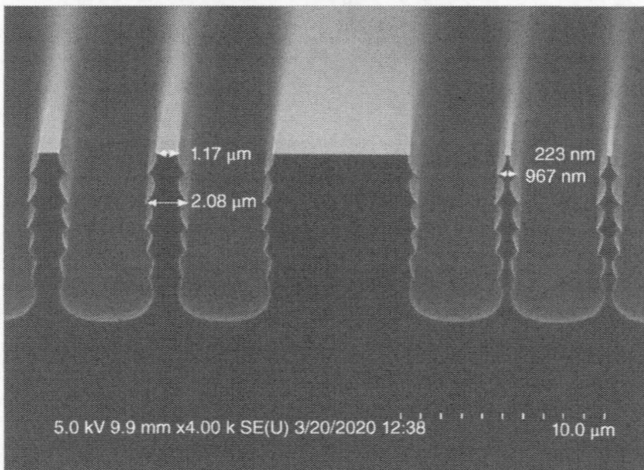

Figure 15.23 Silicon trench, wall, and pillar structures made by Bosch process on ICP RIE.

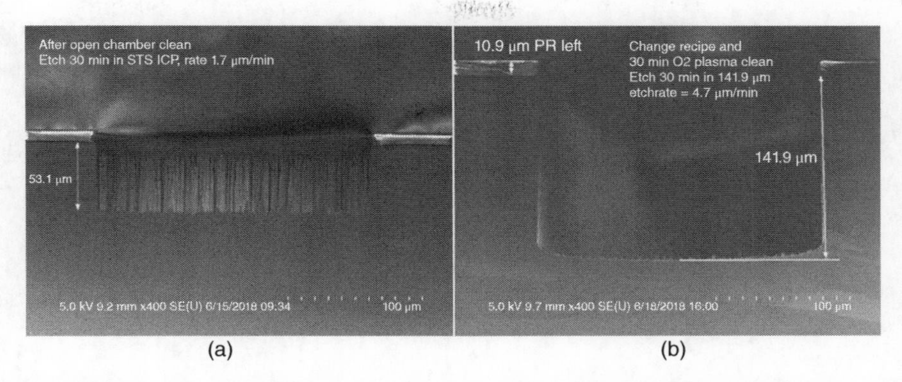

Figure 15.24 Micrograss problem (a) and the structure after the problem is solved (b).

Bosch process. During the oxidation, the reaction rate between the spikes of silicon and oxygen is faster than that of the smooth part. So, after thermal oxidation, the etched sidewall of silicon will become smooth. Meanwhile, the Teflon film on the sidewall will be removed cleanly.

References

1 傅献彩 陈瑞华编. (1979). 高等学校试用教材, 物理化学, 下册, 南京大学物理化学教研室, 高等教育出版社, 1979 修订本, 第 206 页, 第 204 页。.

2 Murarka, S.P., Eizenberg, M., and Sinha, A.K. (2003). *Interlayer Dielectrics for Semiconductor Technologies*, 32. Elsevier Academic Press.

3 Stanley, I.S. (2016). *Chemical, Biochemical, and Engineering Thermodynamics*, 5e, 336. Wiley.

4 NIST Chemistry WebBook, SRD 69.

5 Radzimski, Z.J., Posadowski, W.M., Rossnagel, S.M., and Shingubara, S. (1998). Directional copper deposition using dc magnetron self-sputtering. *Journal of Vacuum Science & Technology B: Microelectronics and Nanometer Structures Processing, Measurement, and Phenomena* 16 (3): 1102–1106.

16

Metal Processes

Metals play important roles in semiconductor microchips. Semiconductors are not conductors, and their resistivity (Chapter 1) is greater than that of metal conductors. Metals mainly have roles in three aspects:

1. Metals are used to make resistors, semiconductor diodes, and transistors (refer to Chapter 6).
2. Metals are used to make ohmic contact.
3. Metals are used to interconnect the devices in an integrated circuit.

In semiconductor processes, three main techniques are used to make metals onto the semiconductor surface:

1. Thermal evaporation,
2. Electron beam evaporation,
3. Sputter deposition.

They are PVD. Metals do not completely cover the whole surface of a device, but some specific areas patterned by photolithography. These patterns are mainly realized by two methods. One is etching, and the other is lift-off. To achieve good contact between a metal and the semiconductor surface, metal annealing (sintering step) is also required. Figure 16.1 is a schematic diagram of the processes of metal etching and lift-off. This is discussed in detail in the subsequent sections.

16.1 Thermal Evaporation Technique

The equipment used for the thermal evaporation process is called thermal evaporator. Thermal evaporation is carried out in the evaporation chamber, using a diffusion pump or a cryopump in combination with a mechanical pump to pump the chamber to 1×10^{-6} Torr. Put the metal to be evaporated (evaporation source or evaporant) on the resistive heat element (heater). The element can be

Semiconductor Microchips and Fabrication: A Practical Guide to Theory and Manufacturing, First Edition. Yaguang Lian.

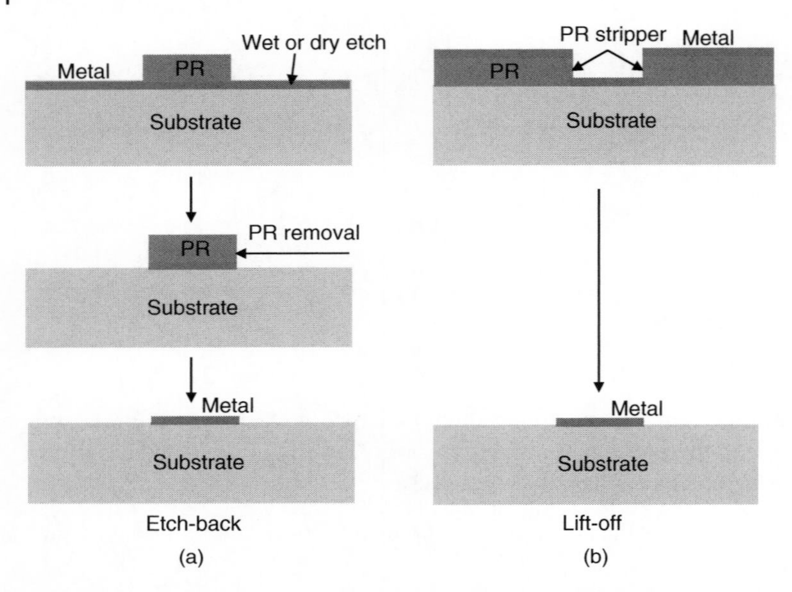

Figure 16.1 The processes of metal etch-back (a) and lift-off (b).

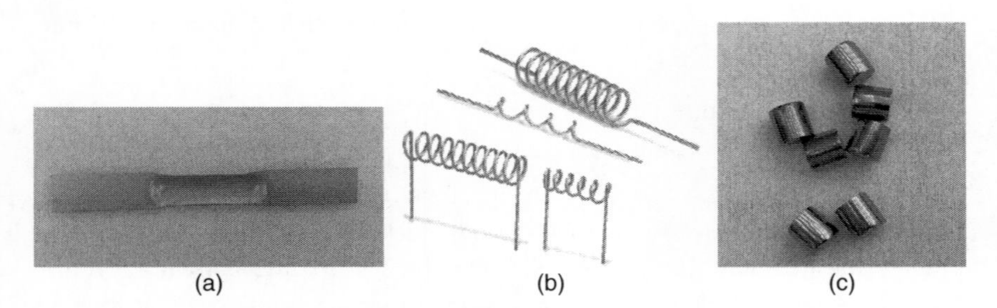

Figure 16.2 Resistive heat element and metal pellets. (a) is boat type heater, (b) is filament type heater, and (c) is titanium pellets. Source: Kurt J. Lesker.

made into boat type or filament type. According to the shape of the filament, the evaporant can be made into different shapes, such as pellet, cane, wire, clip as shown in Figure 16.2. The process chamber has two types: bell-jar type and door-opening type. Figure 16.3 is a bell-jar type thermal evaporator. The material of the heat element is made of tungsten, which has the highest melting point 3400 °C. Tantalum with melting point of 3000 °C and molybdenum with melting point of 2600 °C [1] are also used. Such metals are called high temperature metals or refractory metals. The evaporation source metal changes from a solid state to a gas state directly by the high-temperature heating of the element. Under

(a)

(b)

Figure 16.3 (a) A bell-jar type thermal evaporator which uses diffusion and mechanical pumps, (b) The heater used in the machine is boat type. Source: Cooke Vacuum Products.

1×10^{-6} Torr vacuum, the mean free path of metal gaseous molecules (refer to Figure 15.13) is large enough to deposit on the sample surface without collision.

The main advantages of thermal evaporation technique are (i) simple; (ii) cheap; and (iii) no radiation. The main disadvantages are (i) possible pollution caused by the heater at high temperature; (ii) due to the limited quantity of evaporation source that the heater can be placed, this technique cannot be used for thick film evaporation; (iii) it cannot be used for refractory metal evaporation. These problems can be partially solved in electron beam evaporation.

16.2 Electron Beam Evaporation Technique

The machine used for electron beam evaporation (e-beam evaporation) is called e-beam evaporator. Its structure is basically the same as the thermal evaporator. The process chamber is either bell-jar type or door-opening type. Figure 16.4 is a door-opening e-beam evaporator. The main difference between e-beam and thermal evaporations is that the heater is different. Electron beam evaporation uses an electron gun as the heating source, and the crucible that carries the evaporation source is made into a bowl. Figure 16.5 is a photo of crucibles. This type of crucible can carry more sources and achieve thick film evaporation.

The heater for e-beam evaporation is an electron gun, as shown in Figure 16.6. The electrons emitted from the filament are accelerated by a voltage of about

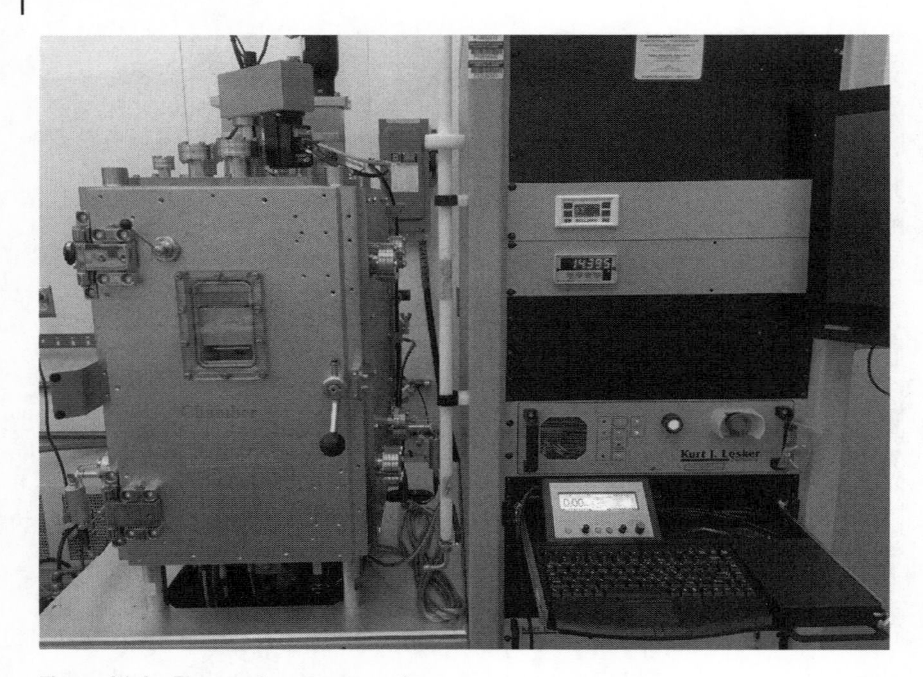

Figure 16.4 Electron beam evaporation.

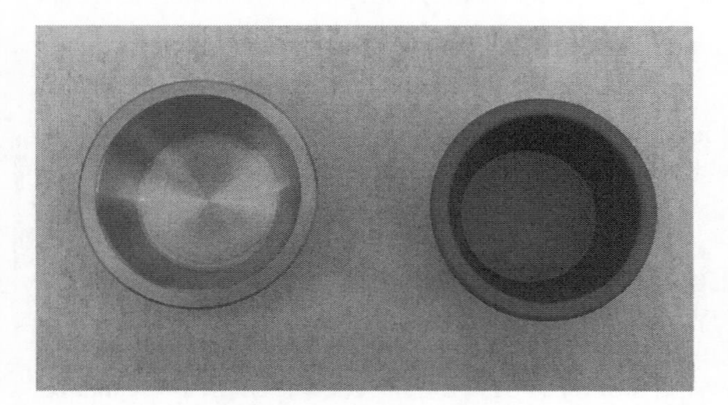

Figure 16.5 Crucibles for e-beam evaporation.

10 kV. They are directed by a magnetic field to the evaporation source placed in the crucible. The huge kinetic energy generated by the acceleration is converted into heat energy and produces a higher temperature than thermal evaporation at the evaporation source. So this technique can evaporate metals including some refractory metals (such as molybdenum) and dielectrics. In addition, because the crucible is placed in a water-cooled seat, its temperature is low, and pollution is

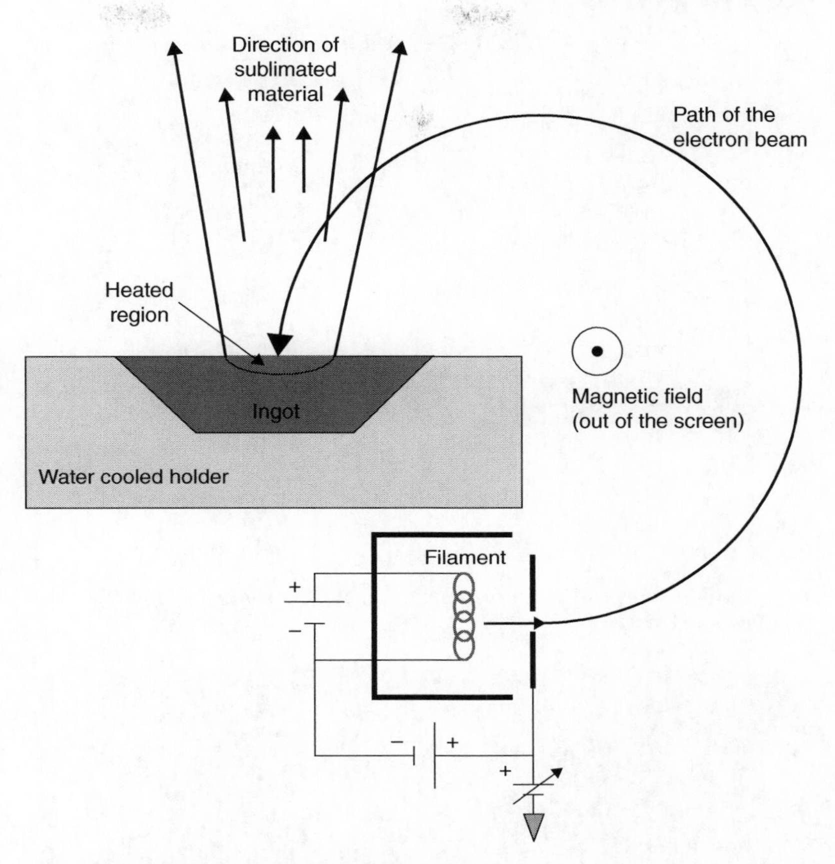

Figure 16.6 The basic structure of electron gun.

small. Its biggest disadvantage is that high-energy electrons can generate X-rays, which can have a bad influence on some devices, especially optoelectronic devices.

Since thermal evaporation and electron beam evaporation have similarities, both are done by heating the evaporation source to make the metal sublimate and deposit on the surface of the sample. So, an electron beam evaporation can also have the function of thermal evaporation. Figure 16.7 is a schematic diagram of the thermal evaporation and electron beam evaporation systems. An e-beam evaporator is shown in Figure 16.4. When the door is opened, the structure inside the process chamber is as shown in the left picture of Figure 16.8. From which we can see the electron gun and the substrate holder. The picture on the right is the e-beam spot generated when the electron beam hits the evaporation source.

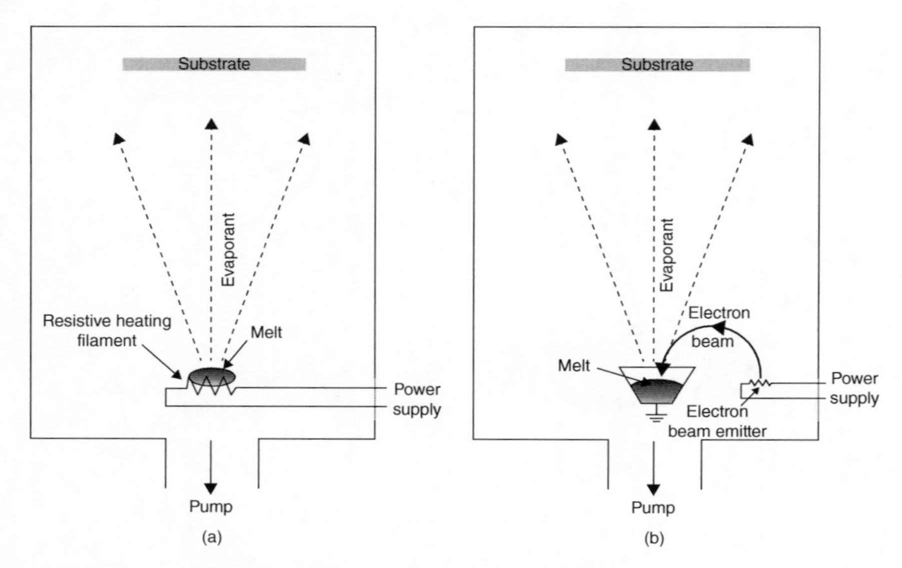

Figure 16.7 The structures of thermal evaporation (a) and E-beam evaporation (b). Source: [2] / Taylor & Francis.

Figure 16.8 The internal structure of e-beam evaporation chamber (a) and e-beam spot (b).

16.3 Magnetron Sputtering Deposition Technique

Although electron beam evaporator can evaporate some refractory metals, it is difficult to evaporate tungsten. Sputtering technique can solve this issue, which was introduced in 15.1. The sputtering process uses Ar^+ to physically bombard the surface of the material to be deposited, the tiny particles are sputtered out from the surface, and these particles are deposited on the surface of the sample to achieve the purpose of the deposition. Metals including tungsten and dielectrics can be deposited by sputtering technique. The material to be deposited is usually made into a circular disc which is called a sputtering target (target). The sputtering process uses DC power or RF power to create the plasma, and the basic structure of the process chamber is CCP. DC power source is used to deposit metals but cannot be used to deposit dielectrics. RF power source can be used to deposit metals and dielectrics. Therefore, in the process, the RF power is more commonly used and is often used in combination with magnetron, which is called "magnetron sputtering" technique. The sputtering equipment that uses this technique is called a magnetron sputtering machine. Here, we are going to discuss this technique.

Figure 16.9 is a control panel of a magnetron-sputtering tool. From here, we can see an RF power supply and a matchwork controller. The process chamber of the tool also adopts the bell-jar type or the door-opening type (see Figure 16.10), the left side of the figure is the bell-jar type, and the structure after the bell-jar is removed; the right picture is the door-opening type, the internal structure behind the door. They all have a wafer holder for placing samples or wafers. There are targets, which can be metal materials or dielectric materials. Below a target is a

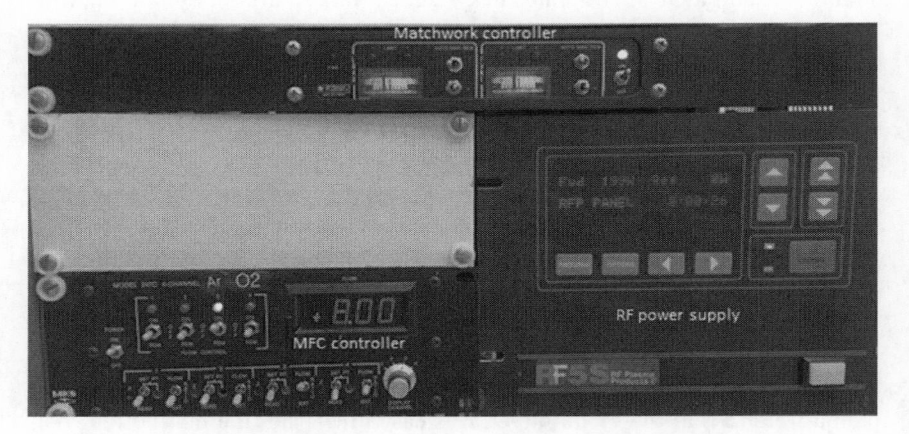

Figure 16.9 The control panel of a magnetron sputtering tool, the photo shows an RF power supply and a match work controller.

Figure 16.10 The internal structures of the magnetron sputtering machine. (a) bell-jar type, (b) door-opening type (Kurt J. Lesker) with shutter covering the gun (two guns). Source: (a) Cooke Vacuum Products.

magnetron, which is used as a cathode. The plasma can be concentrated on the target surface under the action of magnetic field, making a high-density plasma (HDP). Argon is introduced into the chamber, and the ionized Ar^+ is attracted by the cathode, and then is accelerated to sputter the target surface. Combining with HDP, the target can be sputtered out in the form of tiny-particles and deposited on the sample surface. The magnetron adopts a planar structure, ring-shaped, and rectangular. In modern equipment, a ring-shaped planar magnetron is basically used, as shown in Figure 16.11. Due to the action of magnetic field, the intensity of the plasma projected on the target surface is distributed in a circular ring or rectangular ring. This whole set is called target gun, or gun for short. Figure 16.12 shows a working magnetron-sputtering machine with the plasma discharge on top of its gun. Because of the annular distribution of plasma on the target surface, after using for a while, an annular groove will appear on the target surface, as shown in Figure 16.13.

During the sputtering deposition, the plasma is concentrated on the target surface, which generates a lot of heat; so the target gun must be connected to the cooling water. From the metal target, it is easy to transfer the heat to the cooling water. From the dielectric target, it is not easy to transfer the heat to the cooling water. Therefore, the maximum sputtering power for dielectric deposition

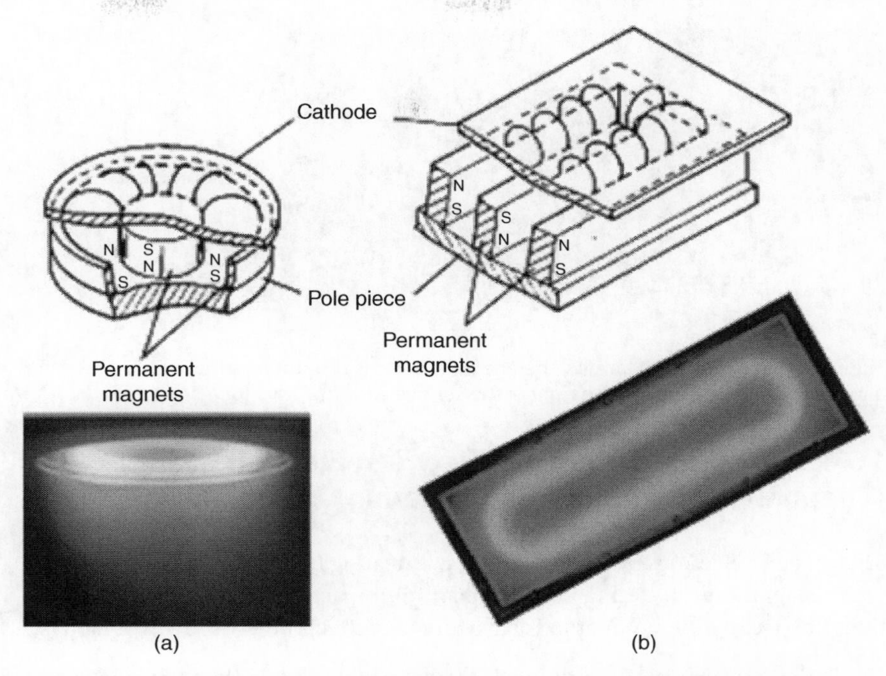

Figure 16.11 The magnetrons with ring-shaped (a) and rectangular (b). Source: [3] Greene et al. 2017 / with permission of AIP Publishing LLC.

Figure 16.12 When the tool is in use, the plasma glow on the top of the target gun.

is much smaller than that for metal deposition. The annular groove only appears on the metal target. The dielectric target does not have obvious groove.

During metal deposition, oxygen can be introduced to the chamber. The ionized oxygen reacts with metal tiny particles to form metal oxides, such as Al_2O_3 and TiO_2. The oxides deposit on the sample surface to create metal oxide films.

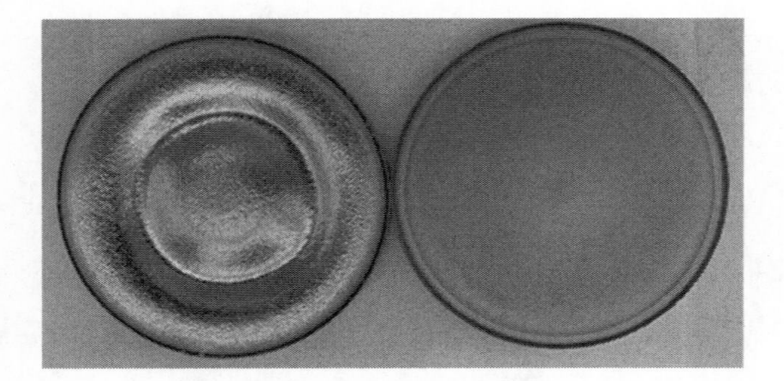

Figure 16.13 Two metal targets, the one on the left is the surface after multiple uses for a long time, and the one on the right is the surface with few uses for a short time.

16.4 The Main Differences Between Electron Beam (Thermal) Evaporation and Sputtering Deposition

Thermal and electron beam evaporation use heating method to evaporate metals, while magnetron sputtering uses momentum impact to complete metal deposition. These two different methods have different effects on the growth of metals:

(1) The step coverage of sputtering deposited metal is better than that of electron beam and thermal, as shown in Figure 16.14. On a substrate with photolithographic patterns, when the metal film is evaporated by thermal and electron beam methods, both methods use heating to sublime the metal from a solid state to a gaseous state and evaporate to the surface of the sample. The kinetic energy of the metal tiny particles is small, and once they touch the surface of the substrate at the bottom of PR opening, the particles will stop. So the coverage of the photo resist step is not good.
The tiny metal particles are produced by sputtering, they are bombarded out by Ar^{+}, and have a lot of momentum. When these particles touch the surface

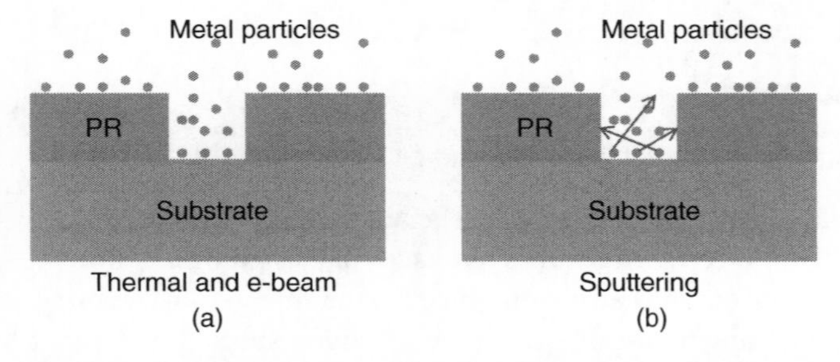

Figure 16.14 Thermal and e-beam evaporation (a), sputtering deposition (b).

Figure 16.15 Schematic diagram of metal connection.

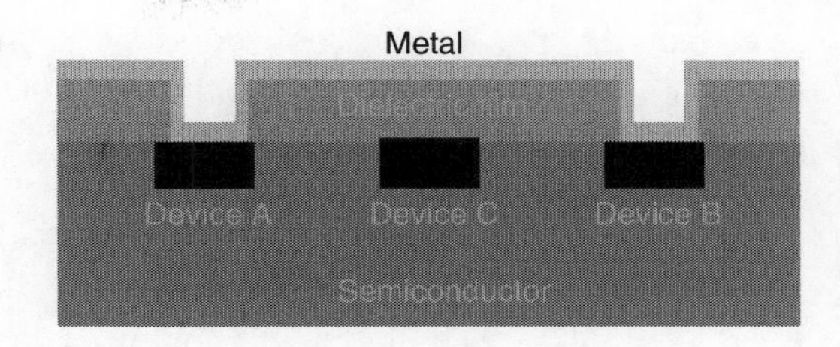

of the substrate at the bottom of PR opening, they will bounce on the surface of the substrate. After the film is created on the surface, the metal particles with high momentum will also re-sputter out the deposited film at the bottom of the PR opening. As a result of the bounce and re-sputtering, the metal particles can better cover the sidewalls of the PR step. The coverage of the metal film is better than that of thermal and electron beam. In the via (Section 11.1) filling process, there is a similar situation.

(2) Because the sputtered metal particles have greater momentum, there is more damage to the surface of the substrate and has an adverse effect on some devices.
(3) In the metal interconnection process (refer to Figure 11.4), to improve the coverage of the dielectric steps of the vias, the sputtering process should be used to complete the metal film deposition, as shown in Figure 16.15.

E-beam and sputtering can also be used for the deposition of dielectric films. The differences between the films done by e-beam and sputtering are the same as metal deposition. In the paper of reference [4], SiO_2 was deposited on GaN as gate dielectric by sputtering. Due to the sputtering-induced surface damage, the electron concentration and mobility of 2-D electron gas were reduced. This phenomenon is consistent with the second difference mentioned above. The author used postannealing treatment to remove the damage.

16.5 Metal Lift-off Process

After the evaporation or deposition of metal is finished, the metal covers the entire surface of the sample. The patterns of metal on the sample surface must be completed by an etch-back or lift-off process (Figure 16.1). As mentioned in Chapter 15, in dry etching, metals are usually etched by chlorine-containing gases. Some metals are not problematic, such as aluminum. The boiling point of the product $AlCl_3$ after reacting with chlorine is 180 °C (refer to Section 15.3). Some metals are more difficult to etch, such as nickel. After reacting with the three commonly used etch gases, oxygen, fluorine, and chlorine, the boiling points of the reactants are all high (over 1000 °C). Metals can also be etched by wet method.

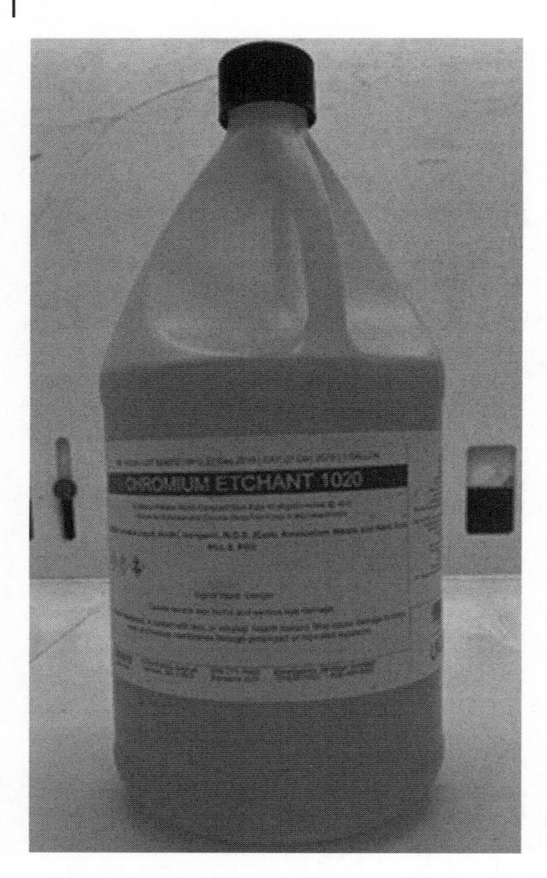

Figure 16.16 Chromium etchant.

Figures 16.16 shows chromium etchant. However, due to the isotropy of wet etching, it is not conducive to the preparation of small-size patterns. Therefore, the lift-off process is the most used technology in semiconductor manufacturing. As mentioned above, in the lift-off process, thermal, or electron beam evaporation should be selected as much as possible to make the metal patterns.

The lift-off process is first to finish the photoresist patterns by photolithography, and after the metal evaporates on the finished patterns, remove the photoresist with a PR stripper or remover. The metal remains in the windows which are not covered by the PR. The metal in other areas, together with the PR, is peeled off from the sample surface by the PR stripper. The lift-off process has three characteristics:

Feature 1. The size of the metal pattern depends on the size of the lithography pattern, so this technology can produce very small metal patterns. By using electron beam lithography, nanosized metal structures can be made.

Feature 2. Because the photoresist is removed to achieve metal lift-off, this technology can be applied to any kind of metal.

Feature 3. As it does not involve metal etching, this technology can easily achieve the evaporation and lift-off of multilayer metals.

From the above discussion, the key to lift-off is that the PR remover can contact the photoresist. If the remover cannot contact the PR, the lift-off cannot be achieved. Photoresist is divided into two types, positive PR and negative PR. The pattern after development is shown in Figures 12.16 and 12.17. From these two figures, we can see that the positive PR development profile is not conducive to lift-off. During the evaporation process, the metal film can easily cover the entire surface of positive PR. The development profile of the negative photoresist is good for lift-off, because the metal film is easy to produce fractures, as shown in Figure 16.17. Mostly, positive PR is used for the lithography, and the lift-off can be realized by using acetone, which is cheap and easy to operate. To realize the positive photoresist lift-off, it is necessary to make the following modifications to its cross-section to make it conducive to the metal lift-off. The purpose of the modifications is to change the cross-section of the PR from the top wide and the bottom narrow to the top narrow and the bottom wide after development. There are three methods to modify the cross-section of positive photoresist after development:

Method 1. PR surface hardening technique

Corresponding to the positive PR AZ 1350, after exposure and before development, immerse the sample in chlorobenzene (C_6H_5Cl) for 1–3 minutes to harden the surface of the resist. After taking the sample out of the chlorobenzene, it is blown dry with a nitrogen gun, and then developed. The cross section obtained is shown in Figure 16.18a [5]. The article I quoted here was published 30 years ago, but it still has reference value.

Method 2. Image reversal technique

With the positive PR AZ 5214E, the positive and negative tones of the PR can be reversed, and the same cross-section as the negative resist can be obtained.

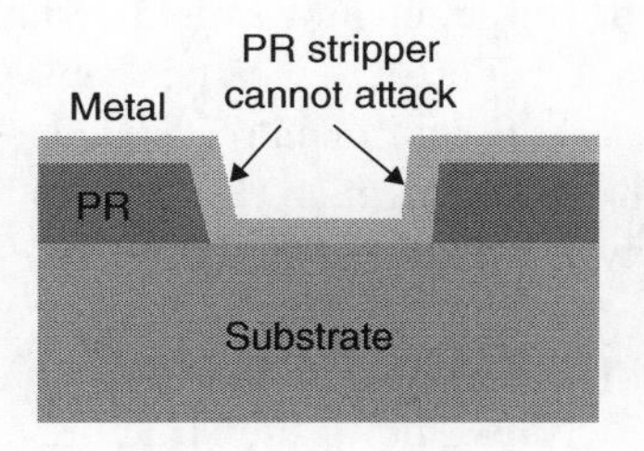

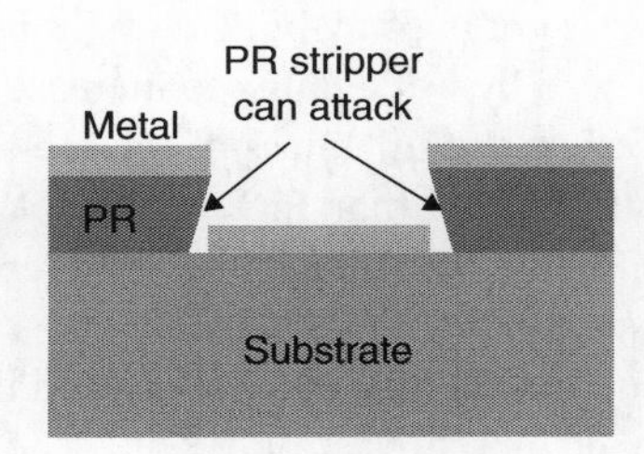

Figure 16.17 PR profile for lift-off. The left is not good (positive), the right is good (negative).

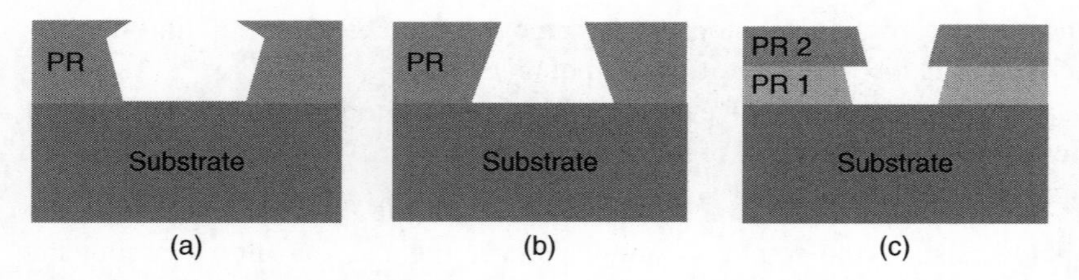

Figure 16.18 Schematic diagram of the profile of positive PR after modification. (a) The PR profile of surface hardening, (b) the PR profile of image reversal, (c) the PR profile of double-layer.

After exposure, put the sample on a hot plate for baking at the same temperature as the soft bake for a few minutes. After baking, the sample is put to the aligner again to do flood exposure (without photomask), and then developed. The PR cross section as shown in Figure 16.18b [6] can be achieved.

Method 3. Double-layer photoresist technique

Before coating ordinary positive PR, spin the first layer of resist, lift-off resist (LOR) or polyamic acid (PA) [5]. This resist layer is prebaked on the hot plate, and then spin the second resist layer-ordinary positive RP. Soft bake the positive PR on the hot plate, the temperature of the prebaking is higher than the temperature of the soft bake. After exposure and development, the developing rate of the first layer of resist is faster than that of the second layer. The cross section of the PR after development is shown in Figure 16.18c.

In these three methods, to do the metal lift-off, the thickness of the metal film should be less than the thickness of the photoresist. The thickness of the metal is less than the thickness of the PR, which is the most common situation in the process. The PVD thickness of metal is generally 100–500 nm, and the thickness of PR is 1–2 μm.

In the double-layer PR process, the first layer of resist can also be replaced by SiO_2. This technique is called SiO_2 assisted lift-off process [5]. At this time, the prebaking step is not needed, but etching is required. The dry etching is used first and then followed by wet etching to increase the undercut of SiO_2, which is beneficial to lift-off. If the thickness of the metal and SiO_2 is controlled so that the metal is slightly thinner than SiO_2, a near-planar structure after metal lift-off can be achieved. Figure 16.19 shows the difference between double-layer PR lift-off and SiO_2 assisted lift-off.

Here we have an example of SiO_2-assisted lift-off. The left photo of Figure 16.20 is the developed photoresist and SiO_2. The photo clearly shows the structure of the PR with a wide top and a narrow bottom, which is more obvious than that in Figure 12.17. The right photo of the figure shows the case where the

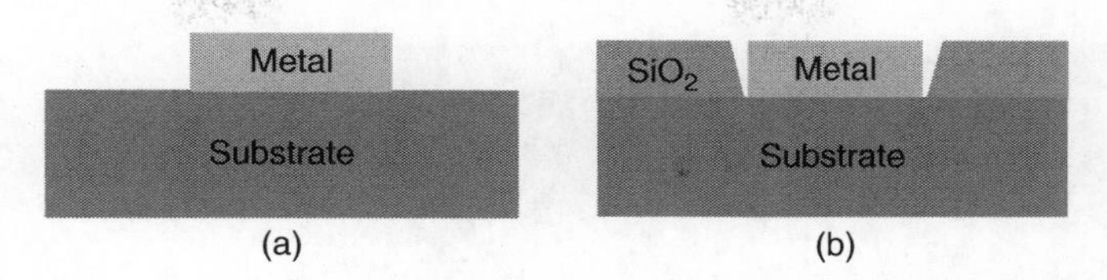

Figure 16.19 The structure after lift-off. (a) The pattern after the double-layer PR lift-off and (b) the pattern after SiO_2 assisted lift-off.

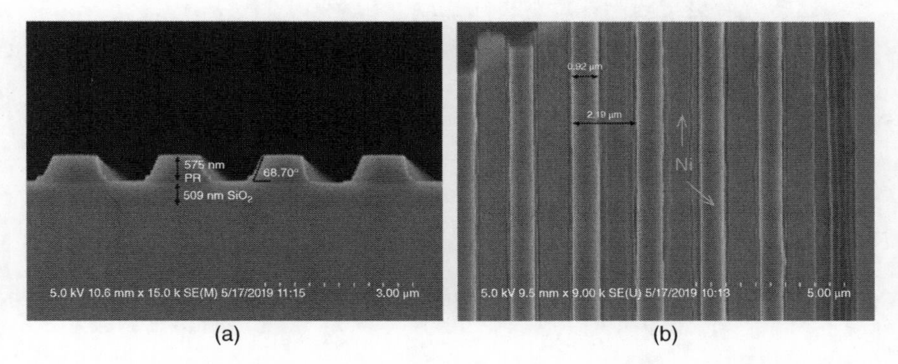

Figure 16.20 The left is the cross-view of the PR and SiO_2 after development. The right is the situation after the Ni lift-off without SiO_2 etching.

metal nickel (Ni) does lift-off without SiO_2 etching. The Ni is not lifted off at all and covered the entire sample surface. After the dry and wet etching of SiO_2, the Ni is easily and smoothly lifted off, and the patterns are perfect, as shown in Figure 16.21.

If we must use dry etching to etch the metal, be careful when removing resist after etching. For positive PR, the commonly used stripper is acetone. In the dry etching process, due to the action of plasma, the surface of photoresist changes, resulting in a hard shell that is not soluble in acetone. The greater power is, the longer etch time is, and the thicker hard shell will be. This phenomenon also occurs when we do dry etching of Si, SiO_2, and Si_3N_4. If we put the sample flat in the liquid, as shown in Figure 16.22a, acetone can remove the PR under the hard shell, but the hard shell remains on the sample surface, as shown in Figure 16.23. Therefore, after dry etching, when placing the sample in acetone, place it upright, as shown in Figure 16.22b, so that the hard-shell falls into the solution instead of remaining on the sample surface. If the etching power is big and etching time is long, use oxygen plasma to etch the resist for several minutes before putting the sample into acetone. If necessary, it should be assisted by ultrasonic vibration.

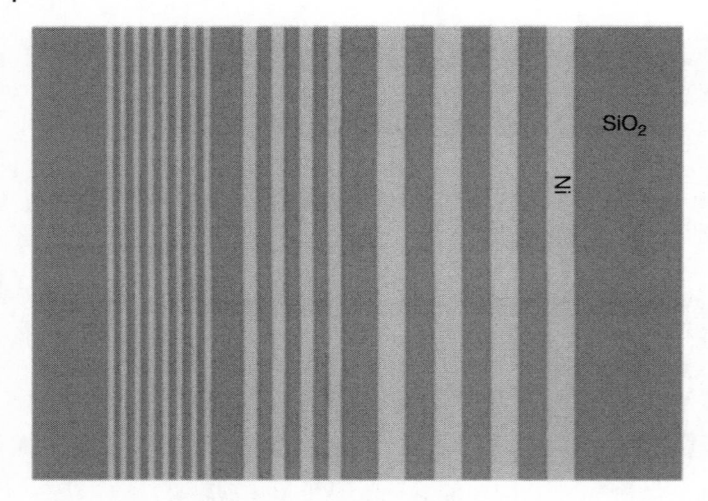

Figure 16.21 After SiO_2 etched, the patterns of Ni lift-off.

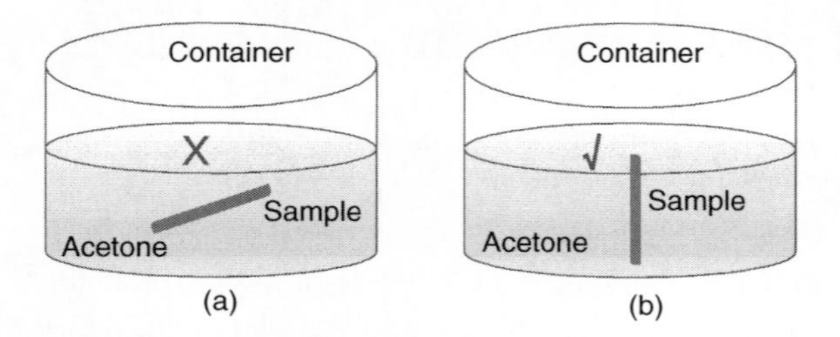

Figure 16.22 Put the sample into acetone after dry etching. The left is wrong, and the right is correct. (a) Don't lay the sample flat in acetone after dry etching, (b) place the sample upright in acetone after dry etching.

Figure 16.23 Hard shell of photoresist left on the sample surface.

16.6 Metal Selection and Annealing Technology

In order to obtain a good Schottky barrier and ohmic contact, the choice of metal must meet certain conditions.

For silicon, when the work function of metal (refer to 6.3) is larger than that of n-type silicon or smaller than that of p-type silicon, the contact between Si and metal can obtain Schottky contact, that is, rectifying contact [7] (refer to 5.3). For ohmic contact, the surface of the substrate must be heavily doped, and the barrier height established by the selected metal should be small (refer to 6.3).

For III–V materials, the situation is more complicated. According to the situation, it can be divided into four basic types: ideal or near ideal Schottky barrier, Bardeen barrier, oxide barrier, and close contact barrier [8]. Ohmic contact is also heavily doped, and the n-type semiconductors are main option. In addition, there are four points that need to be considered:

(1) The adhesion between metals and semiconductors,
(2) The compatibility between metals and semiconductors,
(3) The compatibility between metals,
(4) The appropriate annealing temperature.

We discuss metal selection and annealing as below.

16.6.1 The Selection of Metals

Al is the most used metal in semiconductor processing. Al has low price, low resistivity, simple process, and good adhesion. But in the process, Al also has some problems. Two common problems are electromigration and dissolution of silicon. Electromigration refers to the migration phenomenon of metal under the action of current and temperature, which may break the metal connection and make the chip scrap. This phenomenon is more likely to occur under high-current density and high frequency. Today, the devices that make up microchips are getting smaller and faster. If not handled well, electromigration will have a significant impact on the reliability of the microchips. To overcome this problem, 4–5% Cu is generally added to Al (see Figure 11.4). The dissolution of silicon means that silicon dissolves and diffuses in Al when processed or used at a higher temperature, and corrosion pits appear on the surface of the silicon, which degrades or even fails the device characteristics. To overcome this problem, 1–2% silicon can be added to Al.

Gold is another commonly used metal. The resistivity of gold is lower than that of aluminum. Its electromigration characteristic is better than that of aluminum, and its ductility is good. As it is easy to create a deep-level trap and recombination center in silicon (see Figure 5.6), it is generally avoided in silicon processing. Gold has been widely used in GaAs chips. However, the surface contact characteristics of gold and GaAs are not very good, it cannot meet the specifications of Schottky barrier manufacturing. Therefore, a multilayer-metal process is required, such as the Au/Pt/Ti structure of Schottky barrier. Ti is a barrier metal, Pt is a buffer layer, and Au is conductive layer. In the Au/Ge/Ni structure of ohmic contact, Ni is the adhesion layer, Ge is the buffer layer, and Au is the conductive layer.

In the processing, especially in wire bonding (discuss in Chapter 18), we should avoid the contact between gold and aluminum. Gold-aluminum intermetallic compounds will be created with the contact between two metals, such as Au_5Al_2, Au_2Al, AuAl, $AuAl_2$ [9]. $AuAl_2$ that is called purple plague can cause a big reliability problem in wire bonding.

In addition to the abovementioned metal systems, there are copper interconnection technique, refractory metal tungsten technique, tungsten via plugs (Figure 11.4), and polysilicon gate technique, etc., which will not be introduced in this section.

16.6.2 Metal Annealing

In order to have a good Schottky barrier and ohmic contact, after the metal is evaporated and deposited, the annealing process must be completed. This process can make the metal and the semiconductor contact surface to produce a shallow fusion and achieve the purpose of good contact. Metal annealing is sometimes called sintering. It mainly has two techniques: furnace annealing and rapid thermal annealing (RTA). Furnace annealing uses the tool that has same structure as thermal oxidation furnace. RTA has another name Rapid Thermal Processing (RTP). It uses halogen lamps as the heating source. The structure is shown in Figure 16.24. The furnace annealing has a long temperature ramp-up time, which is suitable for a long time (for example, half an hour or more) annealing process. RTA has a short ramp-up time, which can reach 20–50 °C/s and can achieve annealing in a few seconds to a few minutes. Because of its short temperature ramp-up time and annealing time, so it is called RTA. The metal annealing process mostly uses RTA. The temperature is 300–500 °C and the time is 1–5 minutes. Depending on the device, it may take half to an hour of annealing in a furnace, which requires us to do experiments to determine the best conditions. Sometimes the annealing process cannot be completed at the same time and needs more than one time. In this case, we should complete higher-temperature annealing first, and then do lower temperature one.

The annealing process is mostly protected by introducing nitrogen gas to avoid the reaction of oxygen to the metal at high temperature, which will deteriorate the alloy structure. Sometimes a mixture of 95% N_2 + 5% H_2 is also used. This gas is called forming gas. An example of annealing using forming gas is that when SiO_2 is used as a dielectric film, some oxygen may escape from the film during annealing,

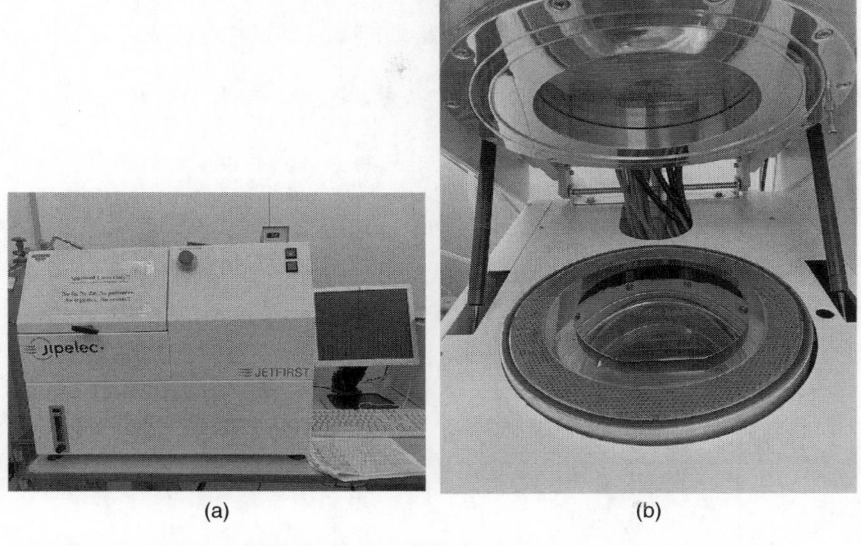

Figure 16.24 RTA equipment (a) and internal structure (b).

and this oxygen may react with the metal film to increase the resistivity of the metal and the surface becomes rough. The phenomenon has a serious impact on aluminum. Using forming gas, the hydrogen in it can react with the escaped oxygen to protect the metal film and improve the alloy properties of the metal.

References

1 饭田修一, 大野和郎, 神前熙等. (1979). 物理学常用数表, [日], 科学出版社, 1979, 198 页.

2 Sarangan, A. (2019). *Nanofabrication Principles to Laboratory Practice*, 54. CRC Press.

3 Greene, J.E. (2017). Review Article: Tracing the recorded history of thin-film sputter deposition: From the 1800s to 2017. *Journal of Vacuum Science & Technology A* 35, 05C204: 25.

4 Pang, L., Lian, Y., Kim, D.S. et al. (2012). AlGaN/GaN MOSHEMT with high-quality gate-SiO_2 achieved by room-temperature radio frequency magnetron sputtering. *IEEE Transactions on Electron Devices* 59 (10): 2650–2655.

5 廉亚光. (1991). 用正性光刻胶进行GaAs IC的非辅助直接剥离技术, 半导体情报, 年, 第 28 卷, 第 3 期, 53 页。

6 MicroChemicals, Basics of Microstructuring, Image Reversal Resists and Their Processing.

7 电子工业生产技术手册 7, 半导体与集成电路卷, 硅器件与集成电路, 581页, 国防工业出版社, 1991年。

8 电子工业生产技术手册 8, 半导体与集成电路卷, 化合物半导体器件, 360 页, 国防工业出版社, 1992年。

9 Galli, E., Majni, G., Nobili, C., and Ottaviani, G. (1980). Gold-aluminum intermetallic compound formation. *Electrocomponent Science and Technology* 6: 147–150.

17

Doping Processes

In the manufacture of semiconductor microchips, it is necessary to change the electrical properties of the original semiconductor substrate material (n-type, p-type, and semi-insulating) to make basic device structures, such as p–n junctions, heavily doped ohmic contact regions, and isolation zones. These structures can be completed by doping processes. In silicon technology, photolithography, oxidation, and doping are the troika of planar technology. The doping processes mainly include two technologies, thermal diffusion and ion implantation. Since there are no thermal oxidation and diffusion process in III–V semiconductors, the doping is generally completed by ion implantation. Let us discuss the processes below.

17.1 Basic Introduction of Doping

Regarding doping, we introduced in Section 5.2, phosphorus doping can make silicon an n-type semiconductor, boron doping can make silicon a p-type semiconductor, and Si doping can make GaAs an n-type semiconductor. In fact, not only these three dopants, but other elements also have the same function. However, these three elements are used the most.

As mentioned above, doping is mainly divided into thermal diffusion and ion implantation. Thermal diffusion is to place the silicon sample at a high temperature of 900–1200 °C. The dopant enters the silicon through diffusion. Thermal diffusion is also referred to as diffusion. 900–1200 °C is obviously too high for III–V materials. The doping technique for these kinds of materials is mainly ion implantation. Ion implantation is to ionize the atoms or molecules to be doped, and the ionized positive ions are injected into the semiconductor substrate or sample in the form of high energy under the acceleration of the electric field. In this sense, the working principle is like RIE and sputtering techniques, but the whole set of equipment is much more complicated and expensive.

Semiconductor Microchips and Fabrication: A Practical Guide to Theory and Manufacturing,
First Edition. Yaguang Lian.

Thermal diffusion is a high-temperature process, and only SiO_2 or Si_3N_4 can be used as a masking film. We use SiO_2 more often. In fact, silicon impurity diffusion is often carried out simultaneously with thermal oxidation. It is suitable for deep doping. Ion implantation is a room-temperature process, so SiO_2, Si_3N_4 can be used as a masking film, and photoresist can also be used as a masking film for implantation, but the photoresist is more difficult to remove from the sample surface than RIE. According to the difference of energy, ion implantation can realize doping from shallow to deep, and it has been widely used in the manufacture of silicon and III–V microchips.

17.2 Basic Principles of Diffusion

Impurity atoms in silicon are mainly diffused in six ways, please see Figure 17.1:

(1) Impurity atoms may reside on a normal lattice site; this is a substitutional impurity.
(2) They may reside in an open space of the lattice; this is an interstitial impurity.
(3) For an impurity atom to behave as a dopant, it must be substitutional so that it can ionize and either donate or accept electrons.
(4) Interstitial impurity atoms can diffuse fairly quickly because they do not need to break any bonds.
(5) Substitutional impurity atoms diffuse fairly slowly because they must break and make bonds in order to propagate.
(6) Moving an impurity atom from an interstitial into a substitutional position is termed dopant activation [1].

In Figure 17.1, cases b, c, e, and f can form substitutional impurity. Boron and phosphorus are substitutional impurities.

The earliest mathematical formula to explain the phenomenon of diffusion is Fick's law, which was proposed by the German scientist Adolf Eugen Fick (September 3, 1829–August 21, 1901) in 1855. It can be divided into Fick's first law and Fick's second law. If it is assumed that the impurity only diffuses in one direction (the direction perpendicular to the sample surface), which is the case of one-dimensional diffusion. The diffusion mode of the impurity in silicon can be obtained by solving Fick's second law:

$$\frac{\partial N}{\partial t} = D\frac{\partial^2 N}{\partial x^2} \tag{17.1}$$

It is a partial differential equation and used to solve one-dimensional diffusion. $N = N(x, t)$ is impurity concentration in silicon, which is a function of diffusion distance (x) and time (t). D is called diffusivity (or diffusion-coefficient), and its

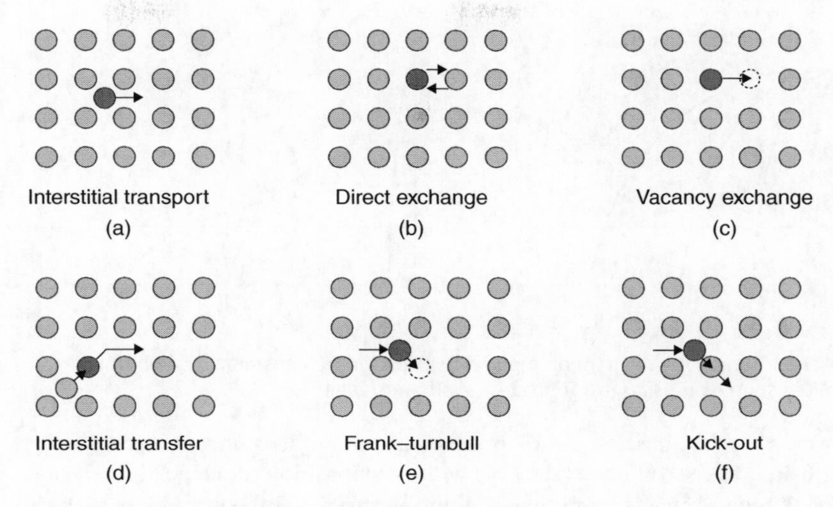

Figure 17.1 Atomic diffusion mechanisms in silicon.

unit is cm^2/s. The higher the value of D, the faster the impurity. D follows Arrhenius Eq. (15.1) and can be rewritten as:

$$D = D_0 \exp(-E/k_B T) \tag{17.2}$$

D_0 is a constant related to the impurity. Typical activation energies (E in the equation) for solid-state diffusion are about 3.3–4.4 eV [1]. In addition, impurities have a maximum solubility in silicon, which is called solid solubility. At 1200 °C, the solid solubility of boron is 5×10^{20} atoms/cm^3; at 1150 °C, the solid solubility of phosphorus is 1.3×10^{21} atoms/cm^3 [2]. For comparison, we can refer to the atomic concentration of major semiconductors listed in Section 6.3.

17.3 Thermal Diffusion

As mentioned above, thermal diffusion is the diffusion of impurities into the silicon crystal at 900–1200 °C. This diffusion is mainly carried out in two ways, constant source diffusion and limited source diffusion, which is also the most used two-step technique in the diffusion process: the first step is predeposition, using constant source diffusion; the second step is redistribution, using limited source diffusion. Although the solving results of Fick's second law are somewhat different from the actual situations, the overall trend is consistent with the process results. By revising this law with other factors, we can get results consistent with reality. In this chapter, we will give a brief introduction to the one-dimensional situation to understand how diffusion works in silicon.

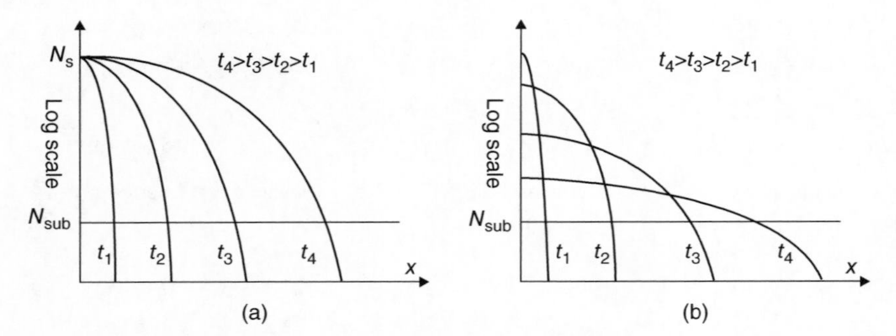

Figure 17.2 Schematic diagram of impurity distribution for constant source diffusion (a) and limited source diffusion (b). Source: [3] Wolf and Tauber.

Constant source diffusion refers to the constant surface concentration during the diffusion process, which is the case with predeposition. Predeposition means that the doping source is first deposited on the surface of the sample for a short time, so it can be considered that it forms a uniformly deposited layer with a small thickness on the silicon surface. If the time is prolonged, the diffusion will advance to the inside of the silicon wafer, but the surface concentration will not change. By solving Fick's second law, the concentration distribution of impurities can be obtained as a complementary error function (erfc) distribution.

Limited source diffusion means that the total amount of impurities is constant during the diffusion process, and this is the case for redistribution. Redistribution means that after the predeposition is completed, the impurity source is removed, and the impurities continue to diffuse at high temperature and advance into the silicon. Since there is no impurity source, the total amount of impurities in the diffusion process is the impurities deposited on the surface during pre-deposition. The redistribution process is sometimes called drive-in diffusion. By solving Fick's second law, the concentration distribution of impurities can be obtained as a Gaussian function. Johann Carl Friedrich Gauss (April 3, 1777–February 23, 1855) was a German mathematician and physicist.

The two distribution curves are shown in Figure 17.2. The "Log Scale" in the figure is an exponential coordinate, x is pointing from the surface of the silicon wafer to the inner direction, t is time, N_s is the surface concentration, its maximum value is the impurity solid solubility at diffusion temperature, and N_{sub} is the substrate background doping concentration.

17.4 Diffusion and Redistribution of Impurities in SiO_2

SiO_2 plays an important role in the doping process. It mainly affects thermal diffusion in two aspects. One is that the redistribution process is often carried out at the same time as the thermal oxidation process. The other is that the thermal oxide film acts as a diffusion mask and is widely used in the process. So it is necessary to discuss the diffusion behavior of impurities in SiO_2 and the redistribution at the Si–SiO_2 interface. In this section, we still use boron and phosphorus as examples. To discuss this issue, we need to introduce segregation coefficient m, which is defined as the ratio of the equilibrium concentration of impurities in silicon N_{Si} to the equilibrium concentration of N_{SiO_2} in silicon dioxide:

$$m = \frac{N_{Si}}{N_{SiO_2}} \tag{17.3}$$

According to the experimental results, the m value of boron is 0.1–0.3, and the m value of phosphorus is about 10 [2]. The segregation coefficient is the first factor that affects the redistribution of impurities in thermal oxidation. Two circumstances are possible: (i) the oxide takes up the impurity ($m < 1$); and (ii) the oxide rejects the impurity ($m > 1$).

Another factor that affects the redistribution is the diffusivity of dopant impurity in the oxide. The larger the diffusivity, the faster the diffusion rate. If the diffusivity in SiO_2 is too large, the impurities will quickly diffuse through the SiO_2 film. In the worst case, the SiO_2 cannot be used as a masking film for impurity diffusion. For the commonly used impurities of boron and phosphorus, their diffusivities in SiO_2 are much smaller than those in Si. The D_0 of B and P in silicon is 10.5 cm^2/s. The D_0 of B in SiO_2 is 2.8×10^{-4} cm^2/s when temperature is lower than 1100 °C and 5.8×10^{-11} cm^2/s when temperature is higher than 1100 °C. The D_0 of P in SiO_2 is 1.59×10^{-11} cm^2/s [4]. So we can use SiO_2 as a masking layer for boron and phosphorus diffusion.

The third factor affecting redistribution is the rate of oxidation. During thermal oxidation, the thickness of the oxide layer continues to increase, and the Si–SiO_2 interface moves with time. Therefore, the rate of the Si/SiO_2 interface movement also has an important influence on the redistribution.

The four cases described in Figure 17.3 use the substrate impurity concentration as an example to illustrate the redistribution of impurities at the Si–SiO_2 interface. Combined with the predeposited impurity distribution, we can estimate the redistribution of the interface when doing the process of thermal oxidation and impurities drive-in. In the figure, a and b indicate that the segregation coefficient is less than 1, and SiO_2 accumulates impurities. Case a is the slow diffusion in the oxide layer, which is the case with boron. Case b is the fast diffusion in the oxide layer, when thermal oxidation contains H_2 (a method of wet oxidation), this is the case with boron. In the figure, c and d are the segregation coefficients greater than 1, and SiO_2 repels impurities. Case c is the slow diffusion in the oxide layer, which is the case for phosphorus. Case d is the fast diffusion in the oxide layer, which is the case for gallium.

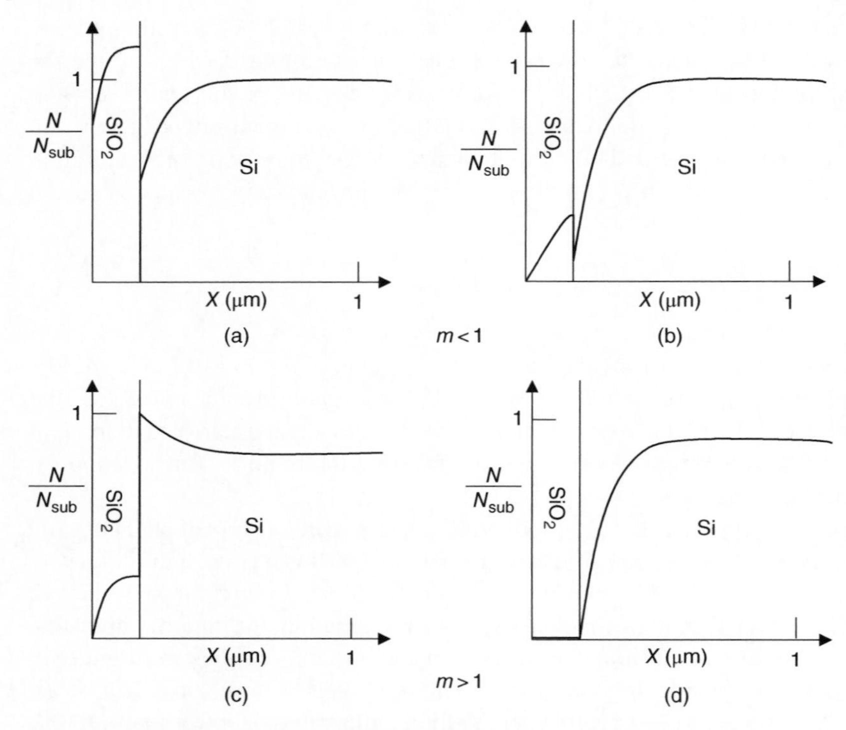

Figure 17.3 Schematic diagram of the redistribution of impurities at the Si–SiO_2 interface. Source: [3] Wolf and Tauber.

17.5 Minimum Thickness of SiO_2 Masking Film

Although the diffusivities of boron and phosphorus in SiO_2 are much smaller than those in Si, impurities are still diffused in SiO_2. Therefore, if SiO_2 is used as a masking film for diffusion, the minimum thickness of SiO_2 that needs to be masked must be considered. In the previous section, the D_0 of B and P in SiO_2 showed that the diffusivity of B is larger than that of P. That means the diffusion rate of B in SiO_2 is larger than the rate of P. The purpose of quoting the data in Section 17.4 is to indicate that the diffusion rates of B and P in SiO_2 are much smaller than those in Si, so that we can use SiO_2 as mask to do B and P diffusion process. Actually, according to the gases passing into the diffusion tube, the diffusion rates of B and P in SiO_2 are different. In some cases, B has faster rate; in some cases, P has faster rate. For more details, please refer to reference [5]. Borosilicate and phosphosilicate (see Section 17.7) are commonly used diffusion techniques. The techniques usually use N_2 during the processing. In this situation, the diffusion rate of B in SiO_2 is smaller than P. If the ratio of substrate surface concentration (N_S) to

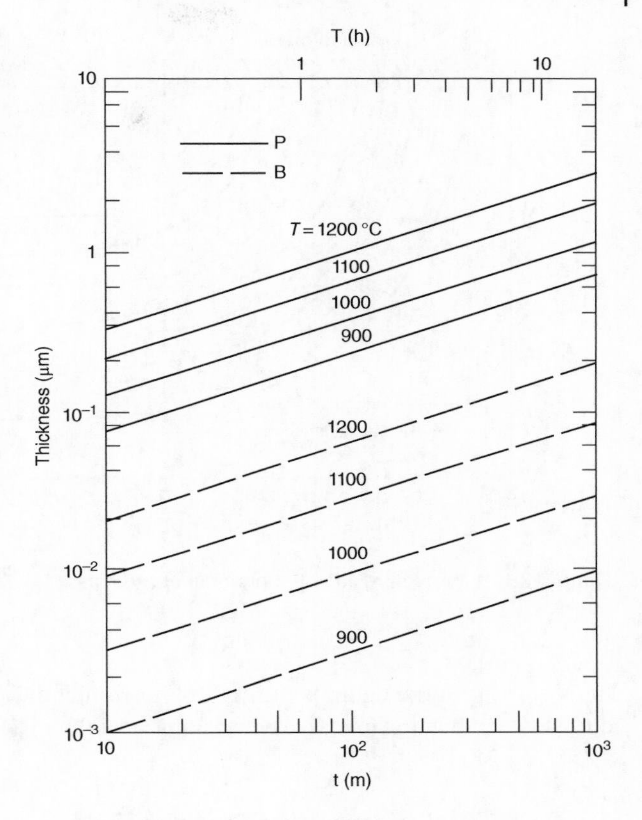

Figure 17.4 The thickness of SiO_2 needed to mask B and P vs. temperature and time [6]. Source: Chinese Technical Books.

substrate background concentration (N_{sub}) is 3×10^3, the required thickness of SiO_2 as a function of time and temperature is shown in Figure 17.4.

17.6 The Distribution of Impurities Under the SiO_2 Masking Film

When impurities diffuse through the masking film, the impurity atoms not only diffuse vertically into the silicon substrate but also diffuse laterally along the interface between the masking film and silicon. Based on the substrate background concentration N_{sub}, if the concentration of the impurity on the surface of the substrate N_s is higher, the depth of impurity diffusion (vertical and horizontal) is greater, as shown in Figure 17.5. The dotted line in the figure is the erfc distribution, and the solid line is the Gaussian distribution. For a given

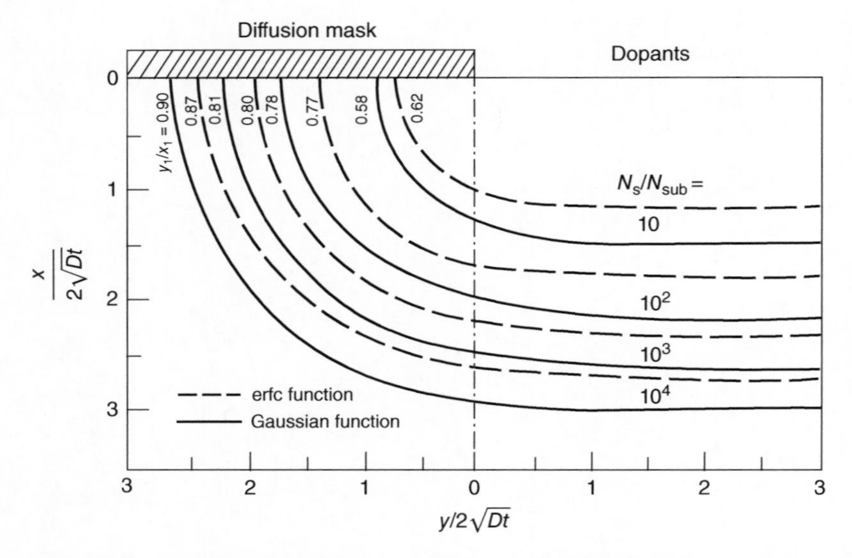

Figure 17.5 Lateral diffusion of impurity under the mask [6]. Source: Chinese Technical Books.

N_s/N_{sub}, y_j/x_j in the figure is the ratio of horizontal diffusion depth and vertical diffusion depth at the interface between mask and semiconductor.

17.7 Diffusion Impurity Sources [2]

The boron and the phosphorus sources used for the diffusion of silicon impurities are divided into solid planar diffusion sources (PDS), portable liquid sources, and coated latex sources. The PDS is made into the same shape as the silicon wafer. It is placed in the quartz boat at the same time as the silicon wafer, put the boat into the quartz tube to be heated to diffuse impurity into the silicon wafer, as shown in Figure 17.6. The portable liquid source is like wet oxidation by passing oxygen to the water bottle and bringing the water vapor into the quartz tube. The process uses nitrogen to pass into the bottle containing the liquid source and carry the source into the quartz tube. The coated latex source is the same as the PR coating process. The source is first coated on the surface of Si wafer through a spinner and then diffused in the tube. Below we will briefly describe each of them.

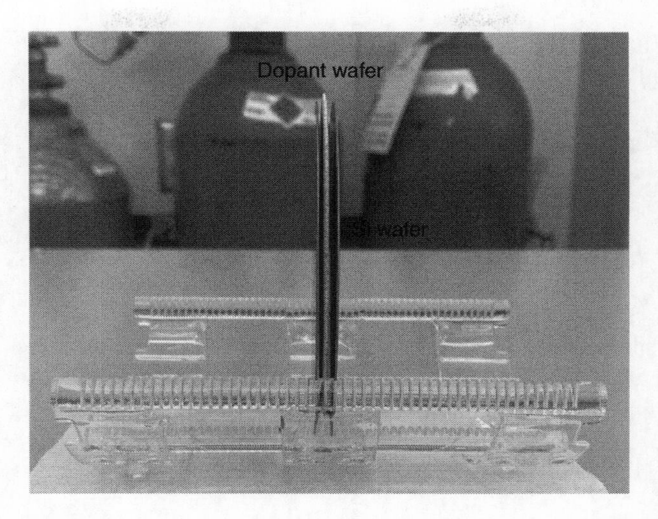

Figure 17.6 Placement of dopant wafer and silicon wafer on a quartz boat.

There are four main sources of boron:

(1) Planar boron nitride (BN) source. The diffusion mechanism of this diffusion source is, when it is heated in an oxygen atmosphere, a layer of boron trioxide is formed on the surface of the source:

$$4BN + 3O_2 \rightarrow 2B_2O_3 + 2N_2 \tag{17.4}$$

This step is called activation. The activation temperature is usually 1000–1200 °C. During predeposition, B_2O_3 reacts with Si, reducing boron atoms to diffuse into silicon to complete p-type doping:

$$2B_2O_3 + 3Si \rightarrow 3SiO_2 + 4B \tag{17.5}$$

The product in Eq. (17.5) includes SiO_2 and B, so this product is called borosilicate glass (BSG). In the process, except for O_2 used in the activation step, N_2 is used in all other steps.

(2) Boron glass–ceramic planar source. The diffusion source is a mixture of B_2O_3 and multiple oxides. It contains 50% B_2O_3, 20–40% SiO_2, 20–30% Al_2O_3, and 5–15% MgO + BaO. Its diffusion mechanism is carried out according to reaction Eq. (17.5). One of the reasons is that B_2O_3 has a low melting point (460 °C) and others have high melting points. During the process, only B_2O_3 escapes from the source in large quantities to finish the B diffusion process. There is almost no diffusion process for other metals Mg and Ba.

(3) Trimethyl borate [$B(OCH_3)_3$] source. It is a colorless and transparent liquid with a boiling point of 67.8 °C and a melting point of −29.2 °C. When in use, it is carried to the quartz tube with nitrogen, and the thermal decomposition reaction occurs when the temperature is higher than 500 °C:

$$B(OCH_3)_3 \rightarrow B_2O_3 + CO_2 + H_2O + C + \cdots \tag{17.6}$$

Then return to reaction Eq. (17.5).

(4) Silica latex boron source. The diffusion source is a kind of latex containing boron. The source is made by dissolving B_2O_3 into silica latex. It is spun on the surface of the silicon wafer by spin coating and prebaking at 300 °C. The solvent in the source is volatilized and undergoes other chemical reactions to form a boron containing SiO_2 layer on the silicon surface. At high temperature (1000–1200 °C), boron diffuses into the silicon. The reaction mechanism also complies with Eq. (17.5).

There are three main sources of phosphorus

(1) Planar phosphorous source. It is a solid sintered from silicon pyrophosphate (SiP_2O_7) and aluminum metaphosphate [$Al(PO_3)_3$] into a wafer shape. At the diffusion temperature, the following reactions occur:

$$SiP_2O_7 \rightarrow SiO_2 + P_2O_5 \tag{17.7}$$

$$Al(PO_3)_3 \rightarrow AlPO_4 + P_2O_5 \tag{17.8}$$

$$2P_2O_5 + 5Si \rightarrow 5SiO_2 + 4P \tag{17.9}$$

The product in Eq. (17.9) includes SiO_2 and P, so this product is called phosphosilicate glass (PSG). The phosphorus atoms in PSG diffuse into silicon to complete n-type doping. The molecular structure of $AlPO_4$ in Eq. (17.8) is very stable, and at the diffusion temperature, aluminum does not be released to diffuse into silicon. N_2 is used in the process.

(2) Phosphorus oxychloride ($POCl_3$) source. It is a colorless and transparent liquid with a melting point of 1.25 °C and a boiling point of 105.3 °C. It is carried into the quartz tube with nitrogen gas and reacts with oxygen and silicon in the tube at high temperature. So besides N_2, O_2 is also used in the process:

$$4POCl_3 + 3O_2 \rightarrow P_4O_{10} + 6Cl_2 \tag{17.10}$$

$$P_4O_{10} + 5Si \rightarrow 5SiO_2 + 4P \tag{17.11}$$

(3) Silica latex phosphorus source. The source is made by dissolving P_2O_5 into silica latex. Its use method is the same as that of silica latex boron source, and the reaction equation is the same as (17.9).

17.8 Parameters of the Diffusion Layer [2]

The parameters of the diffusion layer include structural parameters and electrical parameters. The structural parameter of diffusion refers to the geometric position of the p–n junction formed by diffusion, that is, the distance between the p–n junction and the surface of the diffusion layer. It is called junction depth, which is also represented by the symbol x_j, and its expression is

$$x_j = A\sqrt{Dt} \tag{17.12}$$

In the formula, D is the diffusivity of impurities in silicon, t is time, and A is a constant related to the ratio of surface concentration to substrate concentration.

The electrical parameters of the diffusion junction are expressed by the sheet resistance of the diffusion layer. A semiconductor thin layer that has a square surface, the resistance in the direction of current parallel to one side of the square is called the diffusion layer sheet resistance, as shown in Figure 17.7. The symbol of sheet resistance is R_s, and the unit is Ω:

$$R_s = \frac{1}{\overline{\sigma} x_j} \tag{17.13}$$

In Eq. (17.13), $\overline{\sigma}$ is the average conductivity of the diffusion layer. Its unit is $(\Omega \cdot \text{cm})^{-1} = \frac{1}{\Omega \cdot \text{cm}}$. Conductivity is the reciprocal of resistivity (relationship 1.4), so sheet resistance can also be written as:

$$R_s = \frac{\overline{\rho}}{x_j} \tag{17.14}$$

In the formula, $\overline{\rho}$is the average resistivity of the diffusion layer. The definition of sheet resistance is very useful in the process. The sheet resistance has nothing to do with the specific size l of the square, so once the sheet resistance is measured, the resistance of a diffusion layer can be easily obtained. Please see Figure 17.8, which is a schematic diagram of a strip diffusion layer. The "Diffusion area" in the figure is the area of impurity diffusion layer. The number of squares of this diffusion layer

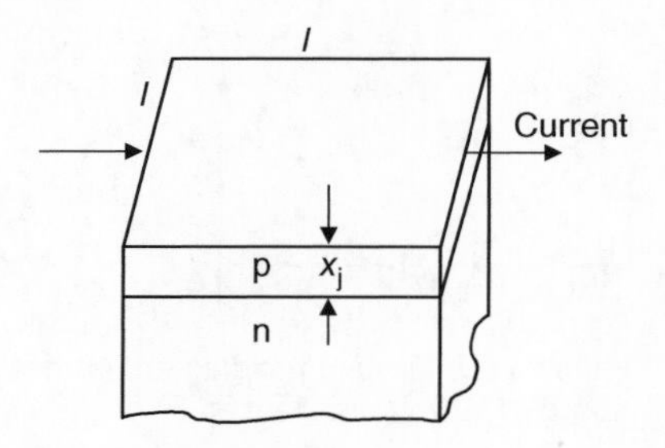

Figure 17.7 Schematic diagram of diffusion layer sheet resistance.

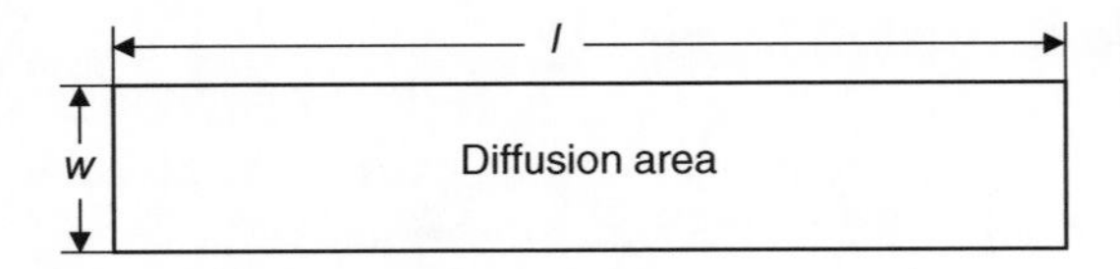

Figure 17.8 Schematic diagram of a strip diffusion layer.

is the ratio of length and width l/w, and the resistance R of the layer is the sheet resistance multiplied by the number of squares:

$$R = R_s \cdot \frac{l}{w} \tag{17.15}$$

17.9 Four-Point Probe Sheet Resistance Measurement

The measurement of sheet resistance mainly adopts the four-point probe method. Figure 17.9 is a schematic diagram of a four-point probe testing a sample. It consists of four probes arranged in a line on the same plane and at equal distances. The distance between the two probes is $s = 1$ mm. When measuring, press the four probes on the surface of the sample to be measured at the same time, the two outer probes are applied with a constant small current, and the inner two probes are used to measure voltage. For rectangular samples, $a/d \geq 4$. If the size of the sample is much larger than the probe pitch, then the sheet resistance R_s is [2]

$$\mathrm{R_s} = \frac{\pi}{\ln 2}\frac{V}{I} = 4.5324\frac{V}{I} \tag{17.16}$$

Figure 17.10 is a photo of a four-point probe. The four-point probe method is not only used to measure the sheet resistance, but also used to measure the resistance of metals after evaporation or sputtering.

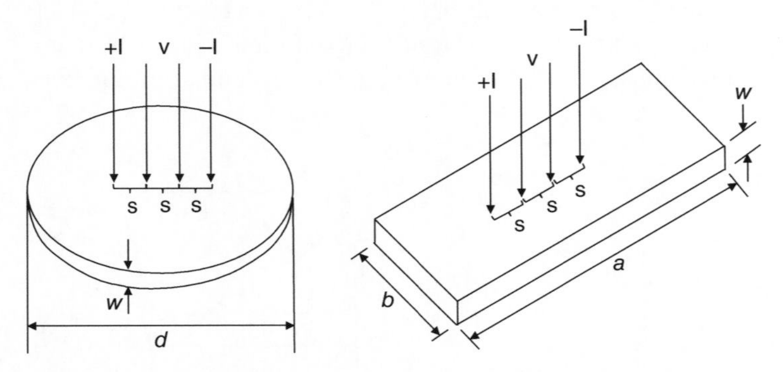

Figure 17.9 Schematic diagram of measuring sheet resistance with four-point probe. Source: [3] Wolf and Tauber.

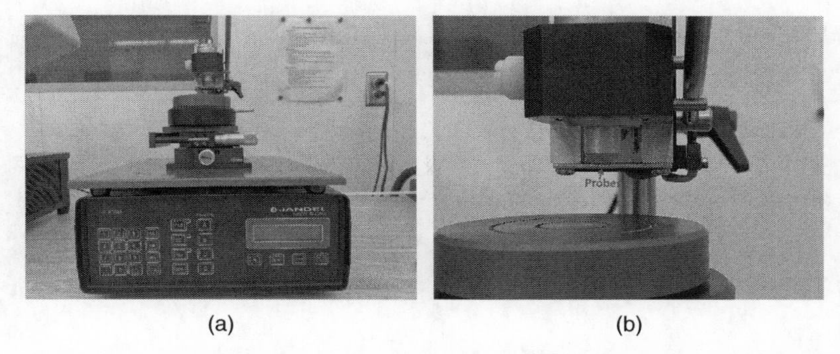

Figure 17.10 A photo of four-point probe (a, JANDEL RM3000) and four probes (b, Arrow point).

17.10 Ion Implantation Process

Ion implantation is also an important means of doping process. In the early semiconductor manufacturing, diffusion was the basic process for making p–n junction. But when the device entered the submicron size, ion implantation became the standard process of doping. Compared with thermal diffusion, ion implantation has the following main advantages:

1. A wide range of doping from shallow to deep can be achieved.
2. It is room-temperature process. So, SiO_2, metal, and photoresist can be used as masking film.
3. In addition to silicon, this technology can also be used for doping of III–V materials.
4. The lateral diffusion is small, which is beneficial to the manufacture of small-sized devices.
5. It can break through the limit of solid solubility of impurities.
6. It can realize some structures' fabrication that cannot be realized by diffusion process.

The main disadvantages are

1. The equipment is complex and expensive.
2. The throughput is lower than diffusion.
3. The semiconductor crystal structure is damaged, and high-temperature annealing is required to repair the crystal structure.

Ion implantation equipment is also called ion implanter. The implanter provides acceleration energy from 0.2 keV to 2 MeV [3] for the ions to be doped. The difference between the minimum energy and the maximum energy is 10 000 times. The commonly used energy unit eV in the microscopic world was introduced in

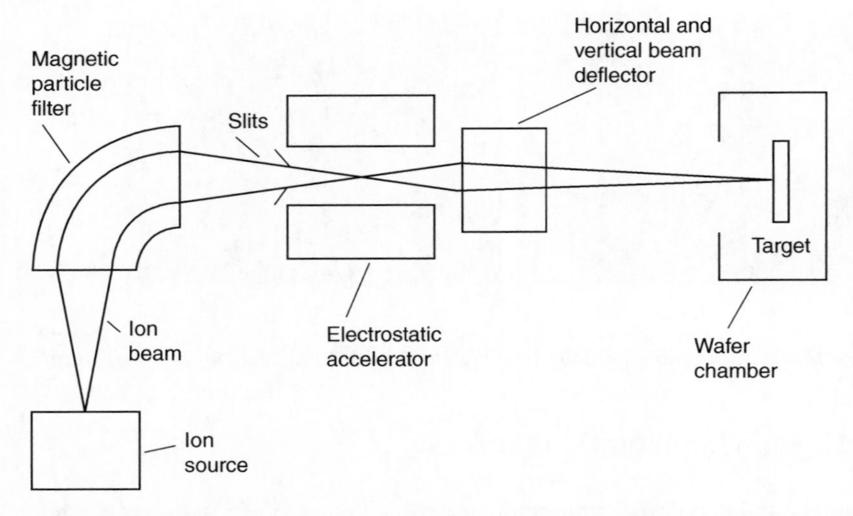

Figure 17.11 Schematic diagram of an implanter. Source: [7] Phelps / IOP Publishing.

Chapter 2. Its physical meaning is the energy obtained by an electron through a potential difference of 1 V. Based on rough calculations, we can think that all ions with a charge can obtain 1 eV of energy when passing through a potential difference of 1 V. To have an intuitive understanding of the energy of an implanter, we use the RIE equipment introduced in Section 15.2 to make a comparison. It is mentioned in this section that when the DC bias voltage of the RIE equipment exceeds 500 V, the system alarms. If 500 V is used as an example, when an ion passes a bias voltage of 500 V, an acceleration energy of 500 eV=0.5 keV can be obtained. The energy provided by an RIE tool to the ion is equivalent to the lowest energy range of the ion implanter. Figure 17.11 is a schematic diagram of an ion implanter.

The implanter uses current to identify the strength of the ion beam. Normally, the current range is 1 μA to 30 mA, and the current of the special purpose implanter is 50–100 mA. Through the current, we define the implantation dose Φ [3]:

$$\Phi = \frac{It}{qA} \tag{17.17}$$

The unit of Φ is the number of implanted ions per square centimeter (atoms/cm^2). In the formula, I is the ion beam current, the unit is ampere; t is the implantation time, the unit is second; q is the charge of an ion (usually equal to an electron charge = 1.6×10^{-19} coulomb); A is the area of the ion beam, the unit is square centimeters.

17.11 Theoretical Analysis of Ion Implantation

The total length of the path taken by an ion from entering the target (sample) surface to the stopping point is called the range, denoted by R. The projection length of R in the incident (X) direction is called the projected range, denoted by R_p. The projection length of R in the plane which is perpendicular to the incident direction is called the projected lateral or transverse range and is represented by R_t. Figure 17.12 is a schematic diagram of these three quantities. We use i to represent the number of ions injected into the target. The R_p of each ion is different, which is represented by R_{pi}. The R_{pi} of all ions incident on a unit area is added up, expressed as $\sum_i R_{pi}$, and then divided by the injection dose Φ, and we get the average projected range $\overline{R_p}$ [8]:

$$\overline{R_p} = \sum_i R_{pi}/\Phi \tag{17.18}$$

Similarly

$$\overline{R_t} = \sum_i R_{ti}/\Phi \tag{17.19}$$

The concentration distribution of impurity in silicon after implantation is well explained with Gaussian function. The quantities mentioned above are closely related to the ion implantation energy. The impurity distribution and junction depth can be obtained by these quantities. When a lot of impurities are implanted into silicon, their distribution is a statistical distribution. In statistics, we use standard deviation to represent the dispersion of a quantity. For the projected range R_p, the standard deviation (projected straggle) is ΔR_p. Its expression is

$$\overline{\Delta R_p} = \left[\sum_i \left(R_{pi} - \overline{R_p}\right)^2/\Phi\right]^{1/2} = \left[\overline{\Delta R_p{}^2}\right]^{1/2} \tag{17.20}$$

The standard deviation of the lateral range (projected lateral straggle) $\overline{\Delta R_t}$ has similar result. We mainly consider the distribution in the incident direction here. The lateral distribution is much smaller than the thermal diffusion, so we will

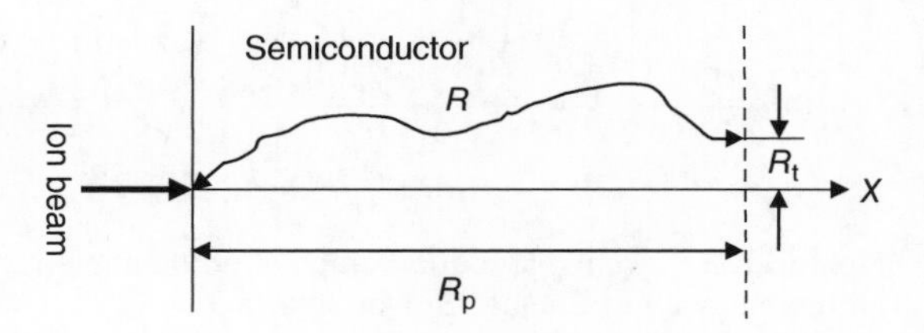

Figure 17.12 Schematic diagram of range, projection range, and lateral component after ion implantation into the sample.

not discuss it. Then the theoretically deduced expression for the distribution of implanted ion concentration N is

$$N(x) = \frac{\Phi}{\sqrt{2\pi}\overline{\Delta R_p}} \exp\left[\frac{-(x - R_p)^2}{2\overline{\Delta R_p}^2}\right] \tag{17.21}$$

The maximum point of concentration is at $x = R_p$, and the expression is

$$N(x) = \frac{\Phi}{\sqrt{2\pi}\overline{\Delta R_p}} \approx \frac{0.4\Phi}{\overline{\Delta R_p}} \tag{17.22}$$

p-type (or n-type) impurity ions are implanted into n-type (or p-type) substrate. When the concentration of implanted impurities is equal to the substrate concentration N_{sub}, the depth is called junction depth x_j. Its expression is [2]

$$x_j = \overline{R_p} + \overline{\Delta R_p}\left(2\ln\frac{N_{max}}{N_{sub}}\right)^{1/2} \tag{17.23}$$

17.12 Impurity Distribution after Implantation

According to formula 17.21, the distribution of ion implantation is different from thermal diffusion. The maximum point of the concentration during diffusion is on the surface, while the peak of the implantation concentration distribution is at R_p. Figure 17.13 shows the results of boron implantation in amorphous silicon with

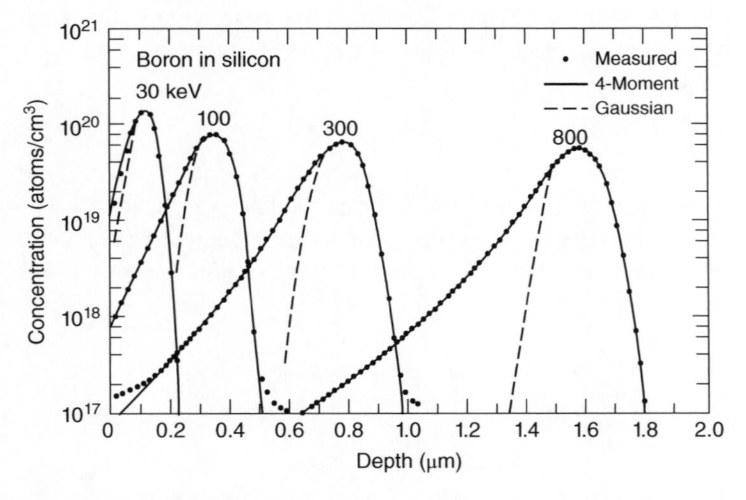

Figure 17.13 Atom distribution graph of boron implanted into amorphous silicon without annealing. Source: [3] Wolf and Tauber.

different energies. The figure shows atomic concentration distribution graph. "4-MOMENT" is a simulation model. For thermal diffusion, the peak of impurity concentration is on the substrate surface (Figure 17.2). For ion implantation, the peak of impurity concentration is below the substrate surface, the higher the implantation power, the deeper the peak of concentration.

The reason why the above figure uses amorphous silicon as an example is because when ions are implanted into single crystal silicon, due to the orderly arrangement of the crystal lattice, it seems that there is "openness" along the crystal direction, and the size of the openness is different in different crystal orientation. After implantation, the depth of ions entering the silicon along the openness is greater than the value predicted by theory and model. This phenomenon is the channeling effect of ion implantation. Figure 17.14 is a schematic diagram of the channeling effect of implantation and silicon crystal openness.

To avoid the channeling effect, during the implantation process, the sample surface and the ion implantation direction are not perpendicular, that is, the ion direction and the openness direction are at an angle of 0°, but a certain angle. Figure 17.15 shows a schematic diagram of the channeling effect of different deflection angles. The "Normalized impurity concentration" in the figure is a method of digital simplification in mathematics. Here, assuming that the

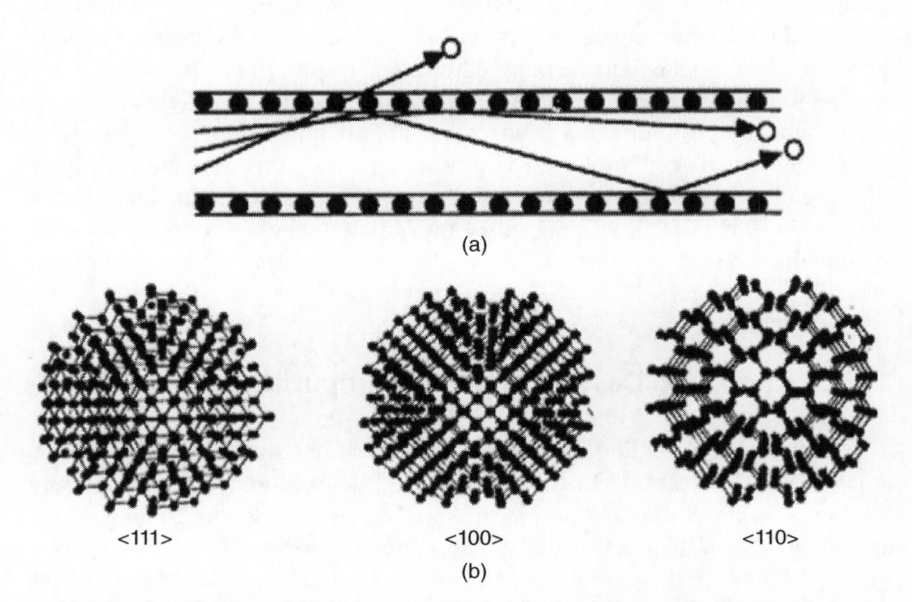

Figure 17.14 Schematic diagrams of: (a) the trajectory of ions entering the "openness" at different angles. (b) the "openness" in different crystal orientation of silicon. Source: [3] Wolf and Tauber.

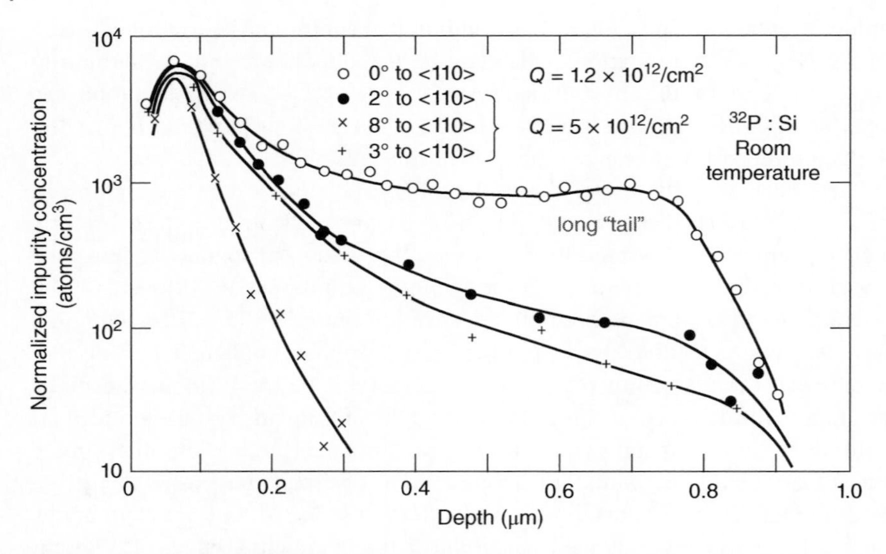

Figure 17.15 Phosphorus impurity profiles for 40 keV ion implantation to silicon at various angles from the ⟨110⟩ axis. Source: [9] Adapted from Bo Cui.

substrate concentration is 1, the implanted concentration is expressed as the ratio of the substrate concentration. Generally the substrate concentration is 10^{16} atoms/cm^3, so the ordinate range of the figure is from 10^{17} to 10^{20} atoms/cm^3. Q is the implantation dose Φ mentioned above, the abscissa is the implant depth, and the implanted impurity is a radioactive isotope of phosphorus, phosphorus-32 (^{32}P). The nucleus of ^{32}P has 15 protons and 17 neutrons, one more neutron than the most common isotope of phosphorus, phosphorus-31. The figure shows that for the silicon ⟨110⟩ axis, when the implantation angle is 8°, the implanted channeling effect is the smallest. "Long 'tail'" is produced by the channeling effect.

17.13 Type and Dose of Implanted Impurities [2]

The impurity implantation energy has been mentioned above, and the range is from 0.2 keV to 2 MeV. In this section, we briefly introduce the commonly used implant impurities and doses. The specific selection should be determined according to the needs of the actual device. For p-type doping, the commonly used implanted ions are B^+ and BF_2^+, and BF_2^+ is used for shallow implantation. For n-type doping, the commonly used implanted ions are P^+, As^+, Sb^+, and PF_2^+. In addition, O^+ and N^+ are used to form an insulating layer, and Ar^+ is implanted to form a high resistance layer. The dose range is from 10^{10} to 10^{18} atoms/cm^2.

17.14 The Minimum Thickness of Masking Film

When doing implantation, we can use different masking films. The most used is SiO_2. Or Si_3N_4, photoresist, and metal can be used as masking films. With the same implantation conditions, the thickness of Si_3N_4 can be thinner (85%) than that of SiO_2, and the thickness of photoresist is thicker (1.8 times) than that of SiO_2 [10]. When implanted, the projected range of ions in Si is close to that in SiO_2, as shown in Figure 17.16. The masking film must meet the minimum thickness requirement: the concentration of implanted ions into the substrate through the masking film is more than two orders of magnitude lower than the substrate concentration. In mathematics, 1 order of magnitude is 10, 2 orders of magnitude are 10^2, 3 orders of magnitude are 10^3, and so on. The minimum thickness of different

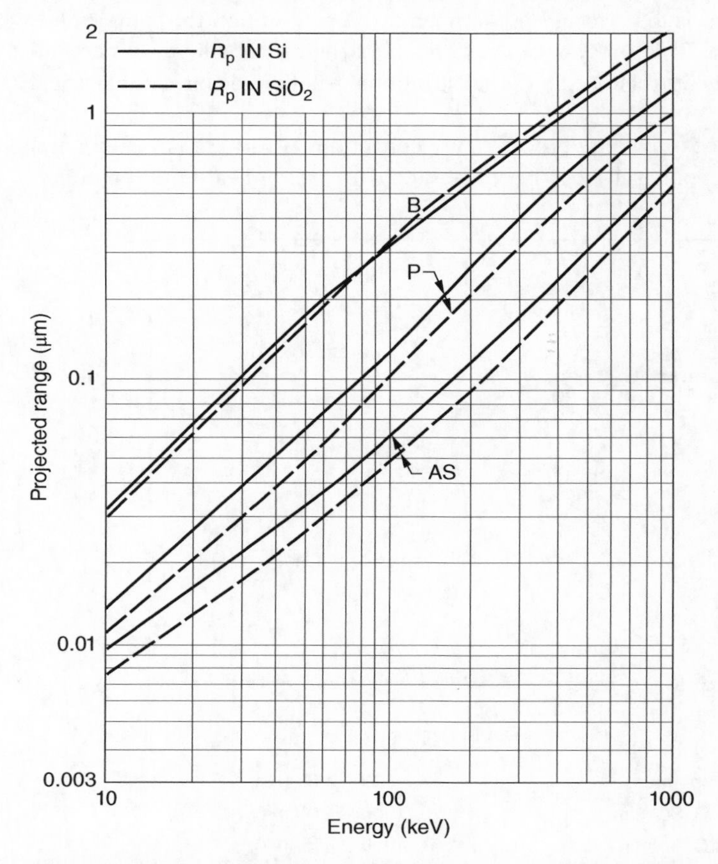

Figure 17.16 The projected range of impurities in Si and SiO_2. Source: [3] Wolf and Tauber.

masking materials is different and should be selected according to the implantation conditions. In the actual implantation process, formula 17.24 can be used to estimate the minimum thickness of the masking film. Note that all parameters in the formula refer to the values in the corresponding masking material [2].

$$d_{\min} = \overline{R_{\mathrm{p}}} + 4\overline{\Delta R_{\mathrm{p}}} \tag{17.24}$$

17.15 Annealing Process

After implantation, the integrity of silicon lattice is destroyed by the impact of high-energy particles, resulting in various defects. The impurity atoms entering the silicon do not reach the substitutional position to activate the electricity. The annealing process is to perform high-temperature treatment on the sample after ion implantation. This process can restore the structure of silicon lattice and push the impurity atoms to reach the substitutional position to generate electrical activation.

Commonly used annealing processes are conventional furnace annealing and rapid thermal annealing (RTA). The conventional furnace annealing equipment is

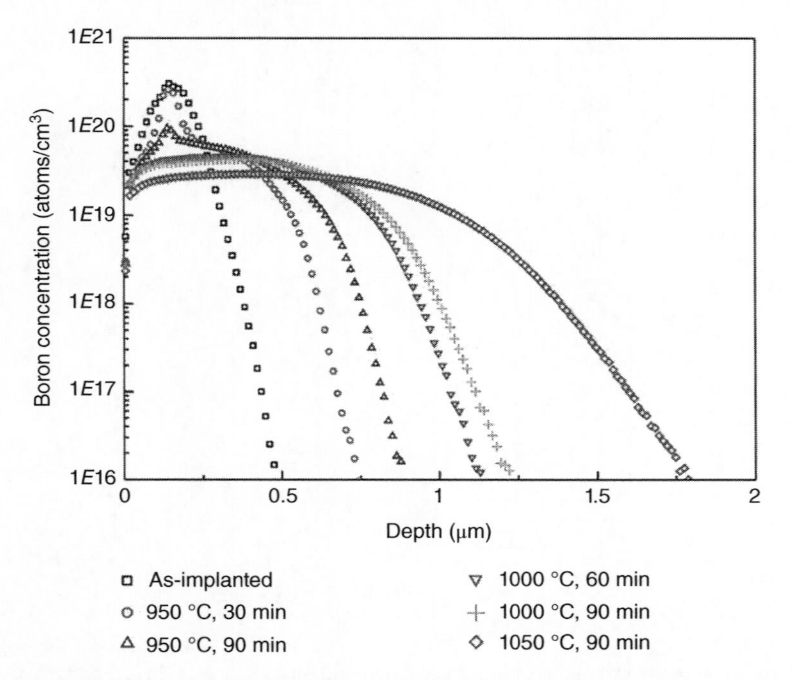

Figure 17.17 The distribution of B impurity concentration for different conventional annealing temperatures and times. Source: [11] Boo et al. / Hindawi Publishing Corporation / CC BY 3.0.

the same as the thermal oxidation furnace. Nitrogen is used to protect the sample during annealing, or using oxygen, so that annealing and oxidation can be performed simultaneously. The annealing time of this process is long, the impurities are pushed deeply into the silicon, and the impurity concentration after annealing is redistributed obviously. It is suitable to make deep junctions. For RTA, the heat treatment time is short, and the change of the impurity concentration distribution after annealing is relatively smaller. For III–V semiconductors, conventional furnace annealing is not suitable, because long-term, high-temperature heat treatment will cause greater damage to these semiconductor structures. Therefore, RTA must be used (refer to 16.6.2). When annealing III–V semiconductors, cover the front and back of the sample with polished silicon wafers to avoid possible material decomposition. The commonly used protective gas for RTA is nitrogen. Helium, argon, and forming gas can also be used.

Figure 17.17 is an impurity distribution diagram with different furnace annealing conditions after boron implantation. "1E16" in the figure is 10^{16}, and so on; "As-implanted" is no annealing. In the case of low temperature and short time, the profile still has peak of concentration. Figure 17.18 is the distribution of boron concentration after RTA. After annealing, the four-point probe is used to determine the activation rate of impurities by measuring the sheet resistance. The results in these two papers show that a short time (10 seconds) of RTA annealing can meet

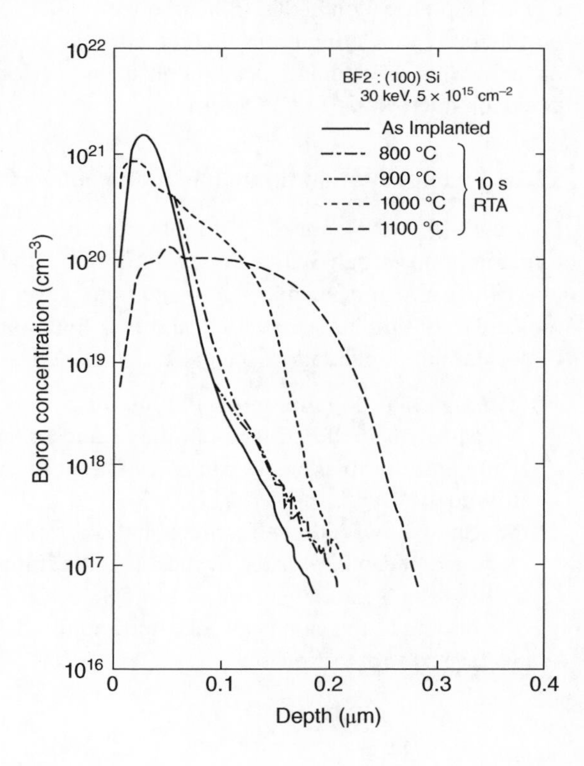

Figure 17.18 Time is 10 seconds, the distribution of B concentration for different RTA temperature. Source: [12] Mikoshiba and Abiko.

the specifications of device manufacturing. However, furnace annealing needs to take long time (30 minutes even more). The junction depth is very shallow in RTA, which is good for the manufacture of small feature size devices.

17.16 Buried Implantation

Figure 17.13 shows that, unlike the diffusion process, the peak of impurity in implantation process is not on the silicon surface. The dose and energy of implantation can be controlled separately. In this way, by controlling the energy and thickness, we can adjust the concentration peak to the surface or any area under the surface of silicon through the masking films, such as SiO_2, Si_3N_4, and photoresist. Plus the dose controlling, the parameters of a device can be precisely controlled. The most common examples are

1. The doping of D, S, and G regions in modern CMOS devices, and ICs is finished by implantation through the pad oxide (Section 13.2).
2. Realize shallow implantation.

The high-energy deep implantation without masking film can realize the buried implantation layer under the silicon surface. The most used field of this technology is the preparation of SOI (Silicon-on-Insulator). SOI means that the silicon is on the top of insulating layer. It is the opposite of the structure produced by the dielectric film growth or deposition on silicon surface. SOI has been widely used in silicon microchips' manufacturing.

17.16.1 Implantation through Masking Film

Figure 17.19 is the impurity concentration distribution diagram of phosphorus implantation through SiO_2 masking film. In the figure, P_2^+ is phosphorus dimer ion. Dimer implants need low dose and high power. This technique is also called molecular ion implantation. BF_2^+ and PF_2^+ mentioned in Section 17.13 are also molecular ion implantation. It can be seen from the figure:

(a) Through masking film implantation, the peak of impurity concentration can be adjusted to the surface of silicon and other semiconductor substrates. Different film thicknesses can use different implantation energies to achieve this goal.
(b) We can use low energy to achieve shallow implantation. If a shallower layer is required, it can be assisted by masking film implantation.
(c) If the SiO_2 is gate dielectric of a MOS device, then through this technique, the doping of the gate channel can be adjusted. This is the precise threshold voltage control technology.

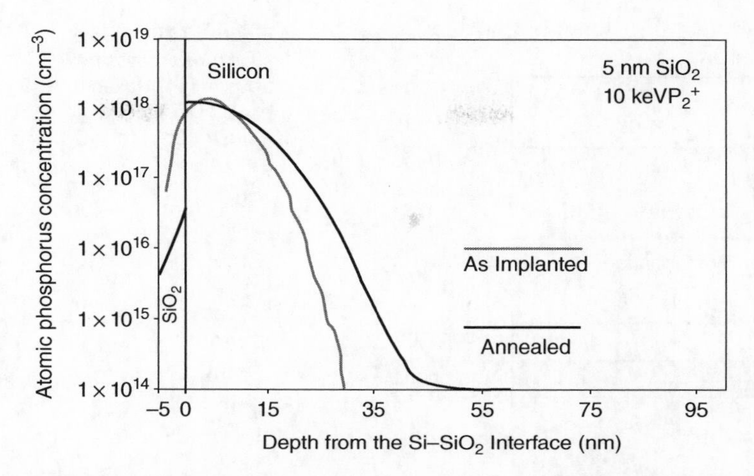

Figure 17.19 Concentration distribution curve of P implanted through SiO_2. Source: [13] Spizzirri et al.

17.16.2 SOI Manufacture

There are three main structures of SOI. [3] The first structure is Silicon-on-Sapphire (SOS), and sapphire is Al_2O_3. The second structure is Separation by IMplanted Oxygen (SIMOX). The third structure is wafer bonding (WB).

The first structure is SOS. SOS technology is limited to fabrication of MOS devices. The structure is epitaxial growth of (100) silicon crystal on the surface of sapphire. Because the lattice constants of these two materials are different, this growth is heteroepitaxial growth. The epitaxial process is the technology of growing a thin layer of crystal on another crystal surface. The growth of homogeneous materials, such as regrowing silicon on the surface of silicon, is homoepitaxial. The growth of different materials is heteroepitaxial. This book does not discuss epitaxial growth. Due to the different lattice constants, the heterogeneously grown silicon material has a high defect density and cannot be used for device fabrication. To improve the situation of defects, a high concentration of silicon is implanted into the epitaxial silicon layer to make silicon at the interface of silicon and sapphire become amorphous. After high-temperature treatment, the amorphous silicon at the interface becomes crystalline silicon. This recrystallization process can greatly reduce the defect density so that the quality of this silicon crystal film can meet the specifications of device manufacturing. Figure 17.20 is a schematic diagram of the SOS manufacturing process flow.

The second structure is SIMOX. Figure 17.21 is an example of a SIMOX manufacturing process flow [3]. This technology is also called Buried

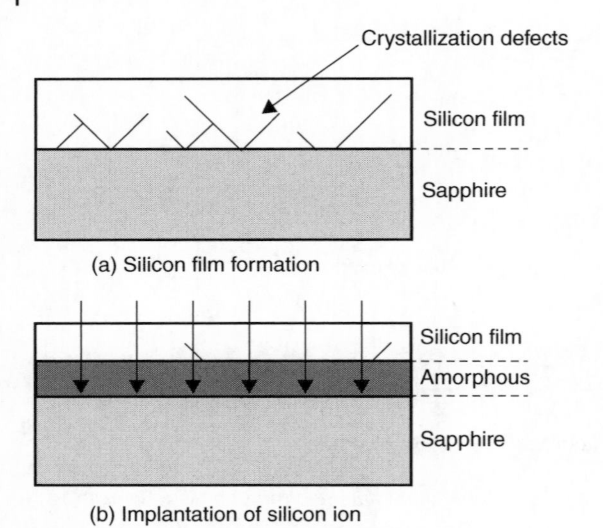

Figure 17.20 SOS structure process flow. Source: [14] Nakamura et al. / Oki Electric Industry Co., Ltd.

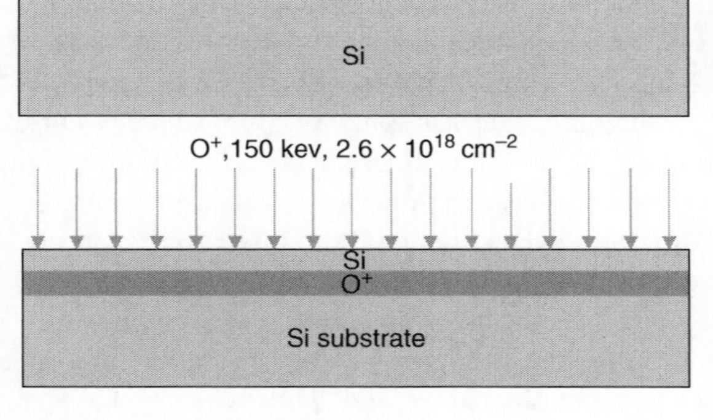

Figure 17.21 Schematic diagram of SIMOX (BOX) process flow.

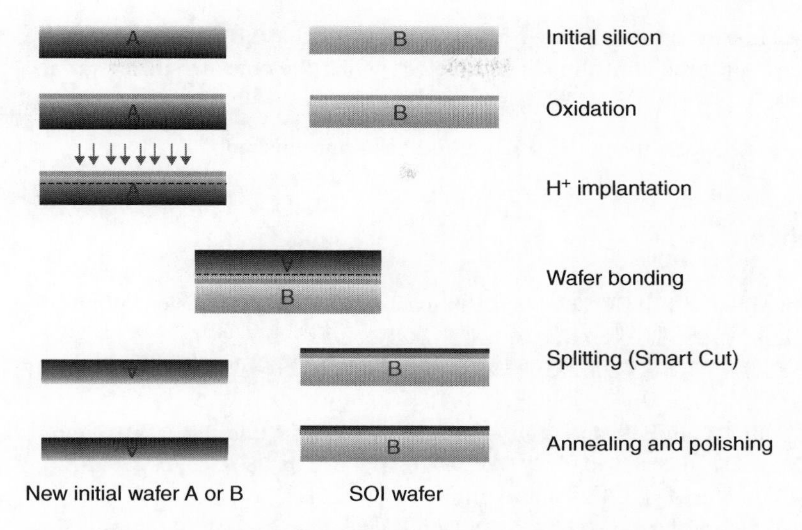

Figure 17.22 Schematic diagram of wafer bonding process flow.

Oxide – BOX – process. If the thickness of the silicon on the BOX is not enough to make device, epitaxial technology can be used to grow thicker silicon crystal layer. Different devices have different requirements for materials. According to needs, the implantation energy and dose of oxygen ions vary within a certain range. If we want to grow an oxide layer on silicon surface, we can anneal the wafer in an oxygen atmosphere.

The third structure is WB. Figure 17.22 is a schematic diagram of the WB process flow, which mainly contains the following six steps:

(1) Prepare two silicon wafers (Initial Silicon) A and B for bonding.
(2) Oxidize A and B to be about 0.5 μm.
(3) Proton H^+ implantation is performed on A, and the implantation depth is about 0.2 μm under the SiO_2 layer. When a hydrogen atom is ionized and loses an electron, it becomes a nucleus with only one positive charge. Since the proton in the nucleus is a positive charge, a hydrogen ion H^+ is often called a proton.
(4) Clean A and B, and then one oxide layer is laminated to another oxide layer. Heat treatment is performed below 400 °C to make the two wafers adhere closely together. This is the wafer bonding.
(5) Subsequently, the two bonded wafers are annealed at ≤600 °C. The bubbles will be generated at the peak of proton implantation of A wafer, causing wafer A to split from the implantation site (Splitting). This technique of cutting wafers is called Smart Cut, and the cut wafers can be reused.

(6) These smartly cut two bonded wafers are annealed at 1100 °C [3] to remove defects. After the annealing is completed, polish the cracked silicon surface (CMP).

So far, an SOI wafer using the WB method has been finished.

References

1 Darling, R.B. (2013). EE-527, MicroFabrication, Solid-State Diffusion, Winter.

2 电子工业生产技术手册 7, "电子工业生产技术手册" 委员会, 半导体与集成电路卷, 硅器件与集成电路 223页, 182页, 259至273页, 244至245页, 305页, 330页, 333–334页, 340页, 国防工业出版社, 1991年。

3 Wolf, S. and Tauber, R.N. *Silicon Processing for the VLSI Era*, Process Technology, 2e, vol. 1, p. 328, 330, p. 297, p. 360, p. 371–372, p. 380–383, p. 256–261.

4 ECE Illinois, ece444, GT10-Silicon Diffusivity Data.

5 Ghezzo, M. and Brown, D.M. (1973). Diffusivity summary of B, P, As, and Sb in SiO_2. *Journal of the Electrochemical Society* 120 (1): 146–148.

6 [美] H. F. 沃尔夫 编. (1975). 半导体工艺数据手册, 天津半导体器件厂 译, 国防工业出版社, 523页, 359页。

7 Phelps, G.J. (2004). Dopant ion implantation simulations in 4H-silicon carbide. *Modelling and Simulation in Materials Science and Engineering* 12: 1139–1146.

8 电子工业生产技术手册 8, "电子工业生产技术手册" 委员会, 半导体与集成电路卷, 化合物半导体器件, 257–258页, 国防工业出版社, 1992年。

9 Bo Cui, E.C.E. Ion implantation, Chapter 8. In: . University of Waterloo, SlidePlayer. https://dokumen.tips/documents/chapter-8-ion-implantation-ii.html

10 Agah, M. Ion implantation, Chapter 8. In: . Virginia Tech, SlideServe. https://pdfslide.net/documents/chapter-8-ion-implantation-instructor-prof-masoud-agah.html?page=16

11 Boo, H., Lee, J.H., Kang, M.G. et al. (2012). Effect of high-temperature annealing on ion-implanted silicon solar cells. *International Journal of Photoenergy* 2012: 921908, 6 pages.

12 Mikoshiba, H. and Abiko, H. (1986). Junction depth versus sheet resistivity in BF_2^+-implanted rapid-thermal-annealed silicon. *IEEE Electron Device Letters* EDL-7 (3): 190–192.

13 Spizzirri PG, Wayne D. Hutchison, Nikolas Stavrias, et al., "ESR studies of ion implanted phosphorus donors near the Si-SiO_2 interface", *ResearchGate*2010.

14 Nakamura, T., Matsuhashi, H., Nagatomo, Y. et al. (2004). Silicon on sapphire (SOS) device technology. *Oki Technical Review* 71 (200), No. 4: 66–69.

18

Process Control Monitor, Packaging, and the Others

So far, we have discussed the main technologies used in semiconductor manufacturing processes. In addition to these, there are other technologies used in the fabrication of microchips, such as epitaxy, molecular-beam epitaxy (MBE), metalorganic chemical vapor deposition (MOCVD), and chemical–mechanical polishing (CMP). These technologies involve more issues on material, chemical, vacuum, and so on. Some machines and processes are complicated. In the university clean room, most users do not use such machines and technologies, so these processes are not discussed in this book.

To make high-quality microchips, we need to carefully control each step of the processes. Process control monitor (PCM) is very important because the details of each step of the process can be obtained through PCM. PCM is to use various patterns to monitor the process. These patterns can be designed together with the device patterns and placed on the same photomask. The patterns can be measured to achieve the purpose of understanding the process quality. In this chapter, we use four examples to illustrate the design of PCM patterns. After that, the chip packaging and some other issues are briefly discussed.

18.1 Dielectric Film Quality Inspection

The dielectrics used in semiconductor microchips are mainly SiO_2 and Si_3N_4 films. There are many parameters to indicate quality of the films. The commonly used ones are stress, refractive index, and breakdown voltage. Among them, the breakdown voltage V_B is the most important parameter. Figure 18.1 is the best structure for measuring V_B, the metal–insulator–metal sandwich structure. To avoid misalignment in lithography causing the two layers of metal to contact each other, the area of the insulator should be slightly larger than the area of the metal layers. The fabrication of the entire structure can be carried out simultaneously with the device fabrication, and the area of the dielectric film (insulator) should

Semiconductor Microchips and Fabrication: A Practical Guide to Theory and Manufacturing, First Edition. Yaguang Lian.

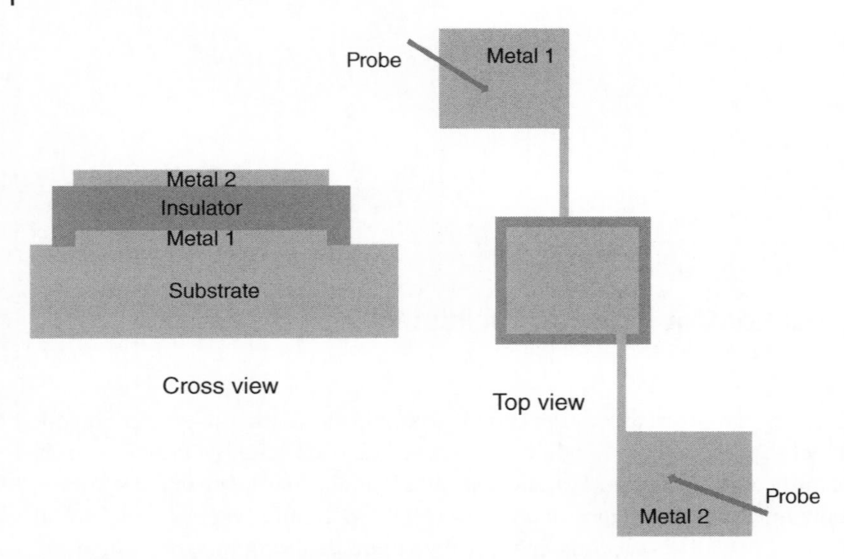

Figure 18.1 Pattern structure for measuring dielectric breakdown voltage.

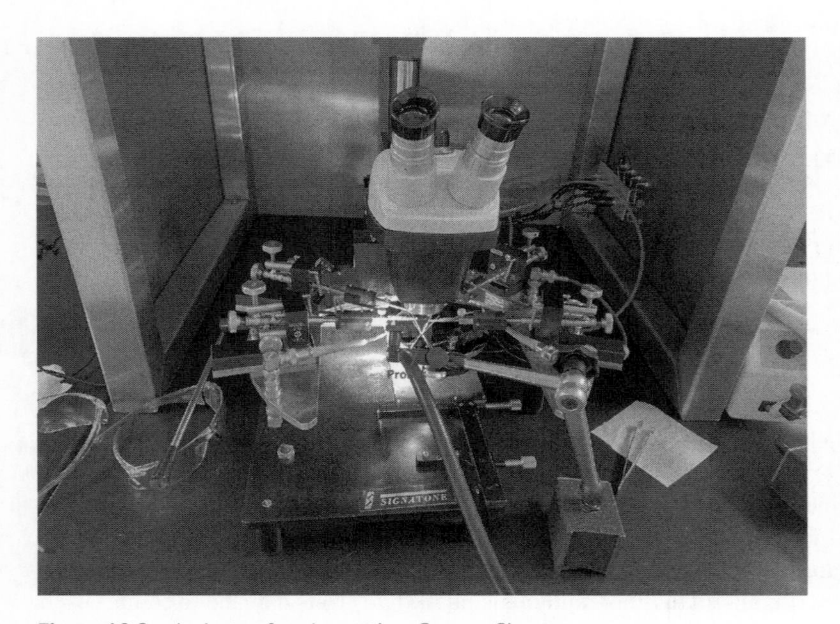

Figure 18.2 A photo of probe station. Source: Signatone.

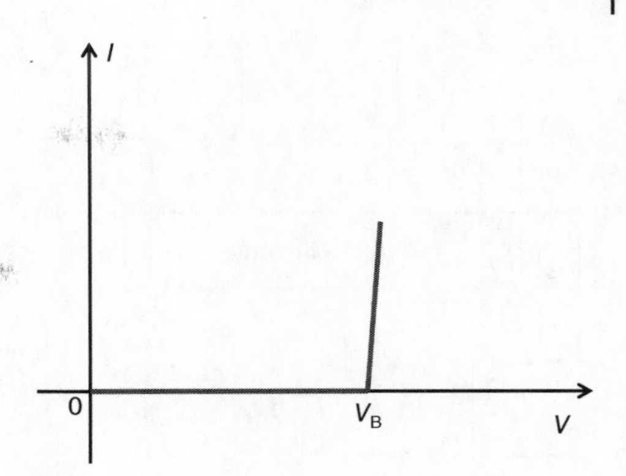

Figure 18.3 The *I*–*V* curve of breakdown voltage.

be as close as possible to the maximum insulation area of the device. Once the structure is completed, use a probe station to measure the V_B of the film. Figure 18.2 is a photo of a probe station. In the photo, the probes are indicated by a arrow. Below the probes is the wafer to be tested. Figure 18.3 is the *I*–*V* curve obtained by measuring V_B of the dielectric film. Before the breakdown occurs, there is no current. Once the breakdown occurs, the current will rise suddenly, and the voltage in this point is V_B. The fewer the number of pinholes in the film (refer to chapter thirteen), the higher the breakdown voltage and the better the quality of the film are; vice versa.

As discussed in Chapter 13, the higher the temperature for dielectric film growth, the lesser the pinholes in the film. So, higher temperature process of dielectric film can provide higher breakdown voltage of the film. If the film is done by lower temperature, we are able to increase the breakdown voltage of the film by higher temperature annealing. For example, plasma-enhanced chemical vapor deposition (PECVD) film may be annealed at the temperature from 400 to 600 °C to increase the breakdown voltage. Due to the stress of film, the annealing temperature cannot be very high. Annealing at very high temperature may cause the film to crack. Therefore, we should optimize the annealing conditions by doing experiments to meet the needs of different device manufacturing.

18.2 Ohmic Contact Test [1]

After the ohmic contact process is completed, the resistance of the contact needs to be measured to observe whether the resistance meets the specifications. There are many methods for measuring contact resistance. Transmission line measurement or transfer length measurement (TLM) is a widely used method. It obtains ohmic contact resistivity by measuring the resistance of the pattern. This resistivity is

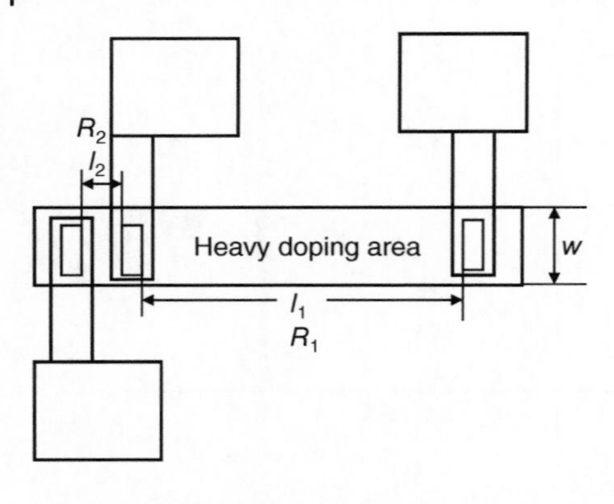

Figure 18.4 Ohmic contact test pattern.

also called specific contact resistivity, and the unit is $\Omega\,cm^2$. The pattern used to test the resistivity is shown in Figure 18.4. l_1 and l_2 are the distances between two near metal contact vias, R_1 and R_2 are the corresponding resistances, and w is the width of the heavy doping area. When measuring, the probes are stuck on three squares of metal film.

The formula of specific contact resistivity is shown below:

$$\rho_c = R_c \cdot A_c \tag{18.1}$$

ρ_c in the formula is the specific contact resistivity, R_c is the contact resistance, and A_c is the contact area. The expression of R_c is as follows:

$$R_c = \frac{R_2 l_1 - R_1 l_2}{2(l_1 - l_2)} \tag{18.2}$$

In general, the specific contact resistivity of metal and Si $\rho_c \approx 10^{-5}$–$10^{-7}\Omega\,cm^2$, metal and metal $\rho_c < 10^{-8}\Omega\,cm^2$ [2].

18.3 Metal-to-Metal Contact

In semiconductor interconnection technology, metal-to-metal (or silicide, Figure 11.4) contact is the basic process, and it is very important to control the quality of metal-to-metal contact in the process flow. The contact between two layers of metal is generally achieved through the via in the dielectric film between the metals. If the via is not well done, for example, after dry or wet etching, there are insulating residuals left on the metal surface of the via, then the next layer of metal cannot have good contact with this layer of metal. The contact resistance increases. So, we need to test the contact between two metals to determine whether the process meets the specifications. Figure 18.5 is the pattern used to measure the contact between metals. The SiO_2 in the figure is called inter-metal

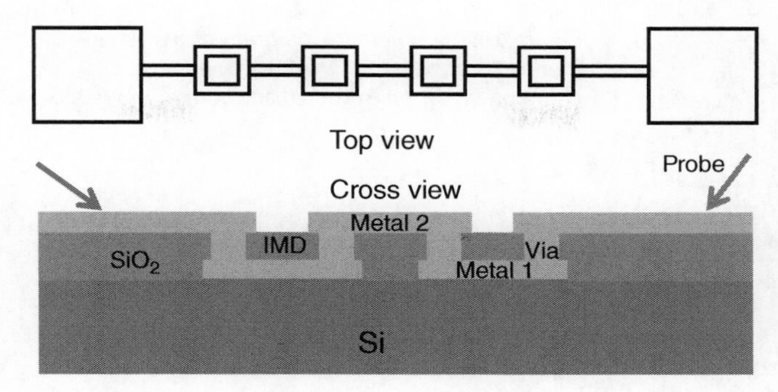

Figure 18.5 The pattern used to test the metal-to-metal contact.

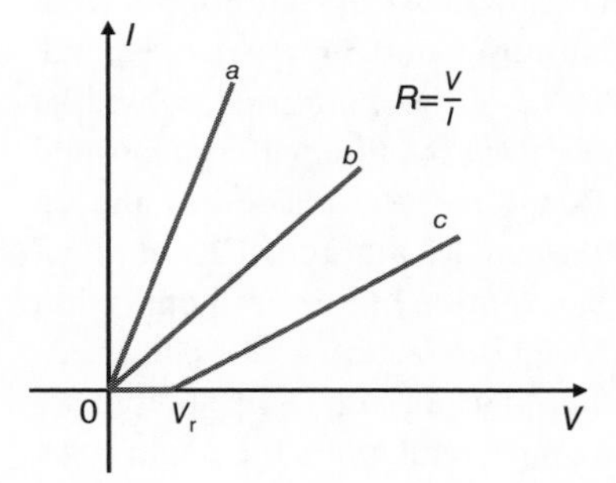

Figure 18.6 The test curve obtained from the pattern in Figure 18.5.

dielectric (IMD). The size of the via should be the same as the smallest size of via in the interconnection.

From the size of the pattern, the thickness of the metal and the resistivity, and according to the formula shown in Figure 1.3, we can calculate the resistance of the test pattern. Perform probe measurement on the pattern shown in Figure 18.5, and the result is shown in Figure 18.6. If the result is curve a, the value is very close to the theoretically calculated value, it means that the contact between the metals is good. If the result is b, the resistance increases, it may be caused by two cases: case 1, during dry etching of the vias, some metal films can be sputtered out, cover the vias' surface and up edge; cases 2, the metal in the vias has voids. If the curve is c, it means that there is a residual of dielectric film between the metals. As shown in Figure 18.7, the residual remains on the contact surface of the two layers of metals. The voltage V_r from the straight line to the rising inflection point is the breakdown voltage of the insulating residual. The thicker the residual, the bigger the V_r.

Let us discuss more about the curve b in Figure 18.6.

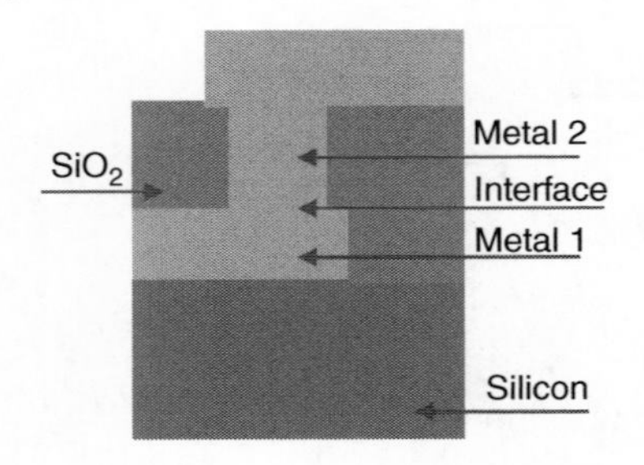

Figure 18.7 Schematic diagram of two-layer metals contact. Metal 2 used as via metal is a common situation in the clean room at university.

Case 1. This problem often occurs when the size of via is greater or equal to 0.5 μm and the aspect ratio is smaller than 0.5. The SiO_2 via is etched by fluorine-containing gases in reactive-ion etching (RIE). But F cannot etch most metals, such as Al. RIE has DC bias and can create physical sputtering during dry etching. If the etching time setting is not accurate and SiO_2 is over etched, the metal will be exposed to F plasma. In this case, the metal surface will be physically bombarded, and some metal particles from the film will be deposited on the sidewall and the up edge of the via. Rough surface of sidewall and up edge of the via makes the connection between metal 2 and metal 1 not good, and the contact resistance becomes bigger. One solution of this problem is that the dry etching of via should be done by under-etching, leave a 10–50 nm SiO_2 film at the bottom of the via, and then use buffered oxide etch (BOE) to remove this thin film. Due to BOE can etch aluminum, be careful when the metal is Al.

Case 2. As the size of the device becomes smaller and smaller, the size of dielectric via also becomes smaller and smaller. It becomes more and more difficult for the metal made by physical vapor deposition (PVD) to fully fill the via. As mentioned in Section 16.4, the step coverage of sputtering is better than thermal and e-beam. However, the sputtered metals still have void issue (refer to Section 13.3). One problem of the voids is to increase the contact resistance of the metals. To overcome the voids, a collimator can be inserted in the space between the target and the wafer to increase the coverage [3]. However, a collimator still cannot solve the problem. So, other techniques are also used. They are planarization in multilevel interconnects, sloped-sidewall vias, and raising the substrate temperature to increase the surface-migration ability of atoms. Even so, when the feature size gets smaller further, sputtered Al still cannot meet the needs of via filling. Several technologies have been investigated for completely filling vertical small vias. The process of chemical vapor deposition (CVD) – tungsten (W) plugs is the most widely used technology in the fabrication for complementary metal-oxide-semiconductor (CMOS) down to about 0.35 μm [3].

Why is tungsten chosen? There is because WF_6, the compound of tungsten and fluorine, has a boiling point of 17.1 °C, which is a gas under the process pressure.

That is also why we can etch tungsten with fluorine-containing gas. In turn, we can use the gas characteristics of WF_6 to achieve CVD deposition of W. CVD-W provides superior via-filling-capabilities. It can be finished by reduction reaction of WF_6 with H_2 or SiH_4 in CVD chamber at the temperatures from 300 to below 450 °C.

The hydrogen–reduction–reaction is given by:

$$WF_6(\text{gas}) + 3H_2(\text{gas}) \rightarrow W(\text{solid}) + 6HF(\text{gas}) \quad (18.3)$$

The silane–reduction–reaction is given by:

$$2WF_6(\text{gas}) + 3SiH_4(\text{gas}) \rightarrow 2W(\text{solid}) + 3SiF_4(\text{gas}) + 6H_2(\text{gas}) \quad (18.4)$$

Since W does not adhere well to SiO_2, a thin TiN adhesion layer is first formed on SiO_2 by sputtering or CVD. Sputtering-TiN is done by introducing N_2 to the chamber as reactive gas during Ti sputtering process to produce TiN film. CVD-TiN is finished by the reaction of $TiCl_4$ (boiling point is 136.4 °C) and NH_3 in CVD chamber. In these two techniques, Si may react with N_2 to create Si_3N_4. So, a thin Ti film should be made before TiN as a contact-forming layer [4]. Ti film is created by sputtering or CVD. During the sputtering process, if N_2 is not input, there will be a Ti film. During the CVD process, if NH_3 is not input, there will be a Ti film. So, the structure of W-plug is Ti–TiN–W. Usually, CVD-W process is finished by blanket process, which means that W, TiN, and Ti films are deposited onto the whole surface of the wafer. After the deposition is done, CMP (Section 10.4) is used to remove excess W, TiN, and Ti from the surface of the wafer and leave them only inside the vias. Right now, some companies (for example, Kurt J. Lesker) supply TiN targets for TiN sputtering process.

The electric resistivity of TiN can be 270 μΩ cm by using sputtering deposition [5]. In addition, the resistivity of tungsten is 4.89 μΩ cm, which is nearly twice that of aluminum (2.5 μΩ cm), and that of titanium is 42 μΩ cm [6]. Therefore, under the premise of meeting the specifications of device manufacturing, the thinner the films of Ti and TiN, the better. Aluminum is still used as the main metal for interconnection. Please refer to Figure 11.4.

18.4 Conductive Channel Control

In the manufacture of metal–semiconductor field effect transistor (MESFET), the thickness (depth) of the gate conductive channel is a parameter that needs to be carefully controlled, refer to Figure 6.10. The gate channel layer can be completed by MBE, MOCVD, or ion implantation. But for different device requirements, the thickness of the formed channel needs to be adjusted, generally thinned. This process is called a recess etching process. To accurately control the thickness to be reduced, we need to design a pattern for process monitoring, Figure 18.8 is

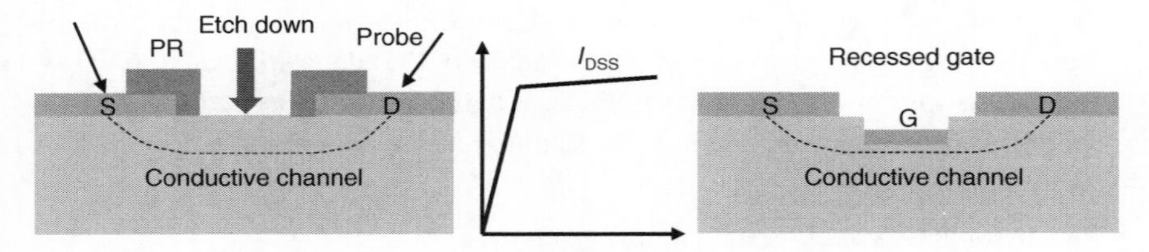

Figure 18.8 Control pattern for recess etching process of MESFET device.

the process monitoring pattern for recess etching. After completing the source (S) and drain (D), perform gate lithography to finish gate pattern, and then dry or wet etching (etch down) on the pattern. After the etching is completed, use a probe station to measure the saturation current I_{DSS} through the gate channel between the source and drain (refer to Figures 6.7 and 6.11). If the I_{DSS} meets the requirements of the device, stop the etching. Otherwise, continue to etch, until the design requirements are met. After that, the gate metal is made, thus completing the manufacture of the recessed gate MESFET.

We have used four examples to discuss how to use PCM patterns to control and monitor the processes. Different processes need to use different PCM patterns. In order to make high-quality microchips, each step of the process requires careful controlling and testing.

18.5 Chip Testing

When the microchips on the wafer are finished, we need to test them. For semiconductor devices and ICs, the probe station is the most important test equipment (Figure 18.2). Diodes and transistors, four probes can meet the requirements. For the testing of integrated circuits, probe cards are needed. Figure 18.9 is a photo of the probe card. If a bad die is found, the machine puts a drop of dye on the bad one to separate it from the good ones.

Figure 18.9 Probe card (a), probe card installed to the probe station and testing ICs (b). Source: Accuprobe Corporation.

18.6 Dicing

After all dies are tested, the wafer must be diced. This process is called dicing and separates each die on the wafer. The equipment used to cut wafers is called a wafer dicing machine. Figure 18.10 is a wafer dicing machine. The core part of the dicing machine is the saw or diamond blade, through which each die on the wafer is separated, as shown in Figure 18.11. After the dies are separated, throw away the one with dye.

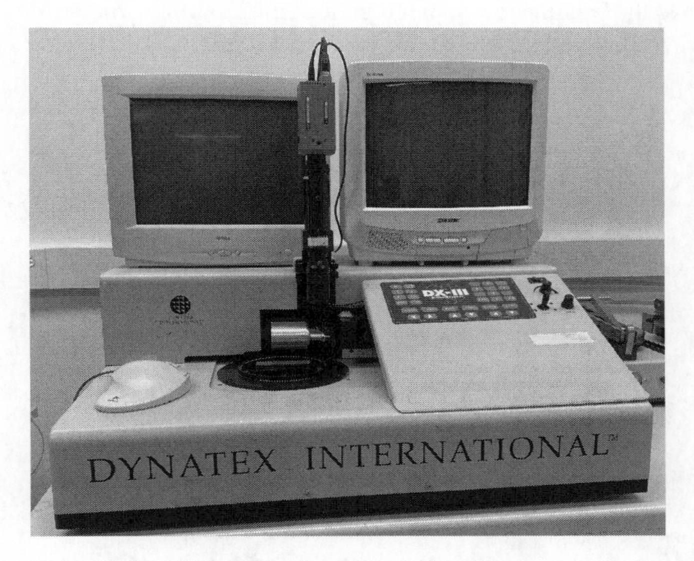

Figure 18.10 Wafer dicing machine. Source: Dynatex International.

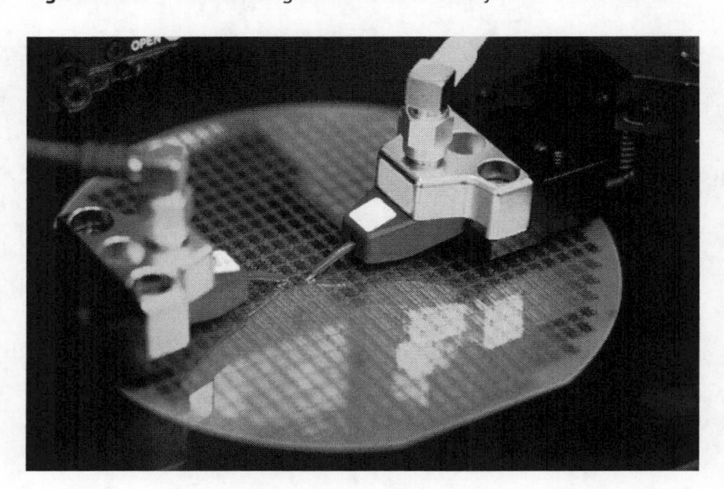

Figure 18.11 The wafer is dicing into dies. Source: Rainer / Adobe Stock.

18.7 Packaging

After a die is separated and selected, it must be packaged, refer to Figures 7.1 and 7.3. Different microchips require different packages, please see Figure 18.12. In most cases, the die and the case are packaged together by wire bonding. The bonding process is to use very thin metal wires, usually aluminum or gold wires, to connect the die and the leads of the case together, as shown in Figure 18.13. Figure 18.14 is a wire bonding photo of an integrated circuit and a simple bonding tool. In addition to the bonding techniques described in this section, there are many other bonding techniques, which will not be described here. At this point, a complete device or IC product is fully made. After final function testing, reliability testing and other special testing if it is needed, the product can be sold to customers.

Figure 18.12 Packages for different types of semiconductor chip.

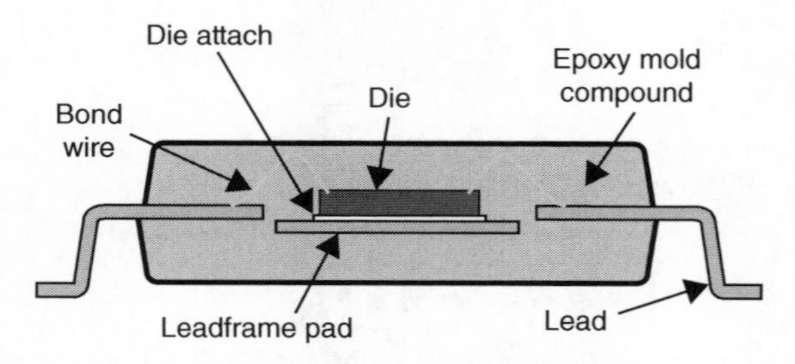

Figure 18.13 Packaging process for semiconductor chip. Source: [7] Joiner.

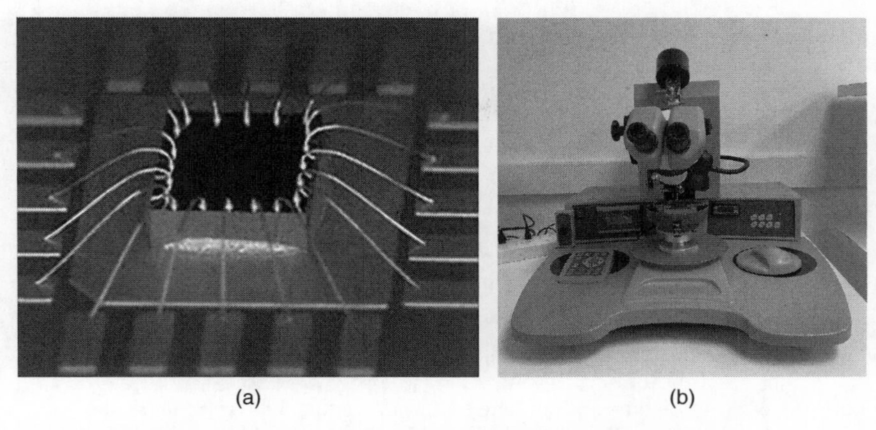

(a) (b)

Figure 18.14 Wire bonding (a) and a simple wire bonding machine (b). Source: (a) Reproduced with permission of First Level Inc.

18.8 Equipment Operation Range

In the manufacturing of semiconductor microchips, a lot of equipments are needed. In these machines, there are many electronic components, such as power supplies and mass flow controllers. When operating the equipment, we cannot use these components at full scale. The range should be controlled between 10% and 90%. For example, an MFC with a nitrogen range of 100 sccm (Figure 14.17). The working range should be 10–90 sccm. To understand this issue, a brief introduction of switch circuit is given. The position of a switch in the circuit is shown in Figure 4.1. The commonly used switch is to turn on and off the circuit by opening or closing two metal sheets, as shown in Figure 18.15. When the switch is off, no current flows through the circuit (the current is zero), and the voltage

Figure 18.15 Commonly used electrical switches. Source: Reproduced with permission of Nidec Copal Electronics Corp.

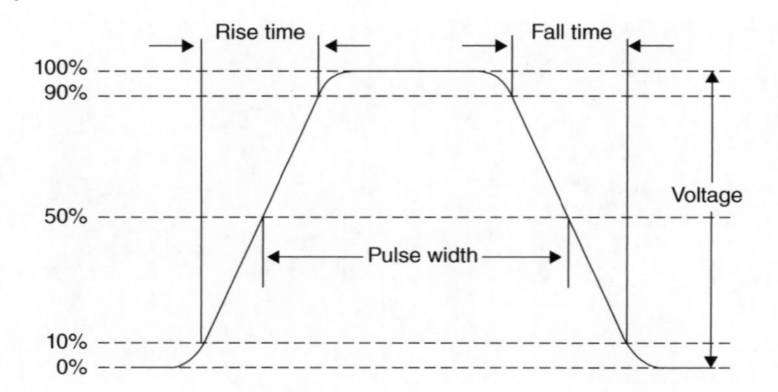

Figure 18.16 The rise time and fall time of a switch circuit. Source: [8] Tektronix.

on both sides of the switch is high (equal to the power supply voltage). When the switch is on, current flows through the circuit (large current), and the voltage on both sides of the switch is low (equal to zero). But in the control circuit composed of microchips, such mechanical switches cannot meet the requirements. We use transistors to achieve similar switching functions. Refer to Figure 6.3, in the application of the transistor in this figure, the "saturation mode" is equivalent to the switch on, because it is a large current and low voltage, the "cut-off mode" is equivalent to the switch off, because it is a small current and a large voltage. But when the transistor changes from saturation to cut-off, it takes a certain amount of time to complete, and vice versa. When such a switch is regularly switched between saturation and cutoff under the control of a periodic signal, the circuit sends out a series of pulse (on and off) signals. The time required for the conversion is the rise time and fall time, as shown in Figure 18.16. In order to have good linear control and avoid the noise fluctuations of the minimum and maximum signals, which is the situation often encountered in the actual operation of a control circuit, the rise time and fall time are selected as 10–90% and 90–10%. 10–90% is a general principle followed by circuit design. So, for all power supplies, mass flow controllers and other electronic components, their operating range should be set between 10% and 90%.

18.9 Low-κ and High-κ Dielectrics

In the production of integrated circuits, the most used material for MOSFET gate dielectric and interlevel dielectric (ILD, refer to Chapter 11) is SiO_2. In order to discuss the κ value, it is necessary to introduce the relative permittivity of the dielectric material. The κ value of a dielectric is its relative permittivity ε_r. The absolute

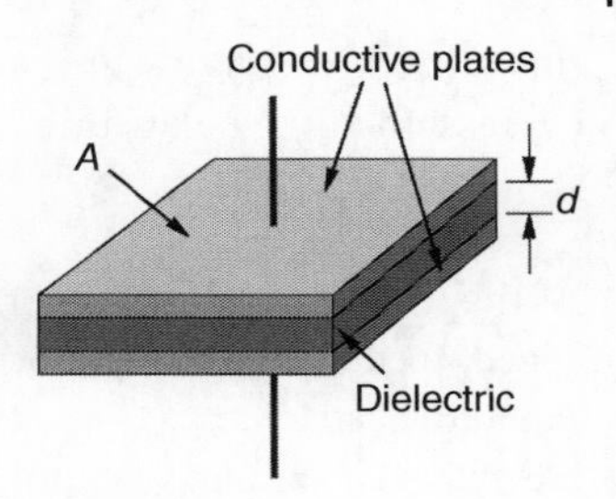

Figure 18.17 The structure of a parallel plate capacitor.

permittivity ε_0 has been introduced in formula (4.3), which refers to the permittivity of vacuum, and its value is:

$$\varepsilon_0 = 8.854 \times 10^{-12}\ \mathrm{F/m} \tag{18.5}$$

The "F" is farad and "m" is meter. The relative permittivity ε_r refers to the ratio of the permittivity of dielectric to that of vacuum. The measurement method is to firstly measure the capacitance C_A of the two plates in air (air permittivity and vacuum permittivity are basically the same). After adding a certain dielectric between the plates, the capacitance C_D is measured. The relative permittivity is:

$$\kappa = \varepsilon_r = C_D/C_A \tag{18.6}$$

The relative permittivity is also called the dielectric constant. In most cases, we use ε_r to represent the relative permittivity of dielectric. The κ value of the most important dielectric SiO_2 is 3.9 [9]. We discussed capacitors in chapter 4. To further discuss the κ, we need to do more about capacitors. Figure 18.17 is a structural diagram of a capacitor.

The capacitance C is given by the following formula:

$$C = \frac{\epsilon_0 \epsilon_r A}{d} = \frac{\epsilon_0 k A}{d} \tag{18.7}$$

As the feature size of ICs becomes smaller and smaller, and the operating speed becomes faster and faster. In the selection of dielectric, with SiO_2 as the standard, there are two directions, low-κ and high-κ. Low-κ means smaller value than SiO_2, high-κ means larger value than SiO_2. Low-κ is used in ILD, and high-κ is used in gate dielectric.

18.9.1 Copper Interconnection and Low-κ Dielectrics

When metal is used as an interconnection "wire" on the dielectric film, the interconnection metal and the metal or conductive layer under the dielectric form a plate capacitor. We assume that the resistance of the metal is R and the capacitance of the capacitor is C. When an electric signal passes through this structure composed of resistors and capacitors, the signal is delayed. The expression of the delay time $\mathcal{T}$ is as follows:

$$\mathcal{T} = RC \tag{18.8}$$

As the IC gets faster and faster, this delay becomes a key factor hindering operation speed. It can be seen from formula (18.8) that in order to reduce $\mathcal{T}$, the resistance R and the capacitance C must be reduced. Since the birth of integrated circuits, Al has played a leading role in the interconnection process due to its low price, easy processing, and good compatibility with silicon and silicon dioxide. The electric wires used in our daily life are made of Al or Cu. The resistivity of Cu (1.55 μΩ cm) is less than that of Al (2.5 μΩ cm), and the price is not expensive. So, we naturally think that if copper is used replacing aluminum in the IC, the delay time of metal interconnection will be reduced. The power consumption also can be reduced. Power consumption is another important factor restricting modern integrated circuits. One more thing is that Cu films exhibit much better electromigration (Section 16.6) than those of Al. The upper limit of current density to prevent electromigration for Cu is 5×10^6 A/cm^2, whereas for Al it is 2×10^6 A/cm^2 [3]. Unfortunately, the processing of copper in semiconductors is much more difficult. There are mainly technical challenges in process listed below:

1. Dry etching cannot be used because the reactants of copper with oxygen, fluorine, and chlorine are not volatile.
2. Copper cannot be in contact with silicon because it is easy to form deep energy level (Figure 5.6).
3. Copper easily diffuses through SiO_2.
4. Copper does not adhere well to SiO_2.
5. Unlike Al, copper does not form a dense, self-passivating oxide on its surface.

Therefore, it was not used in semiconductor manufacture until 1997 when IBM realized the copper interconnection process [10], which was a sensational event in semiconductor field at that time.

Cu interconnects are primarily finished by electrochemical deposition (electroplating and electroless plating) technologies [11]. To overcome the challenges (3) and (4), the barrier layers of Tantalum (Ta) and Tantalum Nitride (TaN) or Ti and TiN are formed before Cu deposition. This process has a special name-Damascene Process. It has two options: single damascene and dual damascene. Due to less steps and lower cost, dual damascene process is widely used in copper interconnects manufacturing. Figure 18.18 is the schematic diagram of dual damascene process flow. As discussed in Section 15.4, we may design a recipe in RIE that it etches SiO_2 only, not Si_3N_4 (or etching rate is very slow). Therefore, Si_3N_4 can be used as etch stop layer. Si_3N_4 capping passivates Cu surface. Figure 18.19 is a photo of copper multilevel interconnects made in IBM. In the figure, we may

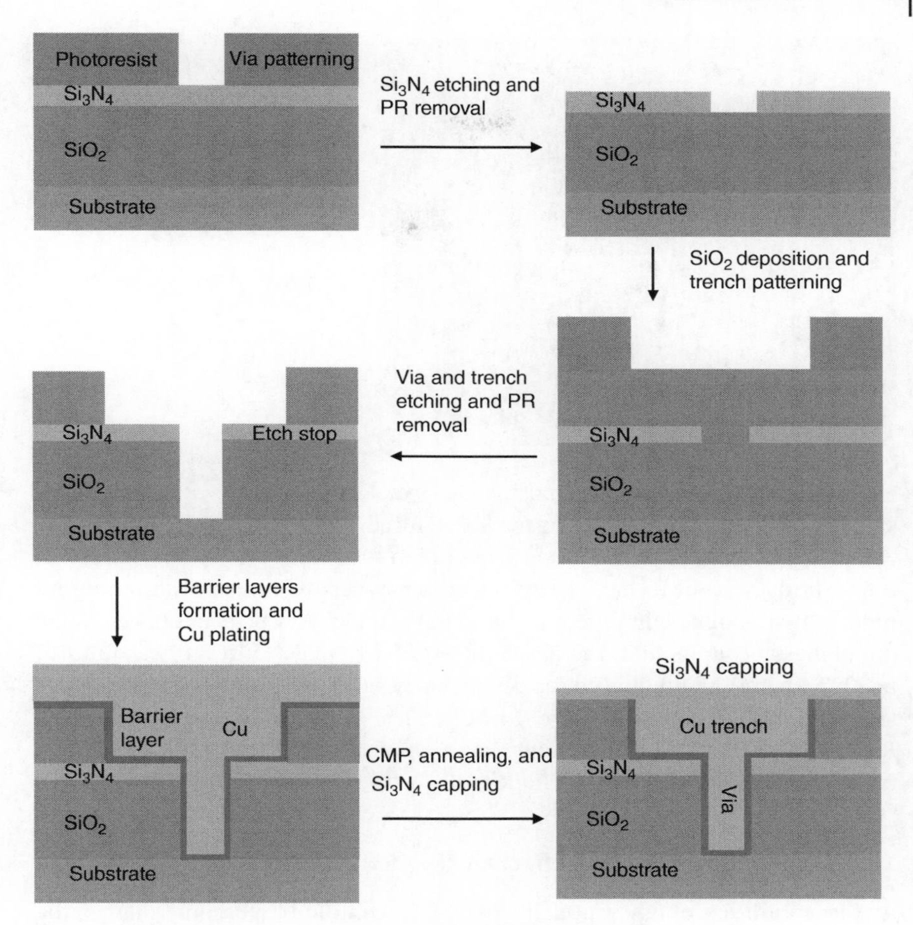

Figure 18.18 A strategy for dual damascene process flow [12].

see the W-plug between the devices (bottom in the figure) and one-level copper interconnect. Other levels don't use W-plug. The vias in one-level have small sizes and need W to fill the openings. Aluminum interconnects have a similar situation.

Copper realizes the low resistance of the interconnection wire, but this is not enough, and the delay time should continue to be reduced. It can be seen from formula (18.8) that in addition to resistance, capacitance can also be reduced. Under a certain process node, the area of interconnection wire and the thickness of ILD have certain requirements, which is equivalent to the fixed values of A and d in the capacitor formula 18.7. ε_0 is also a fixed value. The only thing that can be

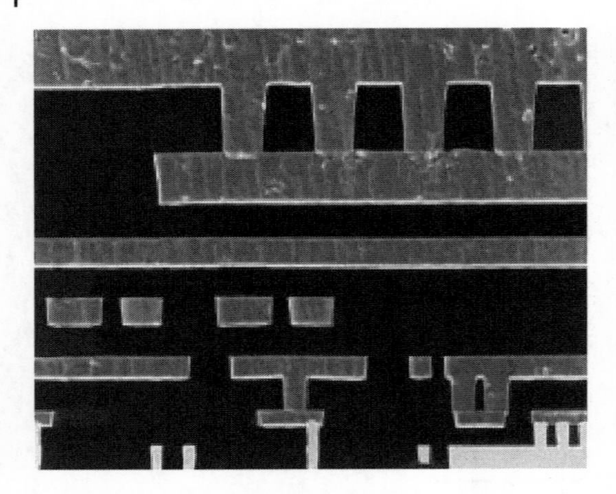

Figure 18.19 Cross section of a six-level copper wiring structure fabricated by IBM. Source: [13] Andricacos et al. 1999 / with permission of IOP Publishing.

changed is k. This shows that by using low-k dielectric, the capacitance becomes smaller and the delay time can be further reduced.

The Cu interconnect was developed by IBM in 1997 using SiO_2 as the ILD at that time. The damascene technology introduces a new set of elements of structure and process that is distinctly different from conventional Al:Cu interconnect. Some difficulties encountered in the development of low k dielectrics. It was fulfilled in 2000 when IBM announced the development of Cu interconnects with SiLK, a low-k dielectric developed by the Dow Chemical Company with k about 2.7 [9]. Right now, there are many available low-k materials and they can be classified into different groups, please see Table 18.1.

18.9.2 Quantum Tunneling Effect and High-k Dielectrics

As the feature size of ICs, especially MOSFETs, continues to become smaller, the thickness of gate dielectric continues to become thinner. The operation of the device will enter the field restricted by quantum mechanics. One of the important issues is the quantum tunneling (tunneling). The tunneling phenomenon is very important in the application and manufacture of semiconductor devices. Numerous devices are based on this phenomenon, including the Zener diode, the Esaki diode, and the resonant tunneling diode. Also, the ohmic contacts made to devices depend upon the tunneling behavior of electrons and holes [15].

For MOSFETs, the basic structures of energy band are shown in Figure 6.14. In this figure, the energy band of SiO_2 in the middle is very wide. So, there are potential barriers between gate metal and SiO_2, and between Si and SiO_2. For Al-SiO_2–Si MOS structure, Ref. [16] gives the measurement results of potential barrier height values at Al-SiO_2 and Si-SiO_2 interfaces. Figure 18.20 is band diagram used in the paper. The measurement results are around 3.65 eV for E_{BG} and 4.40 eV for E_{BS}.

Table 18.1 A classification of low-k dielectrics.

Classification	Material	Dielectric constant (k)
Silica based	SiOF (FSG)	~3.5
	SiCOH	~2.8
Silsesquioxane (SSQ) based	HSQ	~3.0
	MSQ	~2.5
Polymer	Poly(arylene ether) (PAE)	~2.6
	Polyimide	~2.3
	Parylene-*N*(-F)	~2.4
	Teflon (PTFE)	~2.0
Amorphous carbon	C–C (–F)	~2.0
Porous	Porous SiCOH, MSQ, PAE	<2.0
Air gaps	Air	~1.0

Source: [14] Adapted from Cheng et al.

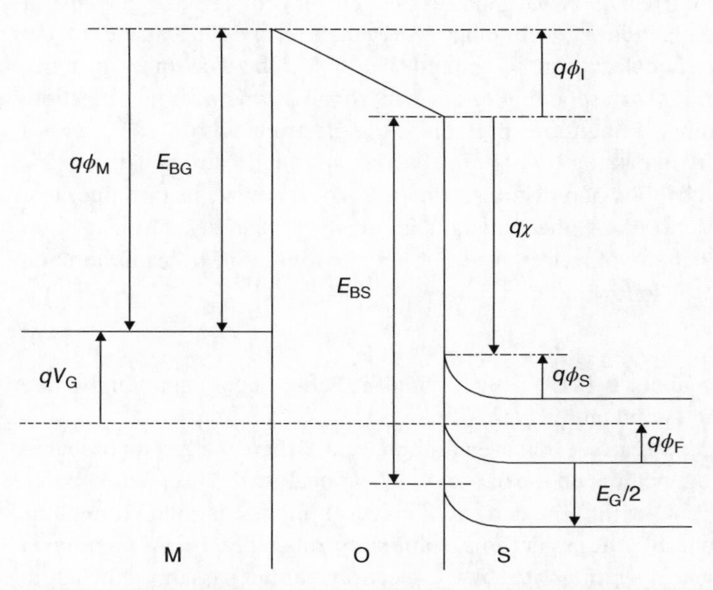

Figure 18.20 Band diagram of the MOS system, at arbitrary gate potential V_G, E_{BG}, and E_{BS} are potential barrier heights at Al-SiO_2 and Si-SiO_2 interfaces, respectively.

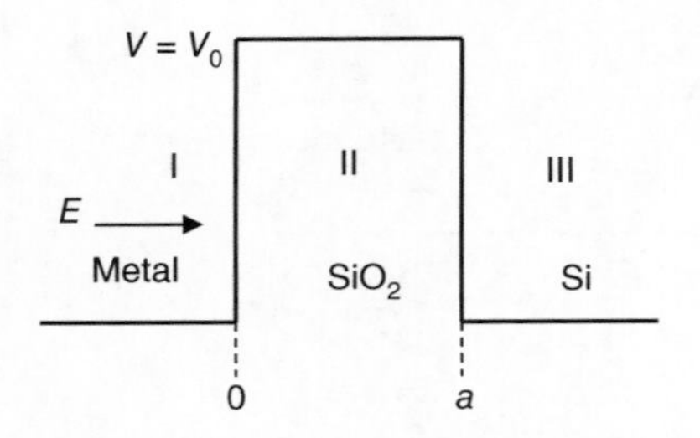

Figure 18.21 A diagram of one-dimensional square potential barrier, the height is V_0 and width is a.

By observing the structures of Figure 6.14 and referring to the above measurement results, we may use a simple diagram in Figure 18.21 to describe a typical MOSFET barrier structure: there is a high potential barrier V_0 between metal and Si. When a voltage is applied to the gate, electrons receive energy E and impinge on the barrier V_0. According to the theory of classical physics, if $E < V_0$, electrons cannot pass through the region II (the electrons and holes in Si have the same situation), which is a necessary condition for a MOSFET to work, because the gate dielectric cannot pass current. But, in actual devices, due to the defects of the dielectric, a very small current may flow through the SiO_2 layer. This current is called gate leakage current, or leakage current for short. Modern semiconductor technology is very mature, and the leakage current caused by dielectric defects can be ignored. If $E > V_0$, the electrons will move across the region II (corresponding to dielectric breakdown) with no reflection. However, according to quantum mechanics, the electrons with $E < V_0$ have a certain probability of passing through the barrier, and the electrons with $E > V_0$ have a certain probability of reflecting from the barrier. Now let us introduce two parameters to describe these phenomena. They are the probability of transmission $|T|^2$ and the probability of reflection $|R|^2$. The probabilities of transmission and reflection satisfy the following condition:

$$|T|^2 + |R|^2 = 1 \tag{18.9}$$

The case we care about is $E < V_0$. By solving the Schrödinger equation, we can get the curves shown in Figure 18.22.

From this figure, we can see that even though $E < V_0$, there is a certain probability that an electron incident on the barrier will be transmitted. This phenomenon, which is inexplicable within the context of classical physics, is called tunneling. The smaller the width a, the larger the transmission probability $|T|^2$. As mentioned above, in the case of operating gate voltage, electrons cannot pass through SiO_2 to generate current in the gate dielectric. But when the SiO_2 of the gate becomes thinner to a certain limit, that is, when a in Figure 18.21 is small enough, the situation changes. At this time, electrons will have sufficient probability to pass through the SiO_2 layer and create current. That means, there will be leakage current

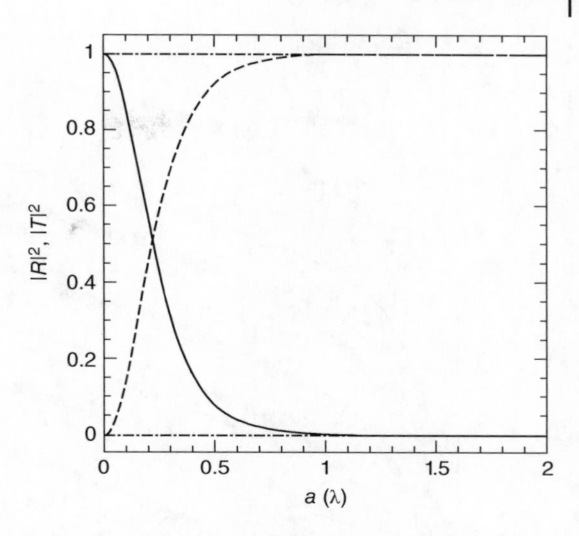

Figure 18.22 Transmission (solid-curve) and reflection (dashed-curve) probabilities for an electron of energy $E = (3/4)\ V_0$ incident on a square potential barrier, as a function of the ratio of the width of the barrier a to the free-space de Broglie wavelength λ (Section 8.2). Source: [17] Fitzpatrick.

flowing in the SiO_2. This current is not caused by the defects in the SiO_2 layer, but by a quantum phenomenon-tunneling effect. So, this leakage current is called tunneling current or tunneling leakage. Once this effect occurs, the device cannot work normally. The tunneling current is closely related to the thickness of the gate dielectric, and is also related to the voltage applied to the gate. The actual MOS situation is much more complicated than the description above. Figure 18.23 is the relationship between the tunneling current and the thickness of SiO_2 and the gate voltage. In the figure, J_g is the current density. From the figure, we understand that the tunnel current includes direct tunnel current and trap-assisted elastic tunnel current. A practical limit of 10–12 Å has been established based on the maximum allowable tunneling current. Transistors with SiO_2 film thickness of 10–12 Å can be used in FETs despite the high leakage current density of 1–10 A/cm^2 [9]. These data are consistent with the results in the Figure 18.23. In such thin layers of SiO_2, leakage current will increase dramatically due to quantum tunneling. In this case, the gate cannot control the channel (Figure 6.13) and "controllability" (section 2.1) fails. When the technology nodes were 90 and 65 nm, the thickness of the gate oxide is close to 1 nm. Therefore, since then, SiO_2 as a gate dielectric had been unable to meet the requirements. To avoid tunneling current, the thickness of the gate dielectric must be increased. Please see the formula (18.7), if we want to increase the thickness d, while keeping the capacitance constant, the k needs to be increased. Which means that a high k dielectric is required. In 2008, the high k dielectric was used for the first time at the 45nm technology node.

There are many high k dielectrics. Figure 18.24 is inorganic dielectrics. Table 18.2 is organic and hybrid dielectrics. For 7-nm CMOS, Hafnium dioxide (HfO_2) is introduced to replace SiO_2 [20].

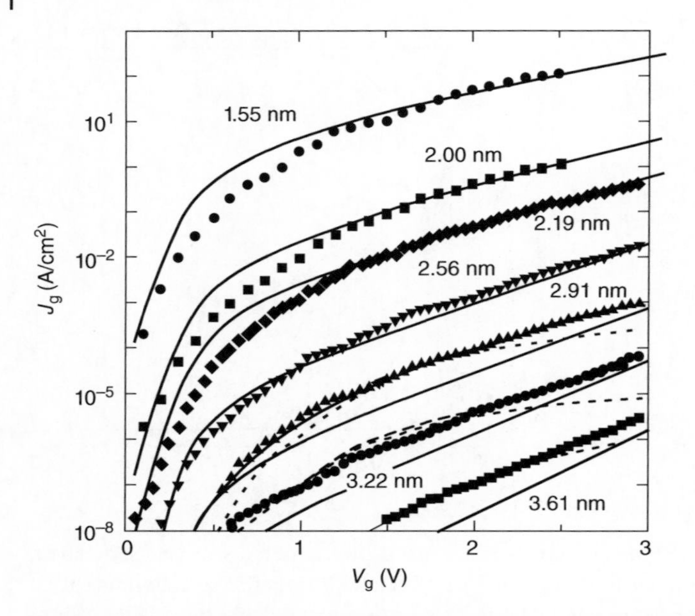

Figure 18.23 Tunneling current density through gate oxides with thickness in the range 1.55–3.61 nm. Solid line: direct tunneling current density. Dotted line: trap-assisted elastic tunneling current density. Dashed line: total tunneling current density. Source: [18] Jiménes-Molinos et al. / AIP Publishing.

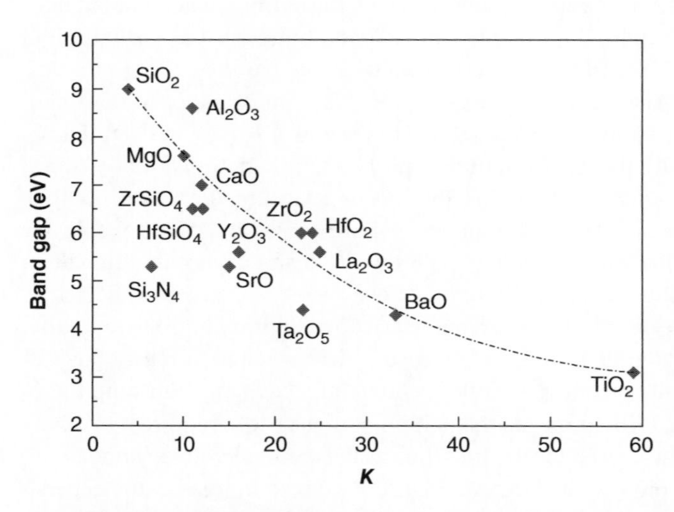

Figure 18.24 Relation between band gap and dielectric constant for some metal dielectrics. Source: [19] Wang et al.

Table 18.2 Summary of fabrication and *k* for representative polymer and hybrid dielectrics.

Material	Fabrication	*k*
PVA	Spin-coating + UV	6.2
Cyanoethylated-PVA	Spin-coating + Annealing	12.6
CEP	Spin-coating + Annealing	13.4
	Spin-coating + Annealing	6.3
P(VDF-TrFE)	Spin-coating + Annealing	10.4
P(VDF-TrFE-CTFE)	Spin-coating + Annealing	52
Acrylate polymer	Spin-coating + UV	5.1
Thiol–ene polymer	Spin-coating + UV	5.1
Cyano-containing polymer	Spin-coating + Annealing	7.2
Zirconium tetraacrylate	Spin-coating + UV	5.5
Zr-SANDs	Spin-coating + Annealing	6.0
Hf-SANDs	Spin-coating + Annealing	9.2
AT/PVP	Spin-coating + Annealing	6.6
Ta_2O_5/PI	Spin-coating + Annealing	5.5
HfO_2/CLHB	Spin-coating + Annealing	6.0
ZrO_2/CEP	Spin-coating + Annealing	16.6
GO/PVDF	Dip-coating + Annealing	15.6

Source: [19] Adapted from Wang et al.

18.10 End

From the above discussions, we can see that semiconductor technology has pushed the industry to the physical limit, from the controllability of Newtonian classical mechanics to the uncontrollability of quantum mechanics. In the technology of CMOS, new structures and dielectrics have been developed to replace conventional MOSFETs and SiO_2 to catch up with Moore's law, please see Table 18.3. As the devices becomes smaller and smaller, the chip size gets smaller and smaller, please see Figure 18.25. The structure of devices and circuits has become complicated, and the manufacturing processes have become more difficult. All these factors have made the design cost, especially the processing cost, to increase rapidly, please see Figure 18.26. Therefore, it is impossible to build an advanced semiconductor industry based on the capabilities of one country alone. This task must be accomplished through the cooperation of the strongest industrial countries in the world.

Table 18.3 Most significant technology nodes over past 21 years.

Technology node	Year of introduction	Key innovations
180 nm	2000	Cu interconnect, MOS options, six metal layers
130 nm	2002	Low-*k* dielectric, eight metal layers
90 nm	2003	SOI substrate
65 nm	2004	Strain silicon
45 nm	2008	Second generation strain, 10 metal layers
32 nm	2010	High-*k* metal gate
20 nm	2013	Replacement metal gate, double patterning, 12 metal layers
14 nm	2015	FinFET
10 nm	2017	FinFET, double patterning
7 nm	2019	FinFET, quadruple patterning
5 nm	2021	Multi-bridge FET

Source: [20] Sicard / Etienne SICARD.

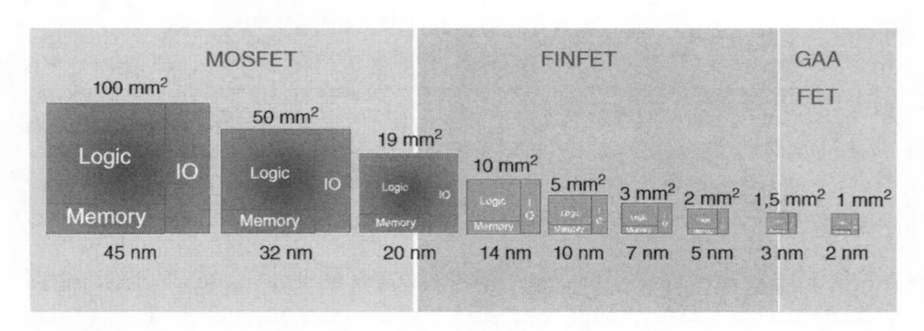

Figure 18.25 The spectacular reduction of feature sizes from 45 to 7 nm [20]. Source: [20] Sicard / Etienne SICARD.

In addition to silicon, germanium and III–V materials, there are other semiconductors. Two commonly used are II–VI materials, such as cadmium selenide (CdSe) and cadmium sulfide (CdS), and IV–IV semiconductors, such as silicon germanium (SiGe) and silicon carbide (SiC). Generally, semiconductor microchips operate at low voltage and low current. However, with the advancement of technology and the development of power electronics, semiconductors are increasingly involved in the field of heavy industry. The HVDC (high-voltage

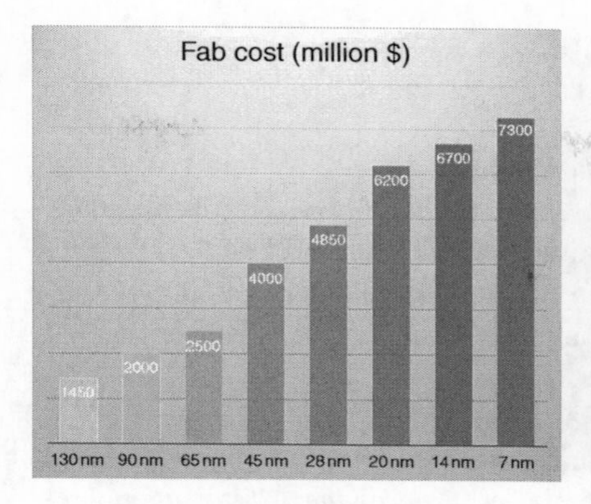

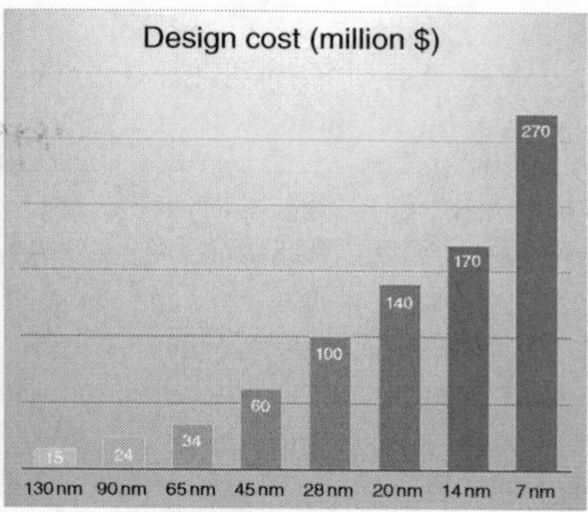

Figure 18.26 The extraordinary increase of the fab cost: more than 7 billion dollars for a 10 to 7-nm process, and associated chip design cost as high as 270 million dollars for a run in 7-nm [20]. Source: [20] Sicard / Etienne SICARD.

direct current) technology based on semiconductor power electronics has made long-distance transmission of direct current possible. In the late 1990s, the silicon insulated-gate bipolar transistors (IGBTs) started to be used for HVDC [22]. Recently, SiC has been involved in this field [23]. We discussed long-distance transmission of alternating current in Chapter 4. There are many differences in high voltage transmission systems between AC and DC. Two things we can clearly observe are: (i) Usually, AC is a three-phase system and uses at least three wires for electricity transmission. HVDC does not need to support three-phase and uses only two wires for electricity transmission. (ii) DC does not use or uses much less transformers than AC. Interestingly, the DC and AC dispute between Thomas Edison and Nikola Tesla (Section 4.4) is reopened today. If long-distance transmission of direct current replaces alternating current, it will have a big impact on the entire industrial system and our daily lives. At least, the electronic products we use every day may not need small transformers (Figure 1.1).

References

1 电子工业生产技术手册 7, 半导体与集成电路卷, 硅器件与集成电路, 653-660页.

2 Ali Javey, EE 143, Jaeger Chapter 7, Section 8: Metallization, UC Berkeley.

3 Wolf, S. and Tauber, R.N. (2000). *Silicon Processing for the VLSI Era, Volume1-Process Technology*, Seconde, P.477-480, P.771-776, P.780. Lattice Press.

4 Wolf, S. *Microchip Manufacturing*, 297–301. Lattice Press.

5 Lima, L.P.B., Diniz, J.A., Doi, I., and Fo, J.G. (2012). Titanium nitride as electrode for MOS technology and Schottky diode: alternative extraction method of titanium nitride work function. *Microelectronic Engineering* 92: 86–90.

6 物理学常用数表 [日] 饭田修一等, 科学出版社, 1979, 134页。.
7 Joiner, B.(2006). Integrated circuit package types and thermal characteristics. *Electronics-Cooling* (1 February).
8 Tektronix (2016). XYZs of oscilloscopes, Primer, P. 48.
9 Murarka, S.P., Eizenberg, M., and Sinha, A.K. (2003). *Interlayer Dielectrics for Semiconductor Technologies*. Elsevier Academic Press, P. 15, P. 38, P. 64, P.329.
10 IBM (1998). Copper Interconnects: The Evolution of Microprocessors.
11 Hasegawa, M. (2007). Fundamental analysis of electrochemical copper deposition for fabrication of submicrometer interconnects. Thesis submitted to Waseda University. March 2007, P. 7.
12 Singer, P. Making the move to dual damascene processing: a look at several different dual damascene processing strategies. *Semiconductor International* 20 (9): 79–82.
13 Andricacos, P.C. (Spring 1999). *Copper On-Chip Interconnects, A Breakthrough in Electrodeposition to Make Better Chips*, 32–37. The Electrochemical Society Interface.
14 Cheng, Y.L., Lee, C.Y. and Haung, C.W. (2018). Plasma Damage on low-*k* Dielectric Materials. *IntechOpen* (5 November).
15 Singh, J. *Quantum Mechanics Fundamentals & Applications to Technology*, 127. A Wiley-Interscience Publication.
16 Piskorski, K. and Przewlocki, H.M. (2006). Distribution of potential barrier height local values at Al-SiO_2 and Si-SiO_2 interfaces of the metal-oxide-semiconductor structures. *Bulletin of the Polish Academy of Sciences, Technical Sciences* 54 (4): 461–468.
17 Fitzpatrick, R. (2010). *Square Potential Barrier*. The University of Texas at Austin.
18 Jiménes-Molinos, F., Gámiz, F., Palma, A. et al. (2002). Direct and trap-assisted elastic tunneling through ultrathin gate oxides. *Journal of Applied Physics* 91 (8): 5116–5124.
19 Wang, B., Huang, W., Chi, L. et al. (2018). High-*k* gate dielectrics for emerging flexible and stretchable electronics. *Chemical Reviews* 118: 5690–5754.
20 Sicard, E. (2017). Introducing 7-nm FinFET technology in Microwind. *HAL* (24 July).
21 Mistry, K. (2017, 2017). *10 nm technology leadership, Technology and Manufacturing Day*. Intel.
22 Fairley, P. (2013). Germany jump-starts the supergrid. *IEEE Spectrum* (May), P. 37–41.
23 Bhattacharya, S. (2017). Transforming the transformer. *IEEE Spectrum* (July), P. 39–43.

Index

Semiconductor Microchips and Fabrication: A Practical Guide to Theory and Manufacturing, First Edition. Yaguang Lian.

g

h

i

j

k

l

m

n

q

r

t

Printed and bound by CPI Group (UK) Ltd, Croydon, CR0 4YY

17/04/2025

14658853-0001